I0891102

Synthesis, Characterizations, and Applications of Ionanofluids

Syam Sundar Lingala
Prince Mohammad Bin Fahd University, Saudi Arabia

Muhammad Waqar Ashraf
Prince Mohammad Bin Fahd University, Saudi Arabia

IGI Global
Scientific Publishing
Publishing Tomorrow's Research Today

Vice President of Editorial Melissa Wagner
Director of Acquisitions Mikaela Felty
Director of Book Development Jocelynn Hessler
Production Manager Mike Brehm
Cover Design Jose Rosado

Published in the United States of America by
IGI Global Scientific Publishing
701 East Chocolate Avenue
Hershey, PA, 17033, USA
Tel: 717-533-8845 | Fax: 717-533-7115
Website: https://www.igi-global.com E-mail: cust@igi-global.com

Copyright © 2026 by IGI Global Scientific Publishing. All rights reserved. No part of this publication may be reproduced, stored or distributed in any form or by any means, electronic or mechanical, including photocopying, without written permission from the publisher. Use of this publication to train generative artificial intelligence (AI) technologies is expressly prohibited. The publisher reserves all rights to license its use for generative AI training and machine learning model development.

Product or company names used in this set are for identification purposes only. Inclusion of the names of the products or companies does not indicate a claim of ownership by IGI Global Scientific Publishing of the trademark or registered trademark.

Library of Congress Cataloging-in-Publication Data

Names: Lingala, Syam Sundar, 1976- editor | Ashraf, Muhammad Waqar, 1967- editor
Title: Synthesis, characterizations, and applications of ionanofluids / edited by Syam Sundar Lingala, Muhammad Waqar Ashraf.
Description: Hershey, PA : IGI Global Scientific Publishing, [2026] | Includes bibliographical references and index. | Summary: "The book covers a broad range of subjects, starting with an in-depth explanation of the fundamental ideas behind ionic liquids and ionanofluids"-- Provided by publisher.
Identifiers: LCCN 2025008262 (print) | LCCN 2025008263 (ebook) | ISBN 9798337312453 h/c | ISBN 9798337312460 s/c | ISBN 9798337312477 ebook
Subjects: LCSH: Ionic solutions--Synthesis | Nanofluids--Synthesis
Classification: LCC QD561 .S96 2025 (print) | LCC QD561 (ebook)
LC record available at https://lccn.loc.gov/2025008262
LC ebook record available at https://lccn.loc.gov/2025008263

British Cataloguing in Publication Data
A Cataloguing in Publication record for this book is available from the British Library.

All work contributed to this book is new, previously-unpublished material.
The views expressed in this book are those of the authors, but not necessarily of the publisher.
This book contains information sourced from authentic and highly regarded references, with reasonable efforts made to ensure the reliability of the data and information presented. The authors, editors, and publisher believe the information in this book to be accurate and true as of the date of publication. Every effort has been made to trace and credit the copyright holders of all materials included. However, the authors, editors, and publisher cannot assume responsibility for the validity of all materials or the consequences of their use. Should any copyright material be found unacknowledged, please inform the publisher so that corrections may be made in future reprints.

Table of Contents

Detailed Table of Contents

This chapter explores the viscosity of ionic liquids (ILs), highlighting its dependence on molecular structure, intermolecular interactions, and external conditions like temperature and pressure. Structural factors, such as cation size, anion nature, and functionalization, significantly impact viscosity, with fluorination, methylation, and branching altering hydrogen-bonding networks. Binary IL mixtures and IL-solvent interactions introduce non-ideal behavior due to solvation and ion-pair dynamics. Temperature and pressure modulate viscosity through ionic diffusion and intermolecular forces. Practical applications include ILs as lubricants, electrolytes, and reaction media, where viscosity is critical. Advances in computational modeling, including machine learning, enhance viscosity prediction and IL design. By integrating experimental and theoretical insights, this chapter provides a comprehensive understanding of IL viscosity and its role in next-generation technologies. Deep eutectic solvents and protic ILs offer new strategies for viscosity modulation, expanding IL applications.

Ionic liquids (ILs) represent a promising new material group showing distinct thermophysical features that lead to significant applications in industrial sectors. The chapter explores ILs through molecular structure examination & classifying mechanisms while assessing their thermophysical features regarding density, viscosity,

Nanostructures, including nanoparticles, are gaining significant interest due to their wide range of applications. Their size, elemental composition, crystal structure, and physical properties—such as reactivity, mechanical strength, conductivity, and optical behavior—are crucial for determining their performance. Smaller particles have greater surface area-to-volume ratios, enhancing reactivity, while elemental composition affects stability and biological compatibility. Crystal structure influences hardness and conductivity, and other factors like shape, surface charge, and modification determine interactions. Various characterization techniques like XRD, SEM, TEM, FTIR, and VSM help analyze these properties. Each technique has its advantages, and combining methods often provides a more comprehensive understanding. Additional techniques, including EDX, UV-Vis, DLS, TGA, and spICP-MS, further enhance analysis by revealing elemental composition, optical properties, stability, and particle size distribution, aiding the development of nanoparticles for advanced applications.

Biodiesel is a promising alternative biofuel that is a subject of much research because of growing demand for sustainable energy sources. In the synthesis of biodiesel, ionic liquids (ILs), a class of organic salts with adjustable physicochemical characteristics, have drawn a lot of interest as effective and eco-friendly substitutes for traditional solvents and catalysts. This chapter examines the various functions of ILs in the production of biodiesel, including as their application as extraction solvents, reaction media, and homogeneous and heterogeneous catalysts. ILs are very useful in transesterification and esterification reactions because of their strong catalytic activity, recyclability, and improved enzyme stability. Particular attention is paid to basic and acidic ILs, their impact on reaction efficiency, and their potential to take the role of conventional catalysts while lowering environmental issues associated with the process.

Muhammad Waqar Ashraf, Prince Mohammad Bin Fahd University, Saudi Arabia
Mohammad Amin Mir, Prince Mohammad Bin Fahd University, Saudi Arabia
Syed Hasnain, Prince Mohammad Bin Fahd University, Saudi Arabia
Kim Andrews, Prince Mohammad Bin Fahd University, Saudi Arabia

Ionic liquids (ILs) are a family of adaptable, adjustable solvents that show a lot of promise for industrial and sustainable chemistry uses. They are excellent candidates to replace traditional volatile organic solvents due to their special physicochemical characteristics, which include low volatility, good thermal stability, and tunable solubility. The structure, production, and several uses of ILs are examined in this work, with a focus on their use in metal ion extraction, organic synthesis, catalysis, and nanoparticle synthesis. ILs' promise as green solvents is highlighted in this critical evaluation of their environmental impact, which also acknowledges issues with toxicity, difficulty of synthesis, and sustainability of the life cycle. This study emphasizes the importance that ILs play in promoting more sustainable industrial processes by incorporating them into chemical education and green chemistry practices.

Harshit Gouri, Department of Chemistry, University of Delhi, Delhi, India

This chapter explores the role of ionanofluids—hybrid materials combining ionic liquids with nanoparticles—in advancing sustainable chemical applications. IoNanofluids offer enhanced thermal stability, tunable viscosity, and superior catalytic properties, making them valuable in green chemistry, catalysis, and industrial processes. The chapter is structured into three key sections: (1) their use in sustainable chemistry, focusing on reducing environmental impact and improving reaction efficiency, (2) their role in catalytic applications, where they enhance reaction rates, selectivity, and catalyst longevity, and (3) their integration into process chemistry, including heat transfer, separation, and electrochemical applications. By addressing both scientific advancements and industrial scalability, this chapter highlights the transformative potential of ionanofluids in modern chemical engineering and concludes with future research directions.

In pharmaceutical applications, ionic liquids (ILs) have become a revolutionary class of materials, especially in drug delivery systems. They are excellent options for boosting transdermal drug transport and boosting bioavailability due to their adjustable physicochemical characteristics, that includes high solubility, thermal stability, and biocompatibility. Each of the four generations of ILs has improved drug solubility, decreased toxicity, and increased biodegradability. By acting as chemical permeation enhancers, they help transport medications that aren't highly soluble in water, increasing therapeutic effectiveness and reducing systemic side effects. Furthermore, ILs have been extensively investigated as environmentally friendly solvents, catalysts, and stabilizers that can replace volatile organic chemicals in drug synthesis, extraction, and purification procedures. Their versatility in the biological sciences is demonstrated by their promise for antibacterial applications, protein stability, and cancer therapy.

Nano-materials have emerged as a promising solution for wastewater treatment, offering enhanced efficiency in pollutant removal through advanced physicochemical mechanisms. Their high surface area, reactivity, and tunable properties enable superior removal of contaminants, including heavy metals, organic pollutants, pharmaceuticals, and microbial pathogens. Key mechanisms such as adsorption, photocatalysis, membrane filtration, electrochemical treatment, and antimicrobial

Preface

INTRODUCTION

The field of ionanofluids—innovative materials that merge the unique properties of ionic liquids with the enhanced functionalities of nanoparticles—has rapidly evolved in recent years, offering remarkable potential across scientific and industrial domains. Despite growing academic interest and a mounting body of experimental research, comprehensive resources that bridge fundamental theory with practical applications remain scarce.

This edited volume, *Synthesis, Characterizations, and Applications of Ionanofluids*, aims to address this gap. It brings together a diverse collection of contributions from leading researchers and practitioners working in the area of ionic liquid-based nanofluids worldwide. The goal is to offer a thorough and accessible reference that spans the synthesis, characterization, and application of ionanofluids, thereby supporting both academic research and industrial innovation.

The book opens with foundational chapters that explore the fundamental principles underlying ionic liquids and their integration with nanoparticles to form ionanofluids. These discussions set the stage for an in-depth examination of the various physical and chemical synthesis strategies employed in the development of these advanced materials. Particular emphasis is placed on the functionalization and stabilization of nanoparticles within ionic liquids—key determinants of stability, performance, and long-term utility.

Characterization techniques represent another core focus of this volume. Detailed chapters explore the use of Fourier-transform infrared spectroscopy (FTIR), atomic force microscopy (AFM), transmission electron microscopy (TEM), scanning electron microscopy (SEM), and differential scanning calorimetry (DSC) in studying ionanofluids. Complementing these are discussions on the measurement and analysis of critical thermophysical properties, such as thermal conductivity, viscosity, density, specific heat, and enthalpy.

Beyond foundational science, this book also highlights the diverse and growing range of real-world applications. Ionanofluids are increasingly being explored for their potential in energy storage systems, thermal management technologies, catalytic processes, biomedical innovations, and environmental solutions. Each application-oriented chapter provides valuable insight into how ionanofluids can contribute to solving global challenges while opening new avenues for research and development.

This volume is envisioned as more than just a reference text—it is intended to serve as a catalyst for collaboration and discovery. By presenting up-to-date research, critical analysis, and practical applications, we hope to inspire new directions in ionanofluid research and foster dialogue among scientists, engineers, and industry stakeholders.

To our knowledge, this is the first book to offer such a comprehensive overview of the field. By consolidating current knowledge and showcasing global expertise, we aim to support the continued growth of ionanofluids research. We are confident that this book will be a valuable resource for graduate students, researchers, and professionals working across materials science, chemical engineering, nanotechnology, and related disciplines.

We extend our sincere thanks to all contributing authors for their insights, dedication, and commitment to advancing this field. It is our hope that this book will serve as both a foundation and a springboard for future innovations in the exciting and still-unfolding landscape of ionanofluids.

ORGANIZATION OF THE BOOK

This volume begins with Chapter 1, which delves into the viscosity of ionic liquids (ILs)—a foundational property that significantly influences their applicability in various chemical and engineering processes. The chapter provides a detailed analysis of how molecular structure, intermolecular interactions, and external variables like temperature and pressure govern IL viscosity. It examines structural modifications such as fluorination and branching, and explores how binary mixtures and solvent interactions introduce non-ideal behavior. The discussion is further enriched with insights into computational modeling and machine learning for viscosity prediction, emphasizing the relevance of this property in practical applications including lubricants and electrolytes.

Chapter 2 offers an in-depth examination of the thermophysical properties of ionic liquids, focusing on key attributes such as density, viscosity, thermal conductivity, and phase behavior. The authors present a comprehensive classification of

ILs based on molecular structure and highlight how these properties can be tuned through careful selection of cation-anion pairs. Applications in energy storage, heat transfer, and pharmaceutical development are explored, alongside critical evaluations of toxicity, scalability, and environmental implications. The chapter concludes with future research directions aimed at overcoming current limitations.

In Chapter 3, the focus shifts to the synthesis, characterization, and wide-ranging applications of ionanofluids—composite systems formed by dispersing nanoparticles in ionic liquids. This chapter highlights the unique capabilities of ILs as tunable, sustainable solvents and their growing utility in catalysis, energy systems, and environmental remediation. Special attention is given to task-specific ILs and the integration of nanomaterials, which amplify properties such as thermal conductivity and catalytic performance. The authors also address pressing challenges such as toxicity and cost, proposing strategies for safer, more effective ionanofluid design.

Chapter 4 provides a critical overview of the techniques used to characterize both single and hybrid nanoparticles, which are essential for the successful development of ionanofluids. The chapter covers a variety of analytical tools, including XRD, SEM, TEM, FTIR, and VSM, each of which offers insights into structure, composition, and functional properties. Additional methods such as EDX, UV-Vis, and DLS are discussed for their ability to provide complementary data on particle size distribution, stability, and optical behavior. The holistic approach presented here underscores the importance of multi-technique analysis in nanoparticle research.

The role of ionic liquids in renewable energy is explored in Chapter 5, which focuses on their function as catalysts and solvents in biodiesel production. This chapter provides a nuanced discussion on the use of both basic and acidic ILs in esterification and transesterification processes, emphasizing their catalytic efficiency, recyclability, and environmental benefits. The multifunctional nature of ILs as reaction media and extraction agents is also considered, illustrating their potential to replace traditional, less sustainable chemical systems in biofuel synthesis.

Chapter 6 extends the sustainability narrative by examining the broader industrial applications of ILs as green solvents. It discusses their unique physicochemical properties—including low volatility and thermal stability—and their advantages in processes such as metal ion extraction, organic synthesis, and nanoparticle fabrication. The chapter also critically evaluates the environmental challenges associated with IL use, including toxicity and life-cycle sustainability, while proposing integration into green chemistry education and industrial best practices.

In Chapter 7, the discussion returns to ionanofluids, this time with a focus on their role in bridging performance and sustainability in modern chemistry. The chapter outlines how ionanofluids enhance catalytic activity, reaction selectivity, and energy efficiency across a spectrum of chemical and electrochemical applications. Organized into three thematic sections—sustainable chemistry, catalysis, and process integration—the chapter provides a strategic perspective on how these hybrid materials are reshaping chemical engineering, while also identifying areas for future innovation.

Chapter 8 explores the growing impact of ionic liquids in pharmaceutical and biomedical fields. With their exceptional solubility, biocompatibility, and tunability, ILs are proving instrumental in drug delivery, especially in enhancing transdermal transport and bioavailability. The chapter traces the evolution of ILs across four generations, each marked by advances in toxicity reduction and biodegradability. Their application extends to drug synthesis, protein stabilization, cancer therapy, and antibacterial treatment—offering a compelling case for their integration into next-generation therapeutics.

The final contribution, Chapter 9, investigates the use of nanomaterials for wastewater treatment—an area of increasing importance amid global environmental concerns. This chapter reviews cutting-edge nanomaterials and the mechanisms through which they remove contaminants such as heavy metals, organic compounds, and pathogens. Techniques including adsorption, photocatalysis, and membrane filtration are discussed, with an emphasis on scalability and future prospects. The findings presented highlight the crucial role of nanotechnology in developing efficient and sustainable water purification systems.

Together, these chapters provide a holistic and interdisciplinary view of ionic liquids and ionanofluids, from fundamental science to practical applications. By compiling the latest research and expert insights, this book serves as a vital resource for scientists, engineers, and industry professionals seeking to harness the full potential of these advanced materials.

CONCLUSION

As the scientific and industrial landscapes continue to evolve toward sustainability, efficiency, and innovation, the need for advanced, adaptable materials has never been more pressing. Ionic liquids and ionanofluids represent a significant leap in this direction—offering tunable properties, environmental compatibility, and multifunctional capabilities that position them at the forefront of next-generation technologies.

This book has been carefully curated to present a comprehensive and multidisciplinary view of these materials, covering everything from their fundamental thermophysical properties to cutting-edge applications in catalysis, energy, biotechnology, and environmental engineering. Each chapter contributes a unique perspective, authored by experts who are actively shaping the future of the field.

Our intent in assembling this volume is not only to inform but also to inspire. By connecting foundational knowledge with emerging research and practical applications, we hope to bridge gaps between disciplines and foster new collaborations among scientists, engineers, and industry professionals. The dynamic nature of ionic liquids and ionanofluids invites continuous exploration, and we believe that the insights shared in this book will serve as a catalyst for further innovation and discovery.

We are deeply grateful to all contributing authors for their commitment, scholarship, and vision. It is our sincere hope that this book will serve as a lasting resource and a starting point for those seeking to explore, understand, and advance the field of ionanofluids.

Chapter 1
The Viscosity of Ionic Liquids

Dipan Kumar Das
https://orcid.org/0000-0003-0224-3295
Centurion University of Technology and Management, India

Padmaja Patnaik
https://orcid.org/0000-0003-4468-7871
Centurion University of Technology and Management, India

Sudip Kumar Das
https://orcid.org/0009-0003-0814-0884
Dr. C.V. Raman University, India

Nibedita Nayak
Centurion University of Technology and Management, India

ABSTRACT

This chapter explores the viscosity of ionic liquids (ILs), highlighting its dependence on molecular structure, intermolecular interactions, and external conditions like temperature and pressure. Structural factors, such as cation size, anion nature, and functionalization, significantly impact viscosity, with fluorination, methylation, and branching altering hydrogen-bonding networks. Binary IL mixtures and IL-solvent interactions introduce non-ideal behavior due to solvation and ion-pair dynamics. Temperature and pressure modulate viscosity through ionic diffusion and intermolecular forces. Practical applications include ILs as lubricants, electrolytes, and reaction media, where viscosity is critical. Advances in computational modeling, including machine learning, enhance viscosity prediction and IL design. By integrating experimental and theoretical insights, this chapter provides a compre-

DOI: 10.4018/979-8-3373-1245-3.ch001

© 2026, IGI Global Scientific Publishing. Copying or distributing in print or electronic forms without written permission of IGI Global Scientific Publishing is prohibited. Use of this chapter to [gene]rative artificial intelligence (AI) technologies is expressly prohibited. The publisher reserves all rights to license its use for generative AI training and machine learning model development.

hensive understanding of IL viscosity and its role in next-generation technologies. Deep eutectic solvents and protic ILs offer new strategies for viscosity modulation, expanding IL applications.

INTRODUCTION

The resistance of a fluid against flowing is indicated by its viscosity, an elementary physical characteristic. Ionic liquids' (ILs) viscosity is an important variable to identify ILs' suitability for use in many industries. ILs form an all-ion series of organic salts with melting points commonly under 100°C. Their novelty, such as their temperature-varying viscosity, excellent thermal stability, and low volatility, makes them desirable for use in many fields of science and trade. Due to the strong ionic interactions and extensive hydrogen bonding in them, ILs have extremely high viscosities in comparison to other molecular solvents. The property has implications on mass transport, diffusion rates, and overall performance of ILs under real conditions. Some study explores the use of hybrid ionic liquids (ILs) solvents as low viscosity alternatives for biogas upgrading. The thermodynamic and thermal properties of these hybrid solvents were measured using polar soft-SAFT. Results showed that all diluents reduced the solvent's viscosity, but increased isobaric heat capacity, vaporization enthalpy, and solubility. Acetone was found to be a good diluent, increasing CH4 and CO2 solubility while decreasing viscosity and heat of absorption (Alkhatib et al., 2022). The strong ionic conductivity and flexibility of poly(ethylene oxide) (PEO)-based polymer electrolytes make them a viable class of materials for application in lithium-ion batteries. For PEO-based polymer electrolytes, the effects of polymer architecture, such as linear, star, and hyperbranched, and salt (lithiumbis-trifluoromethanesulfonyl)imide (LiTFSI)) concentration on the glass transition (Tg), microstructure, phase diagram, free volume, and bulk viscosity, have been thoroughly investigated. These factors all significantly affect the electrolyte's ionic conductivity. The branching of PEO widens the liquid phase towards lower salt concentrations, resulting in decreased crystallization and improved ion coordination. Ion clustering is common at high salt loadings, but cluster size and distribution are architecture-dependent. Ionic conductivity is maximized at a salt concentration of [Li/EO $\approx$ 0.085] for all architectures, with highly branched polymers showing up to three times higher conductivity. However, salt addition in highly branched architectures accelerates monomeric friction coefficients, potentially enhancing battery performance (Bakar et al., 2023). The rose-window instability (RWI) in silicone oil droplets exposed to corona discharge was investigated. It examines the effects of electrode gap, applied voltage, and viscosity on RWI formation. Increasing the electrode gap leads to an enlarged lattice, while higher voltages accelerate the formation and larger inner diameter. Viscosity has little influence on the lattice geometry, but droplets with low viscosity exhibit a more rapid development of instability. The small Ohnesorge number (Oh), influenced by factors like viscosity and surface tension, may play a role in the development of RWI. Surface tension, though not the main focus, is interconnected with Oh and may contribute to the observed results (Chiou et al., 2023). Non-Newtonian nanofluids are very applicable in many fields due to their exceptional conductivity and energy transfer characteristics. In this study, ionic liquids (ILs) functionalize carbon dots (Cdots) to create non-Newtonian nanofluids. Glucose and waste biomass are the basic ingredients for the Cdots synthesis in this straightforward, one-step process that creates the nanofluids (chitin from crab shells). The creation of a novel class of nanofluids, in which the ILs on the surface of the Cdots serve as the base-fluid, is made possible by the utilisation of ILs as both reaction media and functionalization molecules. Here, an unique home-made IL (1-tosylate-3-methyl-imidazolium triflate) [Tmi][Trif] and the well-known benign IL 1-butyl-3-methylimidazolium chloride ([Bmim]Cl) are utilised. The nanofluids derived from both substrates exhibit desirable heat sensitivity behaviour in addition to excellent conductivity and viscosity, light absorption, and good wettability. The surface plasmon resonance effect may be used to increase this heat sensitivity, which is maintained even when put as thin coatings on glass slides. The presented results show that the novel Cdots/IL-based nanofluids represent a flexible and economical path to high-performance thermosensitive non-Newtonian sustainable nanofluids with enormous promise for the heat transfer film systems and energy coatings industry (Gonçalves et al., 2020).

Table 1. Applications of Ionic Liquids (ILs) in various industries

Industry/Application	Description	Key Advantages of ILs
Energy Storage	Electrolytes in lithium-ion batteries, supercapacitors, and fuel cells.	Enhanced electrochemical stability, Improved safety (low flammability), Increased ionic conductivity.
CO_2 Capture	Solvents for capturing and separating carbon dioxide from gas streams.	High CO_2 solubility, Tunable selectivity, Low volatility.
Biomedicine	Drug delivery systems, Pharmaceutical formulations, Active pharmaceutical ingredients(API-ILs)	Improved drug solubility, Enhanced bioavailability, tunable biocompatibility.
Lubrication	Lubricants and additives for reducing friction and wear.	High thermal stability, Low volatility, Excellent lubricity.
Separation and Extraction	Extraction of metal ions and organic compounds, Gas separation.	High selectivity, Tunable miscibility, Environmentally friendly alternatives to traditional organic solvents.

continued on following page

Table 1. Continued

Industry/Application	Description	Key Advantages of ILs
Chemical Synthesis and Catalysis	Solvents and catalysts for various chemical reactions.	Enhanced reaction rates, Improved product selectivity, Recyclability.
Material Synthesis	Synthesis of nanomaterials, polymers, and other advanced materials.	Controlled morphology, Enhanced material properties.
Waste Treatment	Removal of pollutants from wastewater.	Efficient extraction of contaminants.

FACTORS AFFECTING THE VISCOSITY OF IONIC LIQUIDS

Molecular Structure and Ion Interactions

The molecular structure of ILs, such as the size, shape, and charge distribution of cations and anions, has a significant effect on their viscosity. Strong ionic interactions and extensive hydrogen bonding tend to increase viscosity, whereas bulky or asymmetric ions decrease it by disturbing ion packing. Video-rate atomic force microscopy (AFM) was used to study the near-surface nanostructure and dynamics of HMIM FAP and a locally-concentrated IL on highly oriented pyrolytic graphite (HOPG) and Au (111) electrodes. The results showed different nanostructures, with globular aggregates on HOPG and elongated and worm-like nanostructures on Au(111). The 2:1 mixture diffused more rapidly than pure HMIM FAP, providing insights for various IL applications in interface sciences (Li et al., 2024). Na secondary batteries have been studied using the thermal, physical, and electrochemical characteristics of the ionic liquid electrolyte system, Na[FTA]-[C3C1pyrr][FTA], where FTA– = (fluorosulfonyl) (trifluoromethylsulfonyl)amide and C3C1pyrr+ = N-methyl-N-propylpyrrolidinium. Particularly at low x(Na[FTA]) (x(Na[FTA]) = molar fraction of Na[FTA]) and low temperature range, the asymmetric FTA– structure offers a broad liquid-phase temperature range. The only thermal activity in the x(Na[FTA]) range 0.0–0.4 that has been seen is the glass transition at 170–209 K. The Vogel-Tammann-Fulcher equation governs the temperature dependence of viscosity and ionic conductivity, while the fractional Walden rule governs the relationship between molar conductivity and viscosity. The anodic potential limitations at 298 and 363 K are more than 5 V versus. Na+/Na. x(Na[FTA]) has observable impacts on the electrochemical performance of hard carbon and Na metal electrodes. A moderate concentration, x(Na[FTA]), of 0.2–0.3 allows for advantageous charge-discharge behaviour in both situations. At x(Na[FTA]) of 0.3, the hard carbon electrode's discharge capacities at 363 K are 260 and 236 mAh g−1 at current densities of 20 and 200 mA g−1, respectively. x(Na[FTA]) = 0.3 yields the best cycling performance, with high average Coulombic efficiency and adequate capacity retention (Yang et al.,2020). Prostate-specific antigen (PSA) can be detected using a microfluidic device, a biomedical device that extracts and purifies prostate samples, thereby revolutionizing medical diagnostics for prostate cancer, a prevalent global cancer was demonstrated. Buffer salts were combined with ionic liquids (ILs) and polymers, which are known to be ABS components for effective bioseparations. It was shown that PSA can be detected in non-physiological settings using both IL-based and polymer-based ABS. The ABS composed of tetrabutylammonium chloride ([N4444]Cl) and tetrabutylphosphonium bromide ([P4444]Br), both combined with phosphate buffer, and composed of polyethylene glycol with a molecular weight of 1000 g/mol (PEG1000) with citrate buffer were found to be the most effective in extracting PSA from serum. The use of IL-based ABS-mediated extraction in PSA assays led to lower detection limits compared to PBS or human serum, indicating potential for future point-of-care measurements (Flora et al., 2023).

Temperature Dependence of Viscosity

Viscosity with rising temperature decreases as a result of increased thermal motion of the ions, which reduces intermolecular forces. Most ILs display an Arrhenius-type viscosity, where it decreases exponentially as temperature increases. In a flow system containing 1-ethyl-3-methylimidazolium acetate, the influence of ambient temperature on ionic liquid electrospray is experimentally elucidated for the selective absorption of CO2. The liquid temperature has a major impact on the viscosity and surface tension of ionic liquids, among other physical characteristics. Because of the reduction in viscosity, the spray mode transition voltage rises with temperature. Furthermore, at higher liquid temperatures, the threshold voltage for the spray mode transition is raised by the ionic liquid feeding rate. Widely dispersed ultrafine droplets of ionic liquid are continually produced, as evidenced by the ionic liquid electrospray's enhancement of selective CO2 absorption in the flow system and the early decline of CO2 concentration increasing with applied voltage. Because of the decreased solubility at higher temperatures, CO2 absorption in the spraying chamber at steady state is reduced even if the CO2 reaction rate increases with temperature (Takana et al., 2021). In order to properly forecast the viscosity of multicomponent molten slags across a wide range of compositions and temperatures, a new structurally-based viscosity model and database were investigated. Based on the thermodynamic two-sublattice ionic liquid formalism and the CALPHAD (CALculation of PHAse Diagrams) methodology, the model fully describes a multicomponent system using just binary and ternary system information while accounting for the underlying structure of oxide melts. The FeOx-CaO-MgO-Al2O3-SiO2-CaF2-

CrOx-Na2O-MnOx-TiO2-ZrO2-P2O5-Gd2O3-La2O3-V2O5-NiO-CuOx system's model parameters are optimised using the Thermo-Calc software programme. It is demonstrated that the new thermodynamic database TCOX10 may provide extremely dependable computation results for geological and industrial applications, taking into account the derived viscosity model parameters. The current model and database are found to give the smallest deviations to the measurements when compared to previously published modelling work using the same large experimental datasets on the key subsystem CaO-MgO-Al2O3-SiO2 (CMAS). This is also true for the CMAS-based multicomponent glass, which can have a significant impact on the performance of aeroplane turbine engines (ZHANG et al., 2021). In 2009, Nieto de Castro's group presented the idea of IoNanofluids (INFs), which are stable dispersions of nanoparticles in ionic liquids. Exciting characteristics including enhanced heat conductivity, non-flammability, and non-volatility are characterised by INFs. Our research on how the shape and physicochemistry of carbon-based nanomaterials impact the density, viscosity, and thermal conductivity of INFs is continued in this paper. We concentrate on characterising dispersions made of 1-ethyl-3-methylimidazolium thiocyanate and long carboxylic group-functionalized multi-walled carbon nanotubes. KD2 Pro Thermal Properties Analyzer was used to analyse the thermal conductivity of INFs (Decagon Devices Inc., Pullman, WA, USA). The LV DV-II+Pro rotational viscometer was used to examine the viscosity (Brookfield Engineering, Middleboro, MA, USA). An Anton Paar DMA 5000 vibrating tube densimeter was used to determine the density of INFs (Graz, Austria). It was shown that INF made comprised of 1 weight percent long carboxylic group-functionalized multi-walled carbon nanotubes had the highest thermal conductivity increase of 22% (Jóźwiak et al.,2020).

Effect of Water and Organic Solvents

The introduction of organic solvents or water can substantially change IL viscosity by breaking ionic networks and diminishing cohesive forces. Low concentrations of water usually decrease viscosity, while extreme dilution can lead to phase separation or loss of IL properties. The significance of green lubricants and green lubricating additives is becoming more widely recognised these days. This effort produced a high-performance green lubricant by adding [Choline][Proline] ([Cho][Pro]) to glycerol aqueous solution. The impact of environmental factors, such as water and CO2, on the green lubricant was investigated. It was discovered that the water and CO2 contents of the green lubricant might alter its characteristics. The viscosity of the liquid rose when CO2 was absorbed by it, but it decreased as additional water was added. There was a noticeable increase in friction when CO2 was present. Simultaneously, it is discovered that the water content may change the friction. Therefore, by altering the amount of CO2 and water, friction may be controlled (Hua et al., 2021). Chemical and biochemical operations often employ 1-hexyl-3-methylimidazolium acetate's ionic liquid (IL). The promising compounds (N, N-dimethylacetamide, N, N-dimethylformamide, and dimethyl sulfoxide) were chosen as the additives in this work to reduce the viscosity of IL because of its high viscosity. Using a digital vibrating U-tube densimeter and an Ubbelohde capillary viscometer, the thermophysical characteristics of density and viscosity for IL with solvents were measured from 303.15 to 338.15 K at atmospheric pressure (0.0967 MPa), respectively. Quantitative research was done on how the solvents affected the ionic liquid's thermophysical characteristics. The derived parameters of excess molar volumes, thermal expansion coefficient, and energy barrier were also computed based on the measurement values. The findings demonstrated that the composition of the mixture significantly affected the viscosity and excess volume change. The viscosity was replicated using the hard-sphere model. In order to better understand the interactions between IL and the solvents, the chemical structure was studied using infrared spectroscopy (Yang & Feng,2020). Chitosan is a naturally occurring polymer that is highly biocompatible, biodegradable, and low in toxicity. Chitosan's biomedical uses are hampered by its intractable solubility in water and the majority of common solvents. In accordance with the procedure we devised for dissolving chitosan in plain water, chitosan was dissolved in an ionic liquid, frozen overnight at -20°C, and then exchanged with plain water at ambient temperature. In this work, we used a simple water-based chitosan solution to create a drug-carrying chitosan film by solution casting and air-drying. In particular, a plain-water-based solution was used to dissolve brimonidine tartrate (BT), an antiglaucoma medication, and create BT-loaded chitosan film, also known as chitosan-BT film. The end product is a mucoadhesive, translucent, and structurally stable film. In the chitosan film, tiny crystals of the antiglaucoma BT medication develop and are evenly distributed. For possible ocular medication administration, the chitosan-BT film gives BT a high corneal permeability and quick drug release kinetics (Li et al., 2020). The diffusiophoresis of charged hydrophobic nanoparticles (NPs) regulated by an enforced ionic concentration gradient is investigated. The major purpose is to understand the influence of the laterally mobile adsorbed surface ions at the interface on the propulsion of the hydrophobic NPs in diffusiophoresis. Furthermore, the dielectric polarisation resulting from the variation in the dielectric constant between the suspension medium and the NPs is taken into account. The slip velocity condition and the slip length are altered as a result of the friction and electric force produced by the mobile surface ions at the hydrophobic surface. By using a control volume formulation, we are able to obtain an exact numerical solution of the full form of the governing electrokinetic equations. Analytical solutions based on the Debye-HÜckel linearization are used to complement the numerical model. We discover that the surface ions' lateral mobility prevents the coions from diffusing from the side of greater concentration to the side of lower concentration. This creates a repulsive force on the particle, which causes negative mobility to arise. We have demonstrated, using numerical data and analytical solutions, that the diffusiophoresis of a hydrophobic nanoparticle (NP) with a completely mobile surface charge is the same as the diffusiophoresis of a liquid droplet whose viscosity is proportional to the hydrophobic particle's slip length. We demonstrate that a hydrophobic particle's velocity is increased by dielectric polarisation, which may have practical uses (Majhi et al., 2024).

Effect of External Fields

Electric and magnetic fields may also affect the viscosity of ILs by modifying ion alignment, charge distribution, and molecular orientation. These effects are especially applicable in electrorheological and magnetorheological applications. The mechanisms behind fluidic interface deformation modes in liquid-liquid systems, focusing on the presence of droplets that trigger the dimple at the interface was investigated. The researchers replace the pin electrode with a hollow metallic needle and introduce droplets of an aqueous KCl solution into silicone oils with varying viscosities. The voltage between the needle and liquid-liquid interface has a strong influence on the deformation, while the electrolyte flow rate has a small influence. The deflection and current reach limiting values with increasing ion concentration and viscosity, explained by a scaling relationship for droplet size. The researchers use a finite element solver to compute charge transport due to droplets in the oil phase, compute electric potential distribution, flow field, and interface deformation. After calibration with particle tracking velocimetry data, the dimple at the liquid-liquid interface is reproduced. The space charge is identified as the cause of the limiting behaviour of interface deflection and electric current (Dehe & Hardt, 2021). Liquid crystalline hydrogels are an attractive class of soft materials to direct charge transport, mechanical actuation, and cell migration. When such systems contain supramolecular polymers, it is possible in principle to easily shear align nanoscale structures and create bulk anisotropic properties. However, reproducibly fabricating and patterning aligned supramolecular domains in 3D hydrogels remains a challenge using conventional fabrication techniques. Here, a method is reported for 3D printing of ionically crosslinked liquid crystalline hydrogels from aqueous supramolecular polymer inks. Using a combination of experimental techniques and molecular dynamics simulations, it is found that pH and salt concentration govern intermolecular interactions among the self-assembled structures where lower charge densities on the supramolecular polymers and higher charge screening from the electrolyte result in higher viscosity inks. Enhanced hierarchical interactions among assemblies in high viscosity inks increase the printability and ultimately lead to greater nanoscale alignment in extruded macroscopic filaments when using small nozzle diameters and fast print speeds. The use of this approach is demonstrated to create materials with anisotropic ionic and electronic charge transport as well as scaffolds that trigger the macroscopic alignment of cells due to the synergy of supramolecular self-assembly and additive manufacturing (Sather et al., 2021).

EXPERIMENTAL AND COMPUTATIONAL METHODS FOR MEASURING AND PREDICTING IL VISCOSITY

Experimental Techniques

Experimental techniques for viscosity measurement of ionic liquids (ILs) are rheometry, capillary viscometry, and oscillatory shear measurements. These techniques yield accurate values of viscosity at different temperatures, shear rates, and pressures. Ultrasonic rheometry is also used to measure the gas absorption effect on IL viscosity. Tribological experiments also assist in evaluating ILs as lubricants under various frictional regimes. Sophisticated techniques, like microfluidics, allow for small-volume, high-precision measurements of viscosity. Finding the best piezoelectric material for the production of rheological reflectance ultrasonic sensors is the primary objective of this research. The second objective is to evaluate the viscosity changes in an ionic liquid (IL) brought on by gas absorption and incorporate the ultrasonic rheometer into a gas absorption reactor. In order to accomplish the goals, perovskite structures, tungsten bronze, bismuth titanate, lead titanate, lead metaniobate, and lead zirconate titanate materials were constructed on aluminium delay lines and subjected to heat cycling between ambient temperature and 150 °C. The findings demonstrated that the best material for long-term thermal cycling is lead metaniobate in tungsten bronze framework. Consequently, this material was used to build the ultrasonic rheometer, which was then placed in a pressurised reactor to test a reference IL at 50 °C and 80 bar of pressure. Nitrogen and hydrogen gas were both saturated in the reference IL. Whereas the absorption of the gas caused the viscosity value to increase in the nitrogen environment, the viscosity signals stayed constant in the hydrogen atmosphere (Schirru & Adler, 2020).

Table 2. Experimental methods for measuring and predicting IL viscosity

Method	Principle	Advantages	Disadvantages	Applications
Rotational Viscometers (e.g., Couette, cone-plate)	Measures the torque required to rotate an object in the fluid.	Wide range of viscosity measurements, can assess non-Newtonian behavior.	Can be sensitive to temperature control, requires careful calibration.	General IL viscosity measurements, shear-dependent studies.
Capillary Viscometers (e.g., Ubbelohde, Ostwald)	Measures the time it takes for a fluid to flow through a narrow tube.	Relatively simple, accurate for Newtonian fluids.	Limited shear rate control, requires careful cleaning.	Routine IL viscosity measurements, especially for Newtonian ILs.
Falling Sphere Viscometers	Measures the terminal velocity of a sphere falling through the fluid.	Simple principle, can be used for high-viscosity ILs.	Accuracy depends on sphere size and density, limited shear rate information.	High-viscosity ILs, quick estimations.
Vibrational Viscometers	Measures the damping of an oscillating probe within the fluid.	Can be used for in line measurements, and in small volumes.	Can be sensitive to the probe geometry, and calibration.	In line process monitoring, and microfluidic applications.

Computational and Theoretical Models

Computational models for IL viscosity prediction span from molecular dynamics (MD) simulations to thermodynamic equations of state. Hybrid models of ionic liquids for biogas upgrading and CO_2 capture have also been designed, including viscosity prediction. Machine learning is becoming increasingly effective in being able to predict IL viscosity from molecular structures. Even density functional theory (DFT) and quantum chemistry calculations also contribute to insights into molecular interactions that influence viscosity. Equation-of-state models like PC-SAFT give semi-empirical estimations of viscosity for IL-solvent mixtures. This work explores the practical use of Lasso Regression, Kernel Ridge Regression, and K-Nearest Neighbors (KNN) for predicting the viscosity of ionic liquids in a dataset that contains numeric variables (T(K), xIL(mol percent)) and categorical variables (Cation, Anion). In fact, the models took temperature and the mole percentage of ionic liquids as inputs. The models' efficacy is thoroughly evaluated, and K-Nearest Neighbors stands out for having outstanding predictive capabilities. A hyperparameter tuning optimization technique called Tabu Search is applied to increase the accuracy of the model. As seen by the numerical findings, KNN is the best model with the lowest RMSE and an amazing R2 test score of 0.91628. Model performance is further improved by Tabu Search optimization, which highlights the crucial role that hyperparameter tuning plays in producing reliable regression models for forecasting the viscosity of ionic liquid-water mixtures. This study shows measurable gains in predictive accuracy for viscosity forecasts in comparable scenarios, which offers important insights into the optimization process (Ran et al., 2024). A promising novel method for separating CO2 from CH4 is the use of ionic liquids (ILs) encapsulating metal-organic frameworks (MOFs), particularly ZIF-8. The processes behind the method's strong CO2/CH4 selectivity are still unclear, though. Here, we describe the advancements made in comprehending the processes through the use of DFT and MD simulations to study the ZIF-8 aperture configuration change. According to the data, the pristine aperture design performs the best in terms of separation, and the inclusion of ILs stops the apertures from swinging much (i.e. configuration variation). The impact that IL viscosity had on the ZIF-8 configuration modification was then examined. Additionally, MD simulations demonstrate that ILs with high viscosity sustain the clean aperture configuration more (0-87Cp). The aperture stability did not noticeably alter when the IL viscosity was increased over 87Cp (Yu et al., 2021).

Figure 1. Viscosity dependence on the mixture's IL content

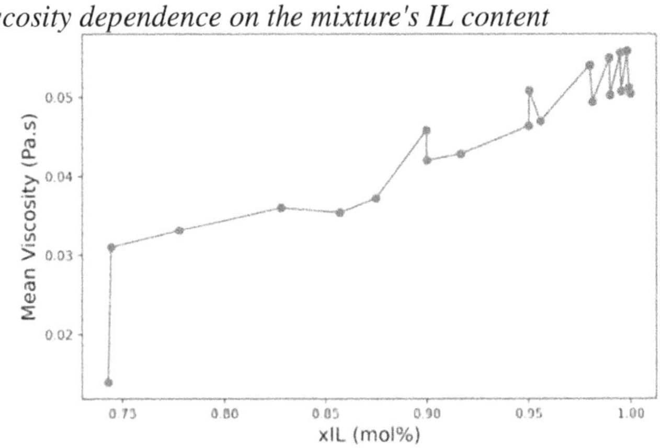

(Ran et al., 2024)

Table 3. Computational methods for measuring and predicting IL viscosity

Method	Principle	Advantages	Disadvantages	Applications
Molecular Dynamics (MD) Simulations	Simulates the motion of individual molecules based on interatomic forces.	Provides detailed molecular-level information, can predict viscosity under various conditions.	Computationally intensive, accuracy depends on force field parameters.	Predicting IL viscosity, understanding structure-viscosity relationships.
Quantitative Structure-Property Relationships (QSPR)	Develops statistical models relating molecular structure to viscosity.	Relatively fast, can be used for screening large libraries of ILs.	Accuracy depends on the quality of the training data, limited to the range of the model.	Predicting IL viscosity from molecular structure, designing new ILs with desired properties.
Density Functional Theory (DFT)	Calculates electronic structure, which can be related to intermolecular interactions and viscosity.	Provides very detailed information of electronic interactions.	Very computationally expensive, and limited to small systems.	Understanding the fundamental interactions that influence IL viscosity.

APPLICATIONS OF IONIC LIQUIDS BASED ON THEIR VISCOSITY

Ionic Liquids in Lubrication

Ionic liquids (ILs) possess superior tribological characteristics owing to their low volatility and adjustable viscosity, and as such, they are well suited for lubrication in mechanical systems. Their capacity to decrease metal contact friction and wear leads to improved gear, engine, and industrial machinery performance. IL lubricants also promise to enhance the durability of low-viscosity axle fluids and high-performance tribological coatings. A new ceramic-phosphonium-based ionic liquids (P-ILs) system has been developed to improve tribological performance without continuous lubricant supply. The system, tested at various temperatures and sliding velocities, showed a low coefficient of friction for phosphonium benzoate and phosphonium salicylate IL at a low temperature of 5°C, but reversed at higher temperatures of 50 and 100°C. The tribological performance of the P-IL+ceramic system is influenced by the thermal response of the ILs (Kasar et al., 2023). A phosphonium phosphate ionic liquid was investigated as a potential anti-wear and anti-pitting ingredient for rear axle lubricant. When this specific IL was initially added to a VHVI8 base oil at a concentration of 2-3 percent, it showed good surface protection against microcracking and wear under rolling-sliding contacts. The encouraging outcomes led to the development of a range of fully formulated gear oils with IL that have low viscosity (about half of SAE 75W-90). Bench-scale rolling-sliding experiments revealed that a standard SAE 75W-90 gear oil was not as effective in mitigating rolling contact fatigue as certain experimental lubricants containing IL. After that, full-scale hub dynamometer testing showed that an IL-containing low-viscosity gear oil could provide more than 3% power output and torque when compared to commercial baselines (Roy et al., 2021). In order to systematically analyse the impacts of cation alkyl chain group length on their physicochemical and tribological characteristics, four new halogen-free borate-based protic ionic liquids were created with distinct ammonium cations but identical organoborate anions. When tested as plain lubricants, the ionic liquids (ILs) under investigation demonstrated minimal wear and up to a 30% reduction in friction when compared to biolubricants. Maximum friction and wear reductions of 25% and 96%, respectively, over the basic oil were achieved by experimenting with blends of 1 weight percent ionic liquid and biolubricant. It's interesting to note that the ionic liquid with the best tribological reaction as a lubricant additive was the one that performed the worst as a neat lubricant because of fluid solidification. The development of a self-assembled adsorbed layer that isolated the frictional surfaces was credited with the tribological advancements (Sierra et al., 2022).

Ionic Liquids in CO_2 Capture and Gas Separation

Viscosity in ILs is of paramount importance for CO_2 capture and separation application, and it influences mass transfer and CO_2 absorption rate. Optimally viscous ILs provide good CO_2 solubility as well as practical operability under industrial processes. The incorporation of ILs into hybrid solvents and membrane applications promotes selectivity and energy saving in carbon capture technology. Specific ionic liquids (ILs) are an attractive alternative to chemical and physical solvents for CO2 capture processes due to their superior properties, such as negligible volatility, high thermal stability, and affinity to capture CO2. However, their high viscosity leads to lower capture rates and higher pumping costs. This study investigated solubility, vapor-liquid equilibria (VLE), and liquid-liquid equilibria for binary and ternary mixtures involving CO2, ILs, water, and methanol using perturbed-chain statistical associating fluid theory (PC-SAFT) equation of state. Both strategies provided accurate correlations in modeling CO2 solubilities in ILs and LLE of binary ILs/water systems. PC-SAFT was found to be a potential practical tool for developing CO2 capture processes with new alternative solvents when sufficient experimental data is available for binary mixtures (Liao et al.,2022). The CO2 absorption capacity of three hybrid solvents based on butyl-3-methylimidazolium acetate ([Bmim][OAc]) and three cosolvents (DEPG250, propylene carbonate, and water) was investigated. The results showed that [Bmim][OAc]-DEPG250 had the highest absorption capacity. The mass ratio and temperature also impacted their absorption capacity, density, and viscosity. The absorption capacities of N2 and CO2 were compared with pure gas. The addition of DEPG250 significantly decreased viscosity, but the absorption capacity remained comparable to pure [Bmim][OAc]. The process also reduced the CO2 capture cost by 11% compared to aqueous amine solution, resulting in a significant decrease in utility cost (Ma et al., 2021). Mixtures of ionic liquids (ILs) with water, amines, and other solvents have drawn interest as potential CO2 absorbent alternatives that might reduce energy consumption in comparison to conventional aqueous MEA procedures and lessen the high cost or viscosity of pure ILs. In this study, CO2 was captured using mixtures of two ionic liquids (1-butyl-3-methylimidazolium acetate [bmim][OAc], 1-ethyl-3-methylimidazolium octylsulphate [emim][OcSO4]) and five amines (1,1,3,3-tetramethylguanidine (TMG), 1,8-diazabicyclo[5.4.0]undec-7-ene (DBU), 1,4-diazabicyclo[2.2.2]octane (DABCO), 1,2′-aminodiethanol (DEA)). Ten model systems in all were examined. Depending on the underlying absorption mechanism, generalised patterns of composition-CO2 absorption connections were found. The chosen systems' viscosity, deterioration, and regeneration capacity were assessed. Regeneration temperatures in DBU-[emim][OcSO4] systems were as low as 60 °C. The addition of water significantly decreased the viscosity of all systems. Non-Newtonian behaviour was seen in some solutions that were high in CO2. 13C NMR was used to validate the mechanism of CO2 absorption in imidazolium ionic liquids with DBU present (Zalewski et al., 2021). Gas-to-liquid mass transfer is frequently thought of as one of the limiting processes in gas fermentations (using O2, CO, H2, CH4, or CO2). Nonetheless, it is well known that the volumetric mass transfer coefficient kLa is greatly influenced by the components of fermentation broths, such as salts, biomass, proteins, antifoam, and organic compounds like acids and alcohols. In three fermentation broths made from syngas fermentation, the impact of ethanol on mass transfer was investigated. Adding ethanol, the anticipated result, to demineralized water doubled kLa in the 0–5 g L−1 range before yielding values of kLa that were almost constant. Ethanol addition considerably raised kLa in the fermentation broths (2–4 times higher than water), and broth salinity had a substantial impact. According to our findings, broth components have the ability to considerably raise kLa, a dynamic parameter in gas fermentation tests (Puiman et al., 2022).

Ionic Liquids in Energy Storage and Electrochemical Devices

Ionic liquids are potential electrolytes in energy storage devices like batteries, supercapacitors, and fuel cells because they possess a broad electrochemical stability and can be made tunable in viscosity. Their regulated viscosity influences ion transport, ionic conductivity, and charge storage, which optimizes the performance of devices. IL-based electrolytes enhance the efficiency and safety of lithium-ion batteries and future energy storage devices. Conductive metal-organic frameworks (c-MOFs) and ionic liquids (ILs) are promising combinations for high-performance supercapacitors. However, nanoconfinement from c-MOFs and high viscosity of ILs slow down the charging process. Solvent can resolve this by adding it. Constant-potential molecular simulations show conditions for a 100% enhancement in capacity and a 6 times increase in charging speed. The solvent acts as an ionophobic agent, enhancing charge storage and creating two distinct ion transport highways (Chen et al., 2024). Researchers have developed a hybrid electrolyte combining a highly fluorinated ionic liquid and a weakly solvating fluorinated ether to achieve the highest energy density in battery chemistry. This hybrid structure allows for the reversible operation of Li0 and LiNiO2 batteries, delivering nearly theoretical capacity for over 300 cycles with 78.6% retention and no unwanted morphological changes. The fluorinated ether's function is more profound than dilution and viscosity reduction, causing drastic changes in Li+-solvation environment, resulting in simultaneous stabilization and interfacing. This study opens more design space for next-generation electrolytes and interphases for these battery chemistries (Liu et al., 2024). It was documented how etheric substituents in imidazolium and ammonium-based ionic liquids impacted the performance of electrochemical double layer capacitors (EDLC) using a reduced graphene oxide (RGO) electrode and gel polymer electrolyte (GPE) (IL). GPEs contain ILs and poly (vinylidene fluoride-hexafluoropropylene) (PVDF-HFP). Ionic liquids (ILs) based on imidazolium and ammonium have different etheric group lengths and concentrations from one another. Electrochemical impedance, cyclic voltammetry, and galvanostatic charge-discharge studies revealed that the longer etheric group substituted {N-methyl-2-(2-methoxyethoxy)-N,N-bis[2- (2-methoxyethoxy)ethyl] Three-allyl-1-[2-(2-methoxyethoxy)ethyl] and ethan-1-aminium bis(trifluoromethanesulfonyl)imide in lieu of ether (AMEt-TFSI) AL3IL-TFSI, or-1H-imidazole-3-ium bis(trifluoromethanesulfonyl)imide, has tender specific capacitances of 250 Fg-1 and 238 Fg-1, respectively, and energy density values of 61.36 wh kg−1 and 61.56 wh kg−1 (Siyahjani et al., 2020). Combining microparticles to create one-dimensional (1D) particle forms that are not produced in a liquid can be challenging. This greatly hinders the development of new materials and possible applications for such structures. Here we provide a simple, efficient, and easy-to-use method for constructing structures that are similar to one particle-thick chains. Electrically conductive particles that are originally dispersed in a nonpolar, weakly conductive liquid are drawn out of the liquid by use of an electric field maintained by capillaries. The effects of capillary and electric interactions, the dispersion phase's viscosity and ionic conductivity, particle size, shape, and density, and pulling rate on our method's overall effectiveness were examined. It has been shown that several micro particle kinds may come together in a matter of seconds to create lengthy chains. It was discovered that capillary forces aid in particle alignment and the formation of single particle-resolution structures. The findings indicate that dispersing liquids with low ionic conductivity and high viscosity are favoured because they lessen the adverse effects of ionic screening. The physical characteristics of the generated beaded structures that we believe are interesting for both basic and practical research are examined in the second section of our study. Finally, we show that electrically conductive micro paths may be designed using 1D particle architectures (Rozynek et al., 2022).

Biochemical and Pharmaceutical Applications

In pharmaceutical and biochemical applications, the rheological properties of ILs determine drug solubility, enzymatic activity, and biocompatibility. ILs find application in drug delivery vehicles, stabilization of proteins, and processing of biomaterials as a result of their tunable rheology. Their applications further include bioseparations, biosensors, and the development of advanced drug-loaded films and gels. Because of their simplicity of use and adaptability, photocrosslinkable bioinks have become more popular in 3D bioprinting. To facilitate photocrosslinking, however, a certain functional group—such as methacrylate or photo-click chemistry—must be present in the polymer backbone. Because of its natural viscosity, methacrylated gellan gum (GGMA) precursor has been shown to have acceptable rheological qualities for an injectable hydrogel. The study uses a two-step crosslinking technique to make GGMA biomaterial ink printable. The ionic crosslinker (Ca2+) transforms GGMA into a weak extrudable hydrogel, followed by photocrosslinking to create a true hydrogel with good shape fidelity. The best printability was found in 2% GGMA with 90 mM calcium chloride, which was then used to print 3D structures. This optimized ink formulation offers good printability and high resolution, making it suitable for various bioprinting applications (Jongprasitkul et al., 2022). Deep eutectic solvents (DESs) and Ionic Liquids (ILs) are used for insulin delivery in various ways, but their high viscosity makes them impractical for pure solution delivery. Water can decrease DESs' viscosity but also affect protein structural stability. This study investigated the effect of DES-water solutions on insulin's structural stability. Results showed that insulin is strongly stabilized in DES and its aqueous solutions compared to water. However, in DES solutions with 50% and 75% water content, malic acid interacts with insulin dimer, causing the separation of monomers. This suggests that aqueous DES is crucial for rapid insulin delivery, as the decomposition of insulin into monomers before reaching the bloodstream can increase pharmacokinetics (Kazhiyev et al., 2023). Sirolimus (SR) is a macrolide with antifungal and antitumor immunosuppressant properties. A gellan gum-based formulation of SR-SUS-ISG was developed, exhibiting stability and minimal irritation. It gels upon contact with ions, increasing drug viscosity and extending retention on the eye surface. It also has favorable shear dilution properties, reducing viscosity and enhancing fluidity. Biocompatibility tests on human corneal epithelial cells and rabbit eyes showed SR-SUS-ISG could be tolerated. Pharmacokinetic studies showed sustained release, improved corneal penetration, and enhanced bioavailability (Liu et al., 2024). Analysis was done on how ZrF4 affected the viscosity of a ZrF4–BeF2 binary salt in order to create BeF2-based molten salts for nuclear transmutation targets. In particular, we used a polarisation ion model to simulate molecular dynamics and assessed the viscosity, static structure, and suitability of the Stokes-Einstein equation for ZrF4–BeF2 in comparison to LiF–BeF2. According to our research, ZrF4–BeF2 has greater viscosity than LiF–BeF2 because the structural relaxation period is prolonged by the longer lifespan of each ionic link. Additionally, it is proposed that the viscosity mechanism of ZrF4–BeF2 is influenced by the Zr[sbnd]F and Be[sbnd]F relaxations. The majority of the elements found in long-lived fission products (LLFPs) are multivalent ions, such Zr, Pd, and Sn, which have a strong affinity for anions and are unlikely to have a significant impact on lowering the viscosity of BeF2 mixed salts. On the other hand, CsF can lessen the viscosity of these salts since Cs is a monovalent ion and also an LLFP. Designing BeF2-based salts for transmutation will include the principle of mixing certain LLFPs with CsF (Shishido & Hashizume,2024). Endovascular embolization is a minimally invasive technique that selectively blocks blood flow to certain areas. The clinical indications for embolization are still growing. Injectable substances known as liquid embolic agents change into solid or semi-solid forms when injected into blood arteries. One important characteristic that sets liquid embolic agents apart is the mechanism that initiates the liquid-to-solid change. GPX is a water-soluble liquid embolic agent made up of inorganic polyphoshate and polyguanidinum, two polyelectrolytes with opposing charges. Ionic strength differences cause the polyelectrolytes to electrostatically condense, which propels in situ solidification. We describe the in vivo efficacy of GPX in acute animal investigations, its material characteristics in vitro, and its potential for chemoembolization. The polyguanidinium MW and/or concentration can be changed to change the viscosity of GPX across a large range. The flow behaviour of GPX was not appreciably altered by formulation with either tantalum microparticles (30 weight percent) or iodinated radiocontrast agents (300 mgI ml−1); the viscosity remained within a therapeutically feasible range (80–160 cP) and was independent of shear rate. The combination of doxorubicin and GPX produced a power law dependence on shear rate and significantly enhanced viscosity at low shear rates. Both retė mirabile and swine kidneys showed high contrast and efficient vascular blockage. Iodinated chemicals produced a transient contrast that vanished in a matter of hours. Over the course of ninety days, the doxorubicin in vitro release profile was linear. The findings show that GPX is a flexible liquid embolic platform that can be made with a variety of viscosities that can be injected at clinically feasible flow rates. It can also be used to deliver doxorubicin to embolized tissues for an extended period of time at zero-order delivery, with either transient or permanent contrast (Stewart et al., 2023). Innovative solvation techniques that effectively disturb the native supramolecular structures of natural polymers—that is, dissolve/gelate—are greatly desired for their modification, derivation, and material creation due to the renewable nature and chemical adaptability of natural biopolymers. Rapid visco analysis (RVA), scanning electron microscopy (SEM), X-ray diffraction (XRD), small-angle X-ray scattering, Fourier-transform infrared (FTIR), and Raman spectroscopy are used in this study to compare the structural disorganisation behaviour at room temperature between waxy maize, high-amylose maize, cassava, and potato starches with the 6:4 (wt/wt) 1-ethyl-3-methylimidazolium acetate ([C2mim][OAc])/water mixture. Both the A-type (cassava and waxy maize) and B-type (potato and high-amylose maize) starches are more vulnerable to structural disruption with the aqueous IL, according to all the viscosity, morphological, and structural alterations. The viscosity of the four starches increases immediately after being ball-milled (for up to five minutes) in RVA. In contrast, native waxy maize and cassava starches that are not ball-milled show a lag in viscosity increase, and high-amylose maize and potato starches show hardly any viscosity change. This suggests that the granule surface compactness regulates the ease of starch gelation by the aqueous IL rather than the amylose concentration (namely, how easily the aqueous IL can reach and disrupt the most vulnerable parts inside the starch granule and subsequently cause the destruction of the rest parts and the breakdown of the whole granule). The results can help create low-energy methods for digesting starch using aqueous ILs (Zhang et al., 2023).

Figure 2. The solvent components' radial distribution function, g(r): (a) choline ion, (b) chloride ion, (c) malic acid, and (d) water, around insulin dimer in pure DES and its aqueous solutions

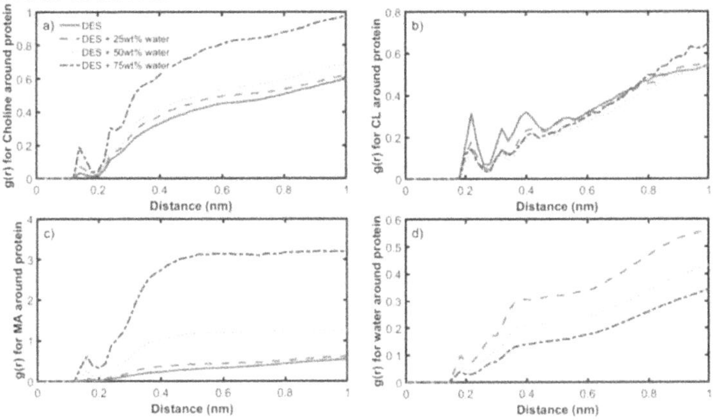

(Kazhiyev et al., 2023)

RECENT ADVANCES AND FUTURE PERSPECTIVES

Recent studies on the viscosity of ionic liquids (ILs) have been aimed at enhancing their tunability for numerous applications, ranging from lubrication, energy storage, to CO_2 capture and biotechnological processes. Improved computational models and experimental methodologies have both clarified IL viscosity as a function of temperature, pressure, and composition. The coupling of ILs with nanomaterials and deep eutectic solvents has further enhanced their feasibility in industrial applications. Future studies will probably concentrate on the development of task-specific ILs with viscosity optimized for specific applications, on environmentally friendly and biodegradable ILs, and on the enhancement of predictive models for their rheological behaviour. Creating more environmentally friendly hydrometallurgical processes means providing substitutes for the traditional solvents used in liquid-liquid extraction (LLE) of metals. In this case, a hydrophobic porous liquid based on silica is suggested as a replacement for the organic phase (PL). Various polyethoxylated fatty amines (EthAs) are added to two distinct sulfonated hollow silica particles (HSPs) to create a canopy that offers the desired liquefying and hydrophobic qualities. According to this study, these characteristics may be adjusted by changing the amount of ethylene oxide units in the EthA. While too short or too long EthA results in solid particles, middle-range molecular weight EthAs enable the formation of a liquid at normal temperature. Additionally, the size and density of the silica spheres affect viscosity: smaller, low-density spheres result in less viscous PLs, whereas bigger spheres (c.a. 200 nm) show less of an effect from density. The synthesis of hydrophobic PLs is accomplished with this method. PLs that are the most hydrophobic can undergo a subsequent phase separation when they come into touch with an aqueous phase. Additionally, preliminary extraction studies on three rare earth elements have demonstrated that in order to see metal extraction, the PL must be functionalized (Ginot et al., 2024). To alter Candida antarctica lipase B, a variety of betaine ionic liquids with varying chain lengths and anions were created (CALB). The findings demonstrated that all modified lipases had improved thermostability and tolerance to organic solvents, as well as improved catalytic activity under various pH and temperature conditions. The modifying impact increased as the length of the ionic liquid chain increased. Overall, CALB modified by [BetaineC16][H2PO4] performed the best; tolerance increased 2.9 times in 50% DMSO and 2.3 times in 30% mercaptoethanol, and the modified CALB enzyme activity increased three times. Thermal stability improved 1.5 times when held at 70°C for 30 minutes. The secondary structure of the CALB enzyme protein and the microenvironment around certain fluorescent groups changed when an ionic liquid was added, according to fluorescence and circular dichroism (CD) spectroscopy research. The structures of CALB changed with [BetaineC4][Cl] and [BetaineC16][Cl] were built, and molecular dynamics simulations were used to examine the stability change processes and enzyme activity of the modified CALB. The results demonstrated that modified CALB had a more stable structure than original CALB, with both the total energy and the root mean square deviation (RMSD) being lower. According to calculations using root mean square fluctuation (RMSF), the improved CALB's stiffness was increased. The modified enzyme-proteins were shown to have better hydrophilicity and hydrophobicity based on solvent accessibility area (SASA) estimations. It was proven by the rise in the radial distribution function (RDF) of water molecules that there were more water molecules surrounding the active sites. Consequently, changed CALB exhibits increased hydrolytic activity and improved structural stability (Xue et al., 2022). Known as green reagents that are excellent for the environment, ionic liquids (ILs) offer a wide range of potential applications. Their high viscosity, however, has prevented them from being widely used. Adding a solvent as an addition is currently an efficient approach. In order to lower the viscosities of the three ILs ([emim]BF4, [bmim]BF4, and [hmim]BF4), working pairs of ethanol with varying molar fractions were selected for this study. The thermal expansion coefficients were calculated by obtaining the densities and viscosities of the three working pairs within the temperature range of 293.15~338.15 K. There is discussion of how ethanol affects the three ILs' volumetric and viscometric properties. Excess mole volume and viscosity variations were added in order to examine the interactions between ethanol and the ILs in more detail. In the working pairs, the interactions between the identical molecules were weaker than those between the different molecules. This is the primary cause of the three ILs' decreased viscosity upon the addition of ethanol. Lastly, the viscosities of the working pairs were predicted using the hard-sphere model. The average absolute relative deviation of the forecast for the three ILs listed above dropped to 4.4 percent, 5.9 percent, and 5.0 percent, respectively, once an adjustable argument was introduced (Yang et al., 2024). Afreen et al. (2022) aimed to develop eight formulations of chlorpheniramine (CPM) niosomes using a 23-factorial design, characterize them through various evaluation tests, optimize them, formulate a gel of optimized dispersion, and conduct in vivo and histopathological studies on rabbits. The selected formulation, N3, showed the highest drug release at pH 6 and followed Korsmeyers-Peppas release kinetics with Fickian diffusion mechanism. After ANOVA and 3D surface plots construction, the gel was evaluated for viscosity, spreadability, extrudibility, drug content, entrapment, stability, and other tests. The niosomal gel successfully developed using non-ionic surfactants and effectively loaded the drug for targeted delivery (Afreen et al., 2022). The diverse variety of uses for flexible sensors in electrical goods is demonstrated. Nevertheless, there are still difficulties in creating sensors with a complicated internal structure and exceptional strength, flexibility, and elasticity. This study produced a prepolymer (PU-PDMS-OH-HEMA) that contains polyurethane (PU) and polysiloxane (PDMS) segments. The prepolymers were mixed with diluents and photoinitiators to create photosensitive resins with low viscosity for 3D printing. The cured photosensitive resins that were produced demonstrated enhanced mechanical qualities. With a resin viscosity of 62 cP and tensile strength and elongation at break of 10.03 MPa and 1046.15 percent, respectively, the PDMS-OH-30 sample has exceptional mechanical qualities. The aforementioned photosensitive polymers were combined with conductive ionic liquids to create flexible sensors that exhibit electrical stability and sensitivity. The work's conclusions provide a way to advance the creation of extremely flexible 3D printing resins for the creation of flexible sensors (Wang et al., 2024).

Figure 3. Experimental viscosity data for the combination of ethanol-[emim]BF4; x is the molar proportion of ethanol

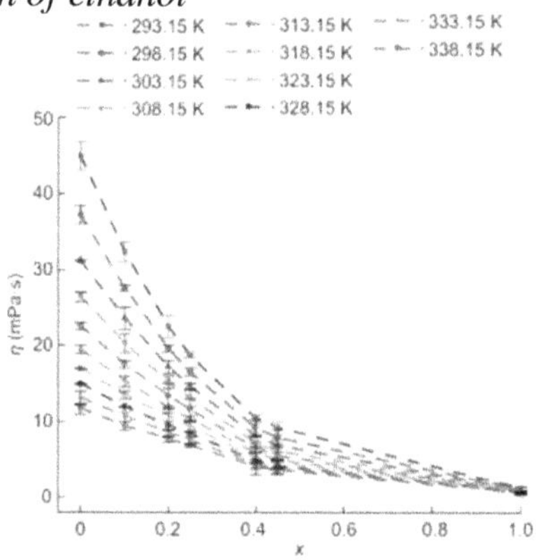

(Yang et al., 2024)

CONCLUSION

Ionic liquid (IL) viscosity study is essential owing to its applications across multiple disciplines. IL viscosity is determined by molecular structure, temperature, solvation, and external fields. Cation-anion interactions and functional group modifications are the primary factors determining viscosity. Both experimental and computational methods have given insights into viscosity behavior, with molecular dynamics simulation and machine learning models proving to be efficient prediction tools. ILs have established their importance in different fields, including lubrication, CO_2 capture, gas separation, electrochemical devices, and biomedical applications. Yet, there are issues with the understanding and control of IL viscosity, including the effects of impurities, long-term stability, and scale-up production problems. Future work must involve the development of ILs with tunable viscosity to address upcoming technologies such as nanofluidics, space exploration, and environmentally friendly processes.

REFERENCES

Afreen, U., Fahelelbom, K. M., Shah, S. N. H., Ashames, A., Almas, U., Khan, S. A., Yameen, M. A., Nisar, N., Asad, M. H. H. B., & Murtaza, G. (2022). Formulation and evaluation of niosomes-based chlorpheniramine gel for the treatment of mild to moderate skin allergy. *Journal of Experimental Nanoscience, 17*(1), 467–495. DOI: 10.1080/17458080.2022.2094915

Alkhatib, I. I. I., Bahamon, D., al Hajaj, A., & Vega, L. F. (2022). Molecular Thermodynamic Modeling of Hybrid Ionic Liquids for Biogas Upgrading. *Industrial & Engineering Chemistry Research, 61*(33), 12190–12207. DOI: 10.1021/acs.iecr.2c00710

Bakar, R., Darvishi, S., Aydemir, U., Yahsi, U., Tav, C., Menceloglu, Y. Z., & Senses, E. (2023). Decoding Polymer Architecture Effect on Ion Clustering, Chain Dynamics, and Ionic Conductivity in Polymer Electrolytes. *ACS Applied Energy Materials, 6*(7), 4053–4064. DOI: 10.1021/acsaem.3c00310 PMID: 37064412

Chen, M., Wu, T., Niu, L., Ye, T., Dai, W., Zeng, L., Kornyshev, A. A., Wang, Z., Liu, Z., & Feng, G. (2024). Organic Solvent Boosts Charge Storage and Charging Dynamics of Conductive MOF Supercapacitors. *Advanced Materials, 36*(30), 2403202. Advance online publication. DOI: 10.1002/adma.202403202 PMID: 38751336

Chiou, Y.-J., Fu, C., & Liao, Y.-H. (2023). A study on rose-window instability in a dielectric droplet exposed to corona discharge. *Journal of Mechanisms, 39*, 491–499. DOI: 10.1093/jom/ufad041

Dehe, S., & Hardt, S. (2021). Deformation modes of an oil-water interface under a local electric field: From Taylor cones to surface dimples. *Physical Review Fluids, 6*(12), 123702. Advance online publication. DOI: 10.1103/PhysRevFluids.6.123702

Flora, F. C., Relvas, S. B., Silva, F. A. E., Freire, M. G., Chu, V., & Conde, J. P. (2023). Combined Use of Ionic Liquid-Based Aqueous Biphasic Systems and Microfluidic Devices for the Detection of Prostate-Specific Antigen. *Biosensors (Basel), 13*(3), 334. Advance online publication. DOI: 10.3390/bios13030334 PMID: 36979546

Ginot, L., el Bakkouche, A., Giusti, F., Dourdain, S., & Pellet-Rostaing, S. (2024). Hydrophobic Porous Liquids with Controlled Cavity Size and Physico-Chemical Properties. *Advancement of Science, 11*(4), 2305906. Advance online publication. DOI: 10.1002/advs.202305906 PMID: 38036426

Gonçalves, H. M. R., Pereira, R. F. P., Lepleux, E., Pacheco, L., Valente, A. J. M., Duarte, A. J., & de Zea Bermudez, V. (2020). Non-Newtonian Thermosensitive Nanofluid Based on Carbon Dots Functionalized with Ionic Liquids. *Small*, *16*(28), 1907661. Advance online publication. DOI: 10.1002/smll.201907661 PMID: 32462808

Hua, J., Björling, M., Larsson, R., & Shi, Y. (2021). Controlling friction in Ionic Liquid/Glycerol Aqueous Solution lubricated contacts by adjusting CO<inf>2</inf> and water content. *Tribology International*, *161*, 107070. Advance online publication. DOI: 10.1016/j.triboint.2021.107070

Jongprasitkul, H., Turunen, S., Parihar, V. S., & Kellomäki, M. (2022). Two-step crosslinking to enhance the printability of methacrylated gellan gum biomaterial ink for extrusion-based 3D bioprinting. *Bioprinting (Amsterdam, Netherlands)*, *25*, e00185. Advance online publication. DOI: 10.1016/j.bprint.2021.e00185

Jóźwiak, B., Dziadosz, J., Golba, A., Cwynar, K., Dzido, G., Zorębski, E., Kolanowska, A., Jędrysiak, R., Gancarz, P., Scheller, Ł., Boncel, S., & Dzida, M. (2020). Thermophysical properties of ionanofluids composed of 1-ethyl-3-methylimidazolium thiocyanate and carboxyl-functionalized long multi-walled carbon nanotubes. *Fluids (Basel, Switzerland)*, *5*(4), 214. Advance online publication. DOI: 10.3390/fluids5040214

Kasar, A. K., Rahman, M. H., D'Souza, B., & Menezes, P. L. (2023). Tribological performance of ionic liquid impregnated porous aluminum borate ceramic. *Tribology International*, *180*, 108219. Advance online publication. DOI: 10.1016/j.triboint.2023.108219

Kazhiyev, S., Kaumbekova, S., & Shah, D. (2023). Malic acid-based deep eutectic solvent and its application in Insulin's structural stability. *Results in Engineering*, *20*, 101529. Advance online publication. DOI: 10.1016/j.rineng.2023.101529

Li, B., Wang, J., Gui, Q., & Yang, H. (2020). Drug-loaded chitosan film prepared via facile solution casting and air-drying of plain water-based chitosan solution for ocular drug delivery. *Bioactive Materials*, *5*(3), 577–583. DOI: 10.1016/j.bioactmat.2020.04.013 PMID: 32405573

Li, H., Wang, J., Warr, G. G., & Atkin, R. (2024). Nanostructure and Dynamics of the Locally Concentrated Ionic Liquid 2:1 (wt:wt) HMIM FAP:TFTFE and HMIM FAP on Graphite and Gold Electrodes as a Function of Potential. *Small*, *20*(46), 2403109. Advance online publication. DOI: 10.1002/smll.202403109 PMID: 39105361

Liao, X., Zheng, K., Wang, G., Yang, Y., Li, Y., & Coppens, M.-O. (2022). Solubility of CO<inf>2</inf>in Ionic Liquids with Additional Water and Methanol: Modeling with PC-SAFT Equation of State. *Industrial & Engineering Chemistry Research, 61*(38), 14364–14373. DOI: 10.1021/acs.iecr.2c02778

Liu, Q., Xu, J., Jiang, W., Gim, J., Tornheim, A. P., Pathak, R., Zhu, Q., Zuo, P., Yang, Z., Pupek, K. Z., Xu, K., & Zhang, Z. (2024). High-Energy LiNiO<inf>2</inf> Li Metal Batteries Enabled by Hybrid Electrolyte Consisting of Ionic Liquid and Weakly Solvating Fluorinated Ether. *Advancement of Science, 11*(46), 2409662. Advance online publication. DOI: 10.1002/advs.202409662 PMID: 39418091

Liu, Y., Chen, X., Chen, X., Chen, J., Zhang, H., Xu, H., Jin, L., Wang, Q., & Tang, Z. (2024). Preparation and in vivo and ex vivo studies of sirolimus nano-in-situ gel ophthalmic formulation. *Journal of Nanobiotechnology, 22*(1), 417. Advance online publication. DOI: 10.1186/s12951-024-02668-1 PMID: 39014353

Ma, C., Wang, N., Ye, N., & Ji, X. (2021). CO<inf>2</inf> capture using ionic liquid-based hybrid solvents from experiment to process evaluation. *Applied Energy, 304*, 117767. Advance online publication. DOI: 10.1016/j.apenergy.2021.117767

Majhi, S., Bhattacharyya, S., & Gopmandal, P. P. (2024). Impact of laterally mobile surface charge on diffusiophoresis of hydrophobic rigid colloids. *Journal of Fluid Mechanics, 997*, A8. Advance online publication. DOI: 10.1017/jfm.2024.778

Puiman, L., Elisiário, M. P., Crasborn, L. M. L., Wagenaar, L. E. C. H., Straathof, A. J. J., & Haringa, C. (2022). Gas mass transfer in syngas fermentation broths is enhanced by ethanol. *Biochemical Engineering Journal, 185*, 108505. Advance online publication. DOI: 10.1016/j.bej.2022.108505

Ran, L., Wang, Z., Yang, B., Amiri-Margavi, A., & Alshahrani, N. (2024). Development of novel computational models based on artificial intelligence technique to predict the viscosity of ionic liquids-water mixtures. *Case Studies in Thermal Engineering, 54*, 104076. Advance online publication. DOI: 10.1016/j.csite.2024.104076

Roy, S., Speed, L.Jr, Viola, M., Luo, H., Leonard, D., & Qu, J. (2021). Oil miscible phosphonium-phosphate ionic liquid as novel antiwear and antipitting additive for low-viscosity rear axle lubricants. *Wear, 466–467*, 203588. Advance online publication. DOI: 10.1016/j.wear.2020.203588

Rozynek, Z., Harkavyi, Y., & Giżyński, K. (2022). Fabrication of 1D particle structures outside a liquid environment using electric and capillary interactions: From fundamentals to applications. *Materials & Design, 223*, 111233. Advance online publication. DOI: 10.1016/j.matdes.2022.111233

Sather, N. A., Sai, H., Sasselli, I. R., Sato, K., Ji, W., Synatschke, C. V., Zambrotta, R. T., Edelbrock, J. F., Kohlmeyer, R. R., Hardin, J. O., Mirau, P., & Stupp, S. I. (2021). 3D Printing of Supramolecular Polymer Hydrogels with Hierarchical Structure. *Small*, *17*(5), 2005743. Advance online publication. DOI: 10.1002/smll.202005743 PMID: 33448102

Schirru, M., & Adler, M. (2020). An ultrasonic rheometer to measure gas absorption in ionic liquids: Design, calibration and testing. *Sensors (Basel)*, *20*(12), 1–20. DOI: 10.3390/s20123544 PMID: 32585862

Shishido, H., & Hashizume, H. (2024). Numerical analysis of the viscosity of ZrF<inf>4</inf>–BeF<inf>2</inf> molten salts for nuclear transmutation using fusion neutron sources. *Fusion Engineering and Design*, *203*, 114456. Advance online publication. DOI: 10.1016/j.fusengdes.2024.114456

Sierra, A., Coleman, M. G., & Iglesias, P. (2022). Tribological Properties of Borate-Based Protic Ionic Liquids as Neat Lubricants and Biolubricant Additives for Steel-Steel Contact. *Lubricants (Basel, Switzerland)*, *10*(10), 269. Advance online publication. DOI: 10.3390/lubricants10100269

Siyahjani, S., Oner, S., Diker, H., Gultekin, B., & Varlikli, C. (2020). Enhanced capacitive behaviour of graphene based electrochemical double layer capacitors by etheric substitution on ionic liquids. *Journal of Power Sources*, *467*, 228353. Advance online publication. DOI: 10.1016/j.jpowsour.2020.228353

Stewart, R. J., Sima, M., Karz, J., & Jones, J. P. (2023). Material characterization of GPX®: A versatile in situ solidifying embolic platform technology. *Frontiers in Bioengineering and Biotechnology*, *11*, 1095148. Advance online publication. DOI: 10.3389/fbioe.2023.1095148 PMID: 36726745

Takana, H., Hara, N., Makino, T., & Kanakubo, M. (2021). Effect of environmental temperature on CO<inf>2</inf> selective absorption characteristics by ionic liquid electrospray in flow system. *Journal of Electrostatics*, *114*, 103634. Advance online publication. DOI: 10.1016/j.elstat.2021.103634

Wang, D., Wang, R., Chen, S., Gao, J., Cai, C., Zheng, Y., Liu, X., Qu, B., Chen, N., & Zhuo, D. (2024). Low viscosity and highly flexible stereolithographic 3D printing resins for flexible sensors. *Materials & Design*, *243*, 113052. Advance online publication. DOI: 10.1016/j.matdes.2024.113052

Xue, Y., Zhang, X.-G., Lu, Z.-P., Xu, C., Xu, H.-J., & Hu, Y. (2022). Enhancing the Catalytic Performance of Candida antarctica Lipase B by Chemical Modification With Alkylated Betaine Ionic Liquids. *Frontiers in Bioengineering and Biotechnology, 10*, 850890. Advance online publication. DOI: 10.3389/fbioe.2022.850890 PMID: 35265607

Yang, F., & Feng, P. (2020). Densities and viscosities of ionic liquid with organic solvents. *Applied Sciences (Basel, Switzerland), 10*(23), 1–18. DOI: 10.3390/app10238342

Yang, F., Zheng, Q., Wang, B., Tan, H., & Wang, X. (2024). Improvement of ionic liquid properties for the absorption refrigeration cycle. *Energy Storage and Saving, 3*(4), 352–361. DOI: 10.1016/j.enss.2024.10.005

Yang, H., Luo, X.-F., Matsumoto, K., Chang, J.-K., & Hagiwara, R. (2020). Physicochemical and electrochemical properties of the (fluorosulfonyl)(trifluoromethylsulfonyl)amide ionic liquid for Na secondary batteries. *Journal of Power Sources, 470*, 228406. Advance online publication. DOI: 10.1016/j.jpowsour.2020.228406

Yu, T., Cai, Q., Lian, G., Bai, Y., Zhang, X., Zhang, X., Liu, L., & Zhang, S. (2021). Mechanisms behind high CO<inf>2</inf>/CH<inf>4</inf> selectivity using ZIF-8 metal organic frameworks with encapsulated ionic liquids: A computational study. *Chemical Engineering Journal, 419*, 129638. Advance online publication. DOI: 10.1016/j.cej.2021.129638

Zalewski, M., Krawczyk, T., Siewniak, A., & Sobolewski, A. (2021). Carbon dioxide capture using water-imidazolium ionic liquids-amines ternary systems. *International Journal of Greenhouse Gas Control, 105*, 103210. Advance online publication. DOI: 10.1016/j.ijggc.2020.103210

Zhang, B., Guo, Y., Lin, L., Qiao, D., & Xie, F. (2023). For efficient treatment of starch using aqueous ionic liquid at room temperature. *Sustainable Materials and Technologies, 36*, e00592. Advance online publication. DOI: 10.1016/j.susmat.2023.e00592

Zhang, R., Hallström, S., Mao, H., Kjellqvist, L., & Chen, Q.ZHANG. (2021). Accurate viscosity prediction for molten slags: A new model and database. *ISIJ International, 61*(5), 1379–1388. DOI: 10.2355/isijinternational.ISIJINT-2020-586

Chapter 2

Thermophysical Properties of Ionic Liquids

Aditya Shrivastav
https://orcid.org/0009-0000-8461-8171
Ajeenkya D.Y. Patil University, India

Sunil Sankathala
https://orcid.org/0009-0004-7809-7647
Ajeenkya D.Y. Patil University, India

Kashvi Chaturvedi
https://orcid.org/0009-0005-3192-3400
Ajeenkya D.Y. Patil University, India

Susanta Das
https://orcid.org/0000-0002-9314-3988
Ajeenkya D.Y. Patil University, India

ABSTRACT

Ionic liquids (ILs) represent a promising new material group showing distinct thermophysical features that lead to significant applications in industrial sectors. The chapter explores ILs through molecular structure examination & classifying mechanisms while assessing their thermophysical features regarding density, viscosity, thermal conductivity, and phase behaviour. IL properties transform in response to specific cation-anion pair selection yet maintains systematic management of material characteristics. It examines IL applications in energy storage & heat transfer operations & their usage in green chemistry processes & pharmaceutical development before showing their better performance than conventional

DOI: 10.4018/979-8-3373-1245-3.ch002

© 2026, IGI Global Scientific Publishing. Copying or distributing in print or electronic forms without written permission of IGI Global Scientific Publishing is prohibited. Use of this chapter to ...rative artificial intelligence (AI) technologies is expressly prohibited. The publisher reserves all rights to license its use for generative AI training and machine learning model development.

solvents. The study discusses how industrial uses and environmental examinations of ILs occur through an assessment of scaling issues and management limitations and toxicity challenges. The chapter concludes with a summary of recommended research approaches to resolve current limitations and improve the use of ILs in developing technological fields.

INTRODUCTION

Ionic liquids (ILs) form when salts maintain a liquid state at or near room temperature, thus having melting points below 100°C. ILs use bulky organic cations like imidazolium and pyridinium together with organic or inorganic anions to form a unique ion-based structure which gives them exceptional physicochemical traits absent in conventional molecular solvents (Shukla et al., 2023). The scientific study of ILs formally started in 1914 when researchers discovered ethyl ammonium nitrate. The advancement of air-stable 1-ethyl-3-methylimidazolium ([EMIM])-based ILs took place during the 1980s under the development work of John Wilkes. The development of room-temperature ILs with adjustable hydrophobicity and acidity properties occurred in 1992 which drove their extensive adoption for various applications. The manufacturing sector embraced ILs as industrial materials during the early 2000s after Chevron implemented its ISOALK™ process which used chloroaluminate IL catalysts to produce alkylate fuels (Greer et al., 2020).

ILs demonstrate distinctive properties during transformation which separate them from conventional solvents. The volatile character of these materials remains below 1 Pa as their vapour pressure makes them highly effective at lowering environmental release and breathing hazards (Shukla et al., 2023). The resistance of ILs to high temperatures can extend to 400°C which makes them suitable for industrial applications (Greer et al., 2020). The pairing ability of specific cations and anions in ILs enables precise control of polarity and acidity and solubility performance characteristics for various applications (Shukla et al., 2023). The main benefits of ILs surpass volatile organic solvents like dimethylformamide (DMF) and acetone because ILs possess reduced vaporisation while being fire-resistant and able to dissolve diverse substances and attain high recycling rates greater than 90% (Greer et al., 2020). The industrial research on ILs includes various applications across green chemistry fields together with pharmaceuticals energy storage and catalysis. A study from 2016 illustrated that Indole alkylation resulted in a 75% IL recovery rate yet yielded products at 62% when compared to traditional method yields of 85% (Greer et al., 2020). The pharmaceutical industry uses Dual Active ILs with medicinal components including lidocaine docusate to improve bioavailability and also enables 98% pure extraction of artemisinin from Artemisia annua (Shukla et al.,

2023). The application of IL-based electrolytes in lithium batteries maintains 95% capacity throughout 500 charge-discharge cycles at 100°C operation temperatures according to Greer et al. (2020). Supercapacitors developed by IoLiTec that utilise [EMIM][BF_4] achieve a capacitance level of 150 F/g (Greer et al., 2020). Industrial catalysis received benefits from ILs in Chevron's ISOALKY™ process which uses [HMIM][Cl-$AlCl_3$] to manufacture 190 kton/year of alkylate fuel while minimising catalyst waste through 60% reduction compared to hydrofluoric acid-based traditional methods (Greer et al., 2020).

The viscosity levels of ILs present between 50-500 cP cause issues with mass transfer throughout industrial processes due to their high viscosity characteristics. Water exhibits a lower viscosity value of 0.3 cP compared to all other liquids. Current pricing models limit widespread industrial adoption because one kilogramme of IL costs between $50 to $500. Modern research efforts involving zwitterionic ILs and deep eutectic solvents aims at solving affordability and biodegradability problems. ILs continue to gain commercial popularity because manufacturers produce more than 1,200 kton/year of processed fuel products from ILs, and they produce over 500 research-grade industrial IL solutions (Greer et al., 2020). Additional research on IL synthesis and design techniques along with government regulatory encouragement along with economic incentives will enhance their uptake in industrial-level applications.

MOLECULAR STRUCTURE AND CLASSIFICATION OF IONIC LIQUIDS

Organic cations together with inorganic or organic anions form ILs which stay in liquid state because of their asymmetric molecular structures prevent crystallisation. Their adjustable properties stem from the different cation-anion pairings which allow developers to create customised solutions for diverse applications according to Flieger et al. (2020). ILs have two components that constitute their molecular structure where imidazolium, pyridinium and phosphonium core cations align with tetrafluoroborate ([BF_4]$^-$) and bis (trifluoromethyl sulfonyl) imide ([TFSI]$^-$) anions (Flieger et al., 2020; Pietro et al., 2020). The classification system for ILs depends on the cation type as well as the anion's range and functionality and their physical status. The research field primarily focuses on imidazolium-based ILs because of their extensive synthetic possibilities and broad applicability (Flieger et al., 2020; Li et al., 2024). IL behaviour changes based on the anion size because Cl^- and Br^- increase viscosity but [TFSI]$^-$ enhances conductivity (Di Pietro et al., 2020; Li et al., 2024). ILs that stay in liquid state at cryogenic temperatures find specific uses within low-temperature operating processes (Li et al., 2024).

The distinctions between ILs depend on their cationic or anionic nature. Cationic ILs feature positively charged ions which attach to negatively charged surfaces but anionic ILs possess negatively charged ions that link up with positive charged surfaces. The different nature of cationic ILs makes them suitable for hair conditioner and disinfectant use and the anionic counterparts are typically designed for detergents and antistatic purposes (Sigma, 2023). Imidazolium-based cations ($[C_{n-}$ $mim]^+$ associated with $[TFSI]^-$ anions display higher fluidity properties compared to the $[BF_4]^-$ version of these ions because weaker intermolecular binding occurs (Di Pietro et al., 2020; Yuan et al., 2018).

The specialised group of ILs called Task-Specific Ionic Liquids (TSILs) incorporates functional groups consisting of thiourea and amines into cationic or anionic structures to generate application-specific properties. Many research applications have proven the usefulness of these ILs in multiple fields of study. The CO_2 capture performance of TSILs increases with thiourea functional groups yet phosphonium-based TSILs conduct ion-exchange mechanisms for neodymium extraction which achieves more than 90% efficiency from mixed ores (Sigma-Aldrich technical document; Do-Thanh et al., 2023). Dicyanamide-based ($[N(CN)_2]^-$) ILs achieve high conductivities above 10 mS/cm at 343K due to their low-viscosity chemical composition which makes them promise for energy storage applications according to Yuan et al. (2018).

The thermal characteristics of ILs strongly result from their molecular structure. The length of cation chains directly affects the viscosity of the substance. The increase in alkyl chain length for imidazolium-based ILs enhances viscosity to the point that $[C1_2mim]$ $[BF_4]$ achieves three times the viscosity compared to $[C_4mim]$ $[BF_4]$ (Li et al., 2024). The Vogel-Fulcher-Tammann model best describes temperature-dependent viscosity variations because this model shows viscosity declines exponentially with temperature increases (Yuan et al., 2018). Conductivity, on the other hand, exhibits an inverse relationship with viscosity. ILs demonstrate a 40% increase in ionic conductivity when researchers replace $[BF_4]^-$ with $[TFSI]^-$ due to diminished intermolecular bond forces per studies conducted by Di Pietro et al. (2020) and Yuan et al. (2018). The unique non-Arrhenius temperature-conductivity behaviour of ILs poses problems during the development of IL-electrolytes for battery applications according to Yuan et al. (2018). ILs exhibit their fundamental thermophysical characteristic in the form of thermal stability. The thermal decomposition points of imidazolium-based ILs containing fluorinated anions reaches 300°C and above thereby providing superior thermal resistance to these ILs (Flieger et al., 2020; Li et al., 2024). Hexafluorophosphate ($[PF_6]^-$) anions present hydrolysis properties that lead to dangerous decomposition end products thus limiting their industrial usage. To deploy ILs in business operations businesses must analyse their stability characteristics at length (Li et al., 2024).

Industrial applications of ILs span across various sectors. Basic ILs developed from Cetyltrimethylammonium bromide act as static charge dispersing agents for textile materials inside the textile industry (Sigma, 2023). Studies show that low-viscosity $[N(CN)_2]^-$ ILs present remarkable potential to boost supercapacitor fast-charging efficiency. TSILs in their chiral form provide pharmaceutical industries with a crucial tool to separate drug enantiomers with enantiomeric purity levels exceeding 99% for enhancing pharmaceutical research and driving drug quality development (Flieger et al., 2020). Thermodynamic system behaviour changes based on which anions researchers select according to modern scientific studies. The change from chloride to iodide anions in $[C_6mim]X$/IPA mixtures causes an opposite change in reaction heat through exothermic behaviour beyond mole fraction 0.4 indicating the critical function of IL components in battery technologies and solvent extraction systems (Li et al., 2024). ILs demonstrate exceptional property modifications that drive research innovation throughout catalytic systems and storage batteries and pharmaceutical applications. Redesign of advanced materials with specific IL pairings and enhanced functionalized IL versions leads to higher performance materials than conventional electrolyte systems through stability optimization along with decreased environmental effects. Studies of molecular structure and thermophysical property relations create vital guidelines for researchers to deploy ILs efficiently in building sustainable high-performance technologies (Li et al., 2024).

THERMOPHYSICAL PROPERTIES OF IONIC LIQUIDS

ILs form a molten salt group which stays liquid at temperatures under 100°C because of their special thermophysical properties that make them ideal for energy storage applications along with catalysis and green chemistry uses (Shukla et al., 2023). Their molecular structure determines these properties and provides control through varying the combinations between cations and anions. Density of ILs originates from the combined effect of the constituent ions molecular weight and van der Waals volume measurement. The densities of ILs usually fall between 1.1 - 1.6 g/cm^3 because of strong Coulombic interactions but they remain less dense than conventional molecular solvents. The molar volume (V_{mol}) measures the space a liquid molecule takes up during expansion and it obeys a temperature-dependent expansion model expressed as:

$$V_{mol} = V_{T_0} + C_1 T$$

The molar volume calculation for absolute zero provides values through V_{T_0} while C_1 serves as the temperature coefficient that determines thermal expansion effects. The relationship between higher temperatures and expanding molar volume reduces the material density according to this equation. ILs made from phosphonium-based cations demonstrate bulkier van der Waals volumes which causes them to possess lower densities than ILs using imidazolium-based cations (Yu & Chen, 2021). The density of ILs decreases with increasing alkyl chain length in the cation since extended nonpolar chains create more free volume which interferes with uniform molecular packing (Markiewicz et al., 2021).

ILs feature viscosity as their core element that shapes their flow properties and diffusion characteristics when applied in lubricating systems and separation methods and electrochemical systems. The viscosity of ILs depends on several elements including alkyl chain length alongside hydrogen bond interactions and anion size and operational temperature. The viscosity of ILs rises as alkyl chain length increases because of stronger van der Waals forces and increased steric hindrance while asymmetric cation types like pyrrolidinium decrease viscosity by disturbing the ordered arrangement of molecules (Silva et al., 2020). ILs behave differently to Arrhenius-dependent molecular liquids due to their non-Arrhenius temperature behaviour that produces excellent

VFT equation results:

$$\eta = \eta_0 exp\left(\frac{B}{T-T_0}\right)$$

where:

- η_0 is the pre-exponential factor,
- B is an empirical constant related to activation energy,
- T is the absolute temperature,
- T_0 is the Vogel temperature, representing the idealized glass transition temperature.

The mathematical expression demonstrates that viscosity of ILs decreases exponentially when temperature rises because this behaviour affects both low-temperature lubrication applications and electrolyte performance (Markiewicz et al., 2021; Aniya & Ikeda, 2019). Anion size plays an important role in viscosity through two opposing effects: TFSI- reduces viscosity because its large size enables charge delocalization and weakens ion-ion interactions while smaller PF6- produces higher viscosity because of stronger hydrogen bonding (Silva et al., 2020). ILs apply their thermal conductivity capacity to power thermal energy storage systems and heat transfer fluid operations. The thermal conductivities of ILs fall between 0.1 and 0.2 W/m·K and provide adequate performance for particular applications even though they exhibit lower values than conventional molecular solvents. Heat transfer in ILs relies on how ions move through the liquid in addition to their molecular weight and the way sound waves scatter energy. The thermal conductivity of imidazolium-based ILs reaches 10-15% higher levels compared to phosphonium-based ILs because their cations and anions remain smaller in size (Yu & Chen, 2021). Molecular dynamics simulations successfully predicted heat capacity values of ILs that matched within 5% of experimental results as reported by Silva et al. (2020). The heat capacity of ILs range from 1.3 to 1.7 J/g·K according to experimental DSC data and results depend on molecular structure alongside ion pairing strength (Yu & Chen, 2021). The phase behaviour of ILs, particularly their melting points and glass transition temperatures (T_g), plays a crucial role in their practical applications. The melting points of ILs span various values because of which anion symmetry interacts with cation size and hydrogen bonding strength together determine the final values. The use of fluorinated anions like $[TFSI]^-$ in imidazolium-based ILs results in reduced melting points compared to ILs with the hexafluorophosphate ($[PF_6]^-$) anion because fluorinated anions form weaker interactions with cations according to Silva et al. (2020).

The glass transition temperature stands as an important property when using ILs in cryogenic processes and solid-state electrolyte technologies. The T_g value of ILs increases when the IL incorporates bulky phosphonium cations because steric hindrance restricts crystallisation whereas imidazolium-based ILs demonstrate lower T_g values to improve their fluidity at cold temperatures (Yu & Chen, 2021). The key characteristic of ILs consists of their low volatility which establishes them as prime substitutes for organic solvents used in green chemistry practises and vacuum technology. Coulombic bond interactions keep ILs below 10^{-7} Pa vapour pressure levels at room temperature thus inhibiting their evaporation. The thermal stability of ILs under TGA analysis reaches 200-400°C which is much higher than the possible evaporation level of molecular solvents. The decomposition point of 1-butyl-3-methylimidazolium bis (trifluoromethyl sulfonyl) imide $[C_4mim]$ [TFSI] occurs at 385°C without detecting any evaporation before the decomposition process happens (Bablee et al., 2025).

APPLICATIONS OF IONIC LIQUIDS BASED ON THEIR THERMOPHYSICAL PROPERTIES

The widespread adoption of ILs results from their adjustable thermophysical properties that support various applications throughout multiple industrial sectors.

ILs show attractive properties including low volatility and high thermal stability combined with customizable cation-anion pair selection that enables their successful deployment in energy storage systems as well as heat transfer appliances and green chemical processes and advanced applications. Energy storage represents a major application area when it comes to IL utilisation because these materials act as electrolytes for lithium-ion batteries and supercapacitors. The electrolytic performance of ILs surpasses conventional electrolytes because of their combined characteristics including high ionic conductivities between 10^{-3}-10^{-2} S cm^{-1} and reaching electrochemical stabilisation ranges of up to 4.6 V Li^+/Li along with their non-flammable properties (Ray & Saruhan, 2021; Pan et al., 2020). Temperature-resistant performance up to 300 degrees Celsius is possible through balanced viscosity and conductivity from imidazolium and pyrrolidinium-based ILs (Ray & Saruhan, 2021). Supercapacitors benefit from 1-ethyl-3-methylimidazolium bis(trifluoromethanesulfonyl)imide ILs because these electrolytes expand the working voltage range from 3.6 - 4.5 V. Hybrid supercapacitors that integrate lithium salts and ILs create improved high-energy-density electrochemical systems because they enhance voltage ranges and thermal stability (Pan et al., 2020). The heat capacity of equimolar 1-butyl-3-methylimidazolium IL binary mixtures reaches 2.08 MJ m-3K-1 values that exceed molten salt capacities (Urzua et al., 2023). The pressure-resistant salt among the group of pyrrolidinium-based ILs maintains structural integrity until 300°C and exhibits low viscosity variation for potential use in concentrated solar power operations. Nanoparticles added to ILs create ionanofluids which produces enhanced thermal conductivity that reaches 15% levels to enhance industrial heat transfer operations (Musiał et al., 2017).

Beyond energy and thermal applications, ILs serve as environmentally friendly solvents and catalysts in chemical synthesis. Their adjustable polarity along with low vapour pressure proves essential for carrying out organic reactions without requiring toxic solvents. Imidazolium-based ILs play a double role by stabilising reaction intermediates which enables higher product yields and enables easy separation. One important use of ILs occurs in 1,2,3-triazole synthesis because these substances both enhance reaction performance and create environmentally friendly production methods (Kumar et al., 2024). The capture of CO_2 shows great promise when using ILs due to their excellent structural control properties combined with low volatility and strong adsorption tendencies toward CO_2 molecules (Numpilai et al., 2024; Shukla et al., 2019). The combination of imidazolium and pyridinium-based ILs with carboxylate or azolate anions uses Lewis's acid-base interactions to improve CO_2 solubility (Shukla et al., 2019; Alguacil & Robla, 2024). The utilisation of 1-butylpyridinium tetrafluoroborate ILs leads to an energy saving of 15% for regeneration while decreasing the overall thermal energy needs by 66% when compared to monoethanolamine (MEA) as a conventional solvent (Shukla et al., 2019; Alguacil & Robla, 2024). Research shows the combination of ILs with membrane or adsorbent materials like silica-supported ILs enhances CO_2 capture processes while improving effectiveness and selectivity making them suitable for industrial applications (Numpilai et al., 2024 and Alguacil & Robla, 2024).

ILs are also an essential part of the lubrication and tribology, where their low level of volatility, their high rate of stability, and their efficiency in boundary lubrication surpass that of the traditional lubricants. Compressor and turbine lubricants, space lubricants, and industrial hydraulic fluids are also used in these ILs (Song, 2021). Based on some of the holdings of phosphonium, ILs are relatively the most environmentally safer when keeping the utilization of zinc dialkyl Di thiophosphate (ZDDP) as the measure to be the same and reducing the toxicity of the environment while maintaining anti-wear properties (Song, 2021). The companies in the pharmaceutical sector have witnessed remarkable potential of ILs in drug solubility improvement as well as bioavailability improvement (Jain et al., 2024; Marrucho et al., 2014). Imidazolium ILs create channels in skin which, according to Jain et al. (2024), increase the treatment of the skin without penetration through the skin that can lead to the impurities of the ion that are not soluble. However, micelles, which are a kind of nano-sized drug particle, have been shown to exhibit lung cancer cells a better result with less side effects (Jain et al., 2024). More interestingly, choline geranate in IL, according to Jain et al. (2024), is capable of dissolving active pharmaceutical ingredients that are hydrophobic thus they are more available to dissolve possibly up to 90%. While still not being able to maintain the therapeutic index of targeted therapy, micelles are better able to minimize cytotoxicity by reducing the exposure of the drug to other tissues (Jain et al., 2024). On the other hand, ILs, like choline geranate, promote the solubility of hydrophobic active pharmaceutical ingredients and the bioavailability in the same proportion as sustainable pharmaceutical development and green chemistry principles. Moreover, amino acid-based ILs are capable of decomposing and are safe to ingest thus aligning with sustainable development and green chemistry (Jain et al., 2024; Marrucho et al., 2014).

The effectiveness of ILs in diverse areas is a direct outcome of their unique thermophysical properties. The less than 350°C decomposition temperatures mean ILs are both thermally stable and of the high thermal resistance and also non-volatile at the ambient atmosphere. As a result, they remain useful for high-temperature applications. Their viscosity and ionic conductivity can be precisely controlled through cation-anion pairing, ensuring the best performance in specific applications (Numpilai et al., 2024). Besides these aspects, their phase behaviour also means they can stay in their liquid phase over an enormous temperature range, a fact that makes them quite adaptable for all kinds of industrial processes (Song, 2021; Marrucho et al., 2014). The Excellent built-in of IL chemical responsiveness, and rapid thermophysical characterizations that accompany them make them revolutionary materials in the energy sector, sustainable engineering, green chemistry, and advanced healthcare. The essence of future research lies in implementing solutions to the deficit of existing IL formulations, to which technological advancements are oriented in the scene of industrial sustainability (Numpilai et al., 2024; Jain et al., 2024).

INDUSTRIAL AND ENVIRONMENTAL IMPACT

ILs are increasingly receiving attention for their industrial use and environmental benefits but need thorough evaluation of their scalability, environmental safety, and regulatory issues. Their broad usage across pharmaceuticals, petrochemicals, and energy storage is because of their thermal stability, low volatility, and adjustable properties (Mota-Martinez et al., 2018). Room-temperature ionic liquids (RTILs) are most commonly used in catalysis and green chemistry, whereas high-temperature derivatives are investigated for energy storage and high-pressure/temperature processes (Mota-Martinez et al., 2018). Still, large-scale application continues to be constrained by high cost of production from intricate synthesis and specialist equipment. Even successful commercial applications, including BASF's BASIL™ process, have not been industrially implemented on a large scale (Mota-Martinez et al., 2018). The extrapolation of laboratory studies to industrial usage is further hindered by technical impediments and lack of scalability (Oskarsson & Wright, 2019). To overcome these challenges, continued advancements in synthesis, such as economically viable precursors and modular design strategies, are being considered to improve the commercial feasibility of ILs. Synergy between industry and academia is crucial in addressing such economic challenges and broadening ILs' application in new fields like carbon capture, renewable energy, and new materials (Wei et al., 2021). Although the industrial applications are promising, ILs environmental safety and biodegradability are still critical issues. Certain ILs have been reported to be ecotoxic, specifically those with certain alkyl chain lengths and cation–anion pairs that are toxic to aquatic and terrestrial life (Flieger & Flieger, 2020). Some short-chain imidazolium-based ILs, for example, resist biodegradation and may accumulate in the environment, which is of concern to long-term environmental pollution (Oskarsson & Wright, 2019). Although their low volatility minimizes air pollution, their high solubility in water and chemical stability lead to long-term soil and water contamination (Wei et al., 2021; Oskarsson & Wright, 2019). Minimizing such hazards calls for the synthesis of biodegradable ILs based on naturally occurring molecules like amino acids, sugars, or choline, which are more environmentally friendly. Research suggests that ILs with longer alkyl chains (C6-C8) break down more effectively than shorter-chain analogues, providing a means to limit environmental persistence (Flieger & Flieger, 2020). Moreover, computational methods like quantitative structure-activity relationships (QSAR) are also being utilized to create less toxic ILs through the improvement of their hydrophobicity and functional groups, complementing initiatives to decrease their ecological footprint (Flieger & Flieger, 2020; Wei et al., 2021).

Regulatory mechanisms for ILs continue to advance, with large regional differences. Chemical safety laws and waste management regulations traditionally have difficulties categorizing ILs as a result of their new structures and physicochemical signatures (Wei et al., 2021). Current environmental laws are not entirely effective in addressing IL-specific hazards, especially regarding persistence and bioaccumulation, making it a requirement for policy to focus on these specific areas (Oskarsson & Wright, 2019). Regulatory directions in the future can be the implementation of more stringent biodegradability requirements to drive the creation of more environmentally friendly ILs, as is being done by the European Union regarding persistent pollutants (Flieger & Flieger, 2020; Oskarsson & Wright, 2019). Furthermore, greater collaboration between regulatory authorities, industry, and academic research is needed to ensure IL innovation is aligned with sustainability goals (Wei et al., 2021). The incorporation of obligatory lifecycle assessments (LCAs) would also permit further assessment of the complete environmental performance of ILs throughout production, use, and disposal, guaranteeing that their industrial deployment is sustainable from a long-term ecological perspective (Wei et al., 2021; Oskarsson & Wright, 2019). As ongoing research continues, a well-balanced strategy linking economic viability, environmental sustainability, and regulatory flexibility (or adaptability) will be instrumental in determining the eventual role of ILs in industry and environmental technologies.

CHALLENGES AND FUTURE PROSPECTS

Industry engagement with ILs has great promise yet several barriers stand in its way because of ILs unique thermophysical properties before their complete potential can be achieved. ILs have high production expenses because researchers face difficulties when seeking appropriate cation-anion pairs from a wide array of possible options (Mota-Martinez et al., 2018; Ovejero-Perez et al., 2024). The multiple problems of unstable thermal behaviour and high viscosity levels prevent many ILs from applying effectively in electrochemical processes and catalysis applications (Mota-Martinez et al., 2018; Khan et al., 2022). The production of pure ionic liquids becomes vital for industrial stability because recycling-related obstacles present economic and sustainability challenges (Khan et al., 2022; Ovejero-Perez et al., 2024). The water hazards of certain IL formulations necessitate environmental research to track down sustainable substitutes since these serve as effective replacements for volatile organic solvents (Ovejero-Perez et al., 2024; Flieger & Flieger, 2020). Researchers focus on developing multiple strategies to manage IL challenges that large-scale operations face. Scientists employ systematic procedures to design proper ionic liquids for specific applications that enhance operational outcomes and yield financial benefits (Mota-Martinez et al., 2018). The field of IL production technology receives active research investigation for developing efficient and scalable manufacturing strategies (Zullo et al., 2021). Research on processes to improve IL recycling continues because the sustainability of these materials depends on effective recycling systems as reported by Khan et al. (2022). The combination of naturally derived anions with cations coming from amino acids or fatty acids enables the development of environmentally friendly and sustainable ILs which produce positive effects on reducing chemical toxicity while generating less environmental impact (Flieger & Flieger, 2020).

The progression of next-generation materials with improved functionality stems from modern trends in IL research which overcomes present material challenges. The research on ILs progresses towards advanced materials applications because their dual features enable customised attributes for energy storage devices alongside biomedical applications (Figueiredo et al., 2024). ILs are used during biomass processing to break down lignocellulosic biomass which provides sustainable solutions for biofuel manufacturing and other bioprocessing operations according to Ovejero-Perez et al. (2024). The electrochemical properties of ILs remain critically important for manufacturers because these solvent substances enhance performance and safety in lithium batteries and supercapacitors and energy storage devices (Greer et al., 2020). The creation of chiral and sugar-based ILs expands their applications in drug formulation and therapeutic solutions as well as chiral synthesis and biomedicine (Khan et al., 2022; Zullo et al., 2021). Despite their difficulty to use researchers work on designing ILs and synthesis methods that will enable broader industrial use. The development of subsequent-generation ILs depends on research efforts that lower costs and improve scalability alongside environmentally friendly optimization and take advantage of ILs distinct thermophysical traits. Sustainable high-performance technologies in future times will need interdisciplinary collaboration with innovative approaches to overcome current IL barriers.

CONCLUSION

ILs research demonstrates that these materials possess extraordinary transformation capabilities within multiple industrial and technological sectors when studying their thermophysical properties. ILs emerge as flexible alternatives to conventional solvents and electrolytes because they possess low volatility, high thermal stability and adjustable physicochemical properties. Values of ionic liquids cover diverse practical applications which include energy storage systems and sustainable chemistry as well as pharmaceuticals and complex material development. The large-scale implementation of ILs depends on the resolution of various technical barriers that scientists must address through continuous research. The high synthesis and purification expenses as well as environmental and toxicity issues create a need to establish green and economical production methods. The understanding of how molecular structural characteristics affect thermophysical behaviours needs deeper research to enhance precise IL utilisation across different uses.

The next phase of research must concentrate on designing ILs rationally to achieve desired properties along with developing green ILs with better biodegradability methods while improving synthesis and scaling-up techniques. Computational modelling should unite with experimental methods to foresee and control IL properties for applications-driven optimization. The evolution of IL regulations demands academic industry and policymaker cooperation to ensure proper development and implementation of responsible IL practises. Investment in this field remains essential because ILs have proven capabilities to advance sustainable technologies along with green chemistry applications. The remarkable thermophysical features of ionic liquids present new possibilities to transform industrial operations across various industries despite ongoing research obstacles. Researchers believe IL technology will experience advanced development which will enable critical changes in the areas of energy sector and environment protection as well as materials science. The development outlook for ionic liquids shows promise since they promise sustainable technological advancement alongside solutions to worldwide problems in forthcoming years.

REFERENCES

Alguacil, F. J., & Robla, J. I. (2024). Using ionic liquids to improve CO2 capture. *Molecules (Basel, Switzerland)*, *29*(22), 5388. DOI: 10.3390/molecules29225388 PMID: 39598774

Aniya, M., & Ikeda, M. (2019). A model for Non-Arrhenius ionic conductivity. *Nanomaterials (Basel, Switzerland)*, *9*(6), 911. DOI: 10.3390/nano9060911 PMID: 31238516

Bablee, A., Amarasekara, A., Gabitto, J., Bhuiyan, A., & Shamim, N. (2025). A Comprehensive Review of the Thermophysical Properties of Energetic Ionic Liquids. *Energies*, *18*(2), 267. DOI: 10.3390/en18020267

Di Pietro, M. E., Castiglione, F., & Mele, A. (2020). Anions as dynamic probes for ionic liquid mixtures. *The Journal of Physical Chemistry B*, *124*(14), 2879–2891. DOI: 10.1021/acs.jpcb.0c00026 PMID: 32186377

Do-Thanh, C., Luo, H., & Dai, S. (2023). A perspective on task-specific ionic liquids for the separation of rare earth elements. *RSC Sustainability*, *1*(5), 1168–1176. DOI: 10.1039/D3SU00007A

Figueiredo, N. M., Voroshylova, I. V., Ferreira, E. S. C., Marques, J. M. C., & Cordeiro, M. N. D. S. (2024). Magnetic Ionic Liquids: Current Achievements and Future Perspectives with a Focus on Computational Approaches. *Chemical Reviews*, *124*(6), 3392–3415. DOI: 10.1021/acs.chemrev.3c00678 PMID: 38466339

Flieger, J., Feder-Kubis, J., & Tatarczak-Michalewska, M. (2020). Chiral ionic liquids: Structural diversity, properties and applications in selected separation techniques. *International Journal of Molecular Sciences*, *21*(12), 4253. DOI: 10.3390/ijms21124253 PMID: 32549300

Flieger, J., & Flieger, M. (2020). Ionic Liquids Toxicity—Benefits and Threats. *International Journal of Molecular Sciences*, *21*(17), 6267. DOI: 10.3390/ijms21176267 PMID: 32872533

Greer, A. J., Jacquemin, J., & Hardacre, C. (2020). Industrial Applications of Ionic Liquids. *Molecules (Basel, Switzerland)*, *25*(21), 5207. DOI: 10.3390/molecules25215207 PMID: 33182328

Jain, A., Shakya, A. K., Prajapati, S. K., Eldesoqui, M., Mody, N., Jain, S. K., Naik, R. R., & Patil, U. K. (2024). An insight into pharmaceutical challenges with ionic liquids: Where do we stand in transdermal delivery? *Frontiers in Bioengineering and Biotechnology*, *12*, 1454247. Advance online publication. DOI: 10.3389/fbioe.2024.1454247 PMID: 39165403

Khan, R. A., Mohammed, H. A., Sulaiman, G. M., Subaiyel, A. A., Karuppaiah, A., Rahman, H., Makhathini, S., Ramburrun, P., & Choonara, Y. E. (2022). Molecule(s) of Interest: I. Ionic Liquids–Gateway to Newer Nanotechnology Applications: Advanced Nanobiotechnical Uses', Current Status, Emerging Trends, Challenges, and Prospects. *International Journal of Molecular Sciences*, *23*(22), 14346. DOI: 10.3390/ijms232214346 PMID: 36430823

Kumar, A., Kumar, V., Singh, P., Tittal, R. K., & Lal, K. (2024). Ionic liquids for the green synthesis of 1,2,3-triazoles: A systematic review. *Green Chemistry*, *26*(7), 3565–3594. DOI: 10.1039/D3GC04898E

Li, F., Zuo, Z., Cao, B., & Ji, X. (2024). Study on the Thermodynamic Properties of Ionic Liquids 1-Hexyl-3-methylimidazolium Halide with Isopropanol Mixtures. *Journal of Chemical & Engineering Data*, *69*(11), 3794–3804. DOI: 10.1021/acs.jced.4c00343

Markiewicz, R., Klimaszyk, A., Jarek, M., Taube, M., Florczak, P., Kempka, M., Fojud, Z., & Jurga, S. (2021). Influence of alkyl chain length on thermal properties, structure, and Self-Diffusion coefficients of Alkyltriethylammonium-Based ionic liquids. *International Journal of Molecular Sciences*, *22*(11), 5935. DOI: 10.3390/ijms22115935 PMID: 34073046

Marrucho, I., Branco, L., & Rebelo, L. (2014). Ionic liquids in pharmaceutical applications. *Annual Review of Chemical and Biomolecular Engineering*, *5*(1), 527–546. DOI: 10.1146/annurev-chembioeng-060713-040024 PMID: 24910920

Mota-Martinez, M. T., Brandl, P., Hallett, J. P., & Mac Dowell, N. (2018). Challenges and opportunities for the utilisation of ionic liquids as solvents for CO2capture. *Molecular Systems Design & Engineering*, *3*(3), 560–571. DOI: 10.1039/C8ME00009C

Musiał, M., Malarz, K., Mrozek-Wilczkiewicz, A., Musiol, R., Zorębski, E., & Dzida, M. (2017). Pyrrolidinium-Based ionic liquids as sustainable media in Heat-Transfer processes. *ACS Sustainable Chemistry & Engineering*, *5*(11), 11024–11033. DOI: 10.1021/acssuschemeng.7b02918

Numpilai, T., Pham, L. K. H., & Wiston, T. (2024). Advances in Ionic Liquid Technologies for CO_2 Capture and Conversion: A Comprehensive review. *Industrial & Engineering Chemistry Research*, *63*(46), 19865–19915. DOI: 10.1021/acs.iecr.4c02072

Oskarsson, A., & Wright, M. C. (2019). Ionic Liquids: New Emerging Pollutants, Similarities with Perfluorinated Alkyl Substances (PFASs). *Environmental Science & Technology, 53*(18), 10539–10541. DOI: 10.1021/acs.est.9b04778 PMID: 31442027

Ovejero-Perez, A., Nakasu, P. Y. S., Hopson, C., Costa, J. M., & Hallett, J. P. (2024). Challenges and opportunities on the utilisation of ionic liquid for biomass pretreatment and valorisation. *Npj Materials Sustainability, 2*(1), 7. Advance online publication. DOI: 10.1038/s44296-024-00015-x

Pan, S., Yao, M., Zhang, J., Li, B., Xing, C., Song, X., Su, P., & Zhang, H. (2020). Recognition of ionic liquids as High-Voltage electrolytes for supercapacitors. *Frontiers in Chemistry, 8*, 261. Advance online publication. DOI: 10.3389/fchem.2020.00261 PMID: 32432074

Ray, A., & Saruhan, B. (2021). Application of ionic liquids for batteries and supercapacitors. *Materials (Basel), 14*(11), 2942. DOI: 10.3390/ma14112942 PMID: 34072536

Shukla, M. K., Tiwari, H., Verma, R., Dong, W., Azizov, S., Kumar, B., Pandey, S., & Kumar, D. (2023). Role and recent advancements of ionic liquids in drug delivery systems. *Pharmaceutics, 15*(2), 702. DOI: 10.3390/pharmaceutics15020702 PMID: 36840024

Shukla, S. K., Khokarale, S. G., Bui, T. Q., & Mikkola, J. T. (2019). Ionic liquids: Potential materials for carbon dioxide capture and utilization. *Frontiers in Materials, 6*, 42. Advance online publication. DOI: 10.3389/fmats.2019.00042

Sigma, V. USA. (2023). *Differences between cationic and anionic - 3V Sigma USA*. 3V Sigma USA. https://3vsigmausa.com/differences-between-cationic-and-anionic/

Silva, W., Zanatta, M., Ferreira, A. S., Corvo, M. C., & Cabrita, E. J. (2020). Revisiting Ionic Liquid Structure-Property Relationship: A Critical analysis. *International Journal of Molecular Sciences, 21*(20), 7745. DOI: 10.3390/ijms21207745 PMID: 33086771

Song, J. (2021). Research Progress of Ionic Liquids as Lubricants. *ACS Omega, 6*(44), 29345–29349. DOI: 10.1021/acsomega.1c04512 PMID: 34778607

Urzúa, J. I., Valenzuela, M. L., Cavieres, J., & Inestrosa-Izurieta, M. J. (2023). Ionic liquid mixtures as energy storage materials: A preliminary and comparative study based on thermal storage density. *RSC Advances, 13*(28), 19412–19419. DOI: 10.1039/D3RA02901H PMID: 37383686

Wei, P., Pan, X., Chen, C., Li, H., Yan, X., Li, C., Chu, Y., & Yan, B. (2021). Emerging impacts of ionic liquids on eco-environmental safety and human health. *Chemical Society Reviews, 50*(24), 13609–13627. DOI: 10.1039/D1CS00946J PMID: 34812453

Yu, Y., & Chen, Y. (2021). Density prediction of ionic liquids at different temperatures using the average free volume model. *ACS Omega, 6*(23), 14869–14874. DOI: 10.1021/acsomega.1c00547 PMID: 34151068

Yuan, W., Yang, X., He, L., Xue, Y., Qin, S., & Tao, G. (2018). Viscosity, conductivity, and electrochemical property of dicyanamide ionic liquids. *Frontiers in Chemistry, 6*, 59. Advance online publication. DOI: 10.3389/fchem.2018.00059 PMID: 29600245

Zullo, V., Iuliano, A., & Guazzelli, L. (2021). Sugar-Based Ionic Liquids: Multifaceted Challenges and Intriguing Potential. *Molecules (Basel, Switzerland), 26*(7), 2052. DOI: 10.3390/molecules26072052 PMID: 33916695

ADDITIONAL READING

Angell, C. A., Ansari, Y., & Zhao, Z. (2011). Ionic Liquids: Past, present and future. *Faraday Discussions, 154*, 9–27. DOI: 10.1039/C1FD00112D PMID: 22455011

Fayer Lab - Ionic Liquids Overview. https://web.stanford.edu/group/fayer/research_il.html

Sharma, P., Sharma, S., & Kumar, H. (2023). Introduction to ionic liquids, applications and micellization behaviour in presence of different additives. *Journal of Molecular Liquids, 393*, 123447. DOI: 10.1016/j.molliq.2023.123447

Silva, W., Zanatta, M., Ferreira, A. S., Corvo, M. C., & Cabrita, E. J. (2020b). Revisiting Ionic Liquid Structure-Property Relationship: A Critical analysis. *International Journal of Molecular Sciences, 21*(20), 7745. DOI: 10.3390/ijms21207745 PMID: 33086771

KEY TERMS AND DEFINITIONS

Catalysis: A chemical process in which a substance (catalyst) accelerates a reaction without being consumed; ILs are widely used as catalysts or catalyst supports in industrial applications due to their tunable properties.

Deep Eutectic Solvents (DESs): A subclass of ILs formed by hydrogen bond interactions between a hydrogen bond donor and acceptor, offering cost-effective and environmentally friendly alternatives to traditional ILs.

Ionic Liquids (ILs): Salts that remain in a liquid state below 100°C due to their large, asymmetric ions, which prevent crystallization, offering unique physicochemical properties like low volatility, high thermal stability, and tunable solubility.

Solvation Power: The capacity of a solvent to dissolve various solutes; ILs can dissolve a wide range of organic, inorganic, and polymeric compounds, often outperforming conventional solvents.

Thermal Stability: The ability of a substance to maintain its chemical structure at elevated temperatures; ILs exhibit high thermal stability, making them suitable for high-temperature industrial processes.

Volatility: A measure of how easily a substance evaporates; ILs are characterized by ultra-low volatility, reducing environmental emissions and inhalation hazards compared to conventional organic solvents.

Chapter 3
Synthesis, Characterizations, and Applications of Ionanofluids

Amruta Prasann Walvekar
https://orcid.org/0000-0001-7262-8380
School of Pharmacy, Indira University, Pune, India

Rutuja Vijay Kamble
https://orcid.org/0000-0002-7919-7197
School of Pharmacy, Indira University, Pune, India

Tejal Nikhil Bele
https://orcid.org/0000-0002-4170-6232
School of Pharmacy, Indira University, Pune, India

Pallavi Singh Chouhan
https://orcid.org/0000-0001-8755-3422
School of Pharmacy, Indira University, Pune, India

Shraddha Devarshi
https://orcid.org/0000-0002-6725-8917
School of Pharmacy, Indira University, Pune, India

ABSTRACT

Ionic liquids (ILs) are a class of versatile, non-volatile, and thermally stable substances that have garnered significant attention for their potential applications in green chemistry, catalysis, and material science. This chapter explores the synthesis,

DOI: 10.4018/979-8-3373-1245-3.ch003

2026, IGI Global Scientific Publishing. Copying or distributing in print or electronic forms without written permission of IGI Global Scientific Publishing is prohibited. Use of this chapter to ative artificial intelligence (AI) technologies is expressly prohibited. The publisher reserves all rights to license its use for generative AI training and machine learning model development.

properties, and diverse applications of ILs, with a focus on their role as sustainable solvents. Their unique ionic composition enables adjustable properties such as viscosity, hydrophobicity, and thermal stability, making them ideal for advanced industrial processes, including energy storage, environmental remediation, and catalysis. Emerging research into task-specific ILs and ionanofluids, combining ILs with nanomaterials, reveals their transformative potential in enhancing thermal conductivity, reaction selectivity, and environmental compatibility. Challenges like toxicity and production costs are also addressed, highlighting future directions in designing safer and more efficient ILs to meet industrial and environmental demands.

INTRODUCTION

The increased use of ionic liquids in various field of chemistry and engineering sciences is closely connected with their characteristics and possibilities of their utilization. These fluids are non-volatile and thermally stable due to ionic nature of these fluids. Their special characteristics enable a wide range of uses. One of the latest developments in this field is the creation of ionanofluids. These materials combine ionic liquids with nano-materials, forming a dispersion of nanoparticles. This combination results into improved performance in several areas of usage. Containing ions, these nano-fluids are particularly useful in heat transfer applications, and therefore in cooling. They also have uses in catalysis, the speeding of chemical reactions by reaction facilitators.

Also, ionanofluids can be employed in solar absorbing surfaces where convection of heat is effective in capturing the sun's energy. They also act as superior emulsifiers and absorb excess moisture that causes wear in the machinery. Other interesting field is in the luminous materials where they participate in operation of lights and displays.

Research on ionanofluids is growing, with many recent studies published in physical and chemical journals. These papers explore the effectiveness and potential of these new materials. With increased discovery, ionic liquids together with their nano-material composites may find potential application across different sectors.

Ionic liquids have been defined as salts with melting point of not more than 100 °C. In some form it also assists in the classification of such substances based on their physical characteristics. However, the term ionic liquid carries different connotations for different persons. Some may also confine it to include other parts or blends or any other preparation. Others may find interest in one of more properties of the substance like ionic conductivity, or solubility, etc. This variation in understanding can create confusion when speaking of ionic liquids in scientific or industrial setting. As such, the freedom definition entails that the key feature should be low melting point and ionic properties. These characteristics set ionic liquids apart from solvents and salts and thus make it possible for them to use in multiple purposes. (Plechkova et al., 2009; Karadas, Atilhan, & Aparicio, 2010; Welton, 1999; Freemantle, Welton, & Rogers, 2009).

Asymmetrical ions maintained in their respective positions by forces that are exerted on them. Ionic liquids or ILs have received much attention due to the presence of numerous advantages. Such properties as low melting point by which the substance remains in liquid state when the environment temperature is low and low vapor pressure by which they do not easily evaporate. They also mentioned that such ILs are easier to work with and are more stable in different circumstances than standard ILs. They also other desirable attributes such as high thermal stability which means the material can undergo high heat without causing it to degrade. Another benefit being they are fire resistant; this means that they are able to burn and cause the fire hazards easily.

The use and application of ionic liquids cuts across many scientific disciplines. Since their discovery up to 1914, the substances have been in use in physical chemistry, biological analysis, inorganic chemistry and organic chemistry. Due to their flexibility they are now widely used in many filed of research and industry. The current research emphasis the soluble ability of ILs to dissolve and selectively recover several components. This involves Organic compounds that play many roles in chemical reactions, Biomolecules such as proteins and nucleic acids. The separation of metal ions is also possible using ionic liquids whether they are transition metals or inner-transition metals.

Science has also searched for ways on how ILs may come into contact with supramolecular assemblies such as cell organelles, which are extremely significant parts of living cells. This versatility of material application makes ionic liquids significant in the contemporary field of chemistry and material science. Due to their distinct characteristics, they are capable of executing jobs that normal solvents cannot do thus causing enhanced and new approaches to problems in research and industry. Further research on ionic liquids, focused on them is likely to expand more as a useful tool.

The term ionic liquids, or low-temperature salts, or molten salts is usually attributed to a rather broad class of quite distinct materials. They can be easily altered as far as their construction is concerned in order to get various uses out of it. These usually have low vapour pressures which would indicate that its volume cannot easily evaporate into environment. This feature is particularly useful in as many industrial processes as is imaginable. Also, ionic liquids are characterized by high thermal stability; they do not decompose at high temperatures. They also have relatively heard liquid state ranges meaning that they only freeze at high temperatures. These are the characteristics that have placed the ionic liquids in high demand for many uses. (Wilkes 2002, Wasserscheid & Keim, 2000).They are gradually implemented in such sectors as storage of energy which benefits from stability and low volatility. The representatives of ionic liquids are widely used in green chemistry as solvents that do not present a threat to the environment, thus minimizing the amount of waste released. They ability to implement themselves as catalytic agents is also discussed to enhance the rate of the several reactions. As indicated by various features of ionic liquids, there is a large potential for application in various scientific and industrial branches, (Plechkova & Seddon, 2008).

Ionic liquids (ILs) characterize by practically unlimited miscibility with other solvents and high solubility of a vast amount of compounds. This property improves their electrochemical stability hence a wide application in different scientific and technological areas. Their diverse structure allows for multiple foreseeable uses in fields including power, catalysis, and material science (Brennecke & Maginn, 2001). It is essential to comprehend how to synthesize these ionic liquids derived from the class ions cited above. A better synthesis allows the utilization of more of them and creates new applications in various fields.

In recent decades there has been tremendous advancement in the chemistry specially known as green chemistry. As this interest has grown, so has the amount of research aiming at generating materials that present less of a threat to human health or the environment. Researchers are now trying their best to synthesize new chemical compounds which reduce adverse impacts. Thus, the synthesis of ionic liquids has received higher attention. Technological and industrial development is putting pressure for demands of cleaner and enhanced material. Due to the extensive research going on in the synthesis pathways of ionic liquids, its uses are expected to enhance the sustainability.

STRUCTURE OF IONIC LIQUIDS

Ionic liquids are chemically composed of two varieties of ions known as cations and anions. The anions generally contains inorganic structures and are relatively smaller than that of the cations. Cations have a positive charge and are generally, bigger in size, consist of organic substances. This size disparity typically results in reduced attractive forces between the two forms of ions enabling ionic liquids to exist in a liquid phase at temperatures less than 100 degrees centigrade.

On the other hand, other salts have more stable physical structures. They possess a characteristic crystal structure and normally starts to soften only at much higher temperatures; roughly about 800 degrees centigrade. The salts possess a relatively high stability due to the bonds established between the soft-scaled and like charged cations and anions. Because they have near perfect size and charge compatibility, they bond more closely and form a solid instead of a liquid. Therefore, despite ionic liquids being definite structures resembling salts, their physical characteristics are rather different mainly due to the type of interaction between ions.

Ionic liquids are considered as certain kind of salts which have low melting points and are usually below 100 °C. This make them different from molten liquids which are in liquid state only at higher temperature above 100 °C. Room temperature ionic liquids (RTILs) are special type of ILS that remain liquid at the standard temperature and hence are highly suitable for a number of uses(Mutelet et al 2005)

Ionic liquids' fundamental constituents consist of two groups of chemical structures. Among them there are imidazole, containing $C_3H_4N_2$ and pyridinium containing C_5H_5N . Cations of this type are considered to be one of the most important for defining properties and behaviour of the ionic liquids. These structural features and molecular organization directly influence the behaviour of ionic liquids with other materials. A comparison of the structural characteristics of these two main categories of ionic liquids is depicted in Figure 2. Awareness of these components is crucial to the ability to consider the potential applications of ionic liquids in areas such as chemistry and materials science.

Ions liquids are really contemplating material that can be optimized for several purposes given by the changing components. These liquids consist of cations and anions, these are positively and negatively charged ions respectively. The properties of these ions mean that through choosing specific pairs of these ions, one can engineer the ionic liquids to possess the required physical and chemical characteristics for the particular application.

For example, frequent anions employed in these mixtures are bromide (Br-), chloride (Cl-), tetrafluoroborate (BF4-) and hexafluorophosphate (BF_6-). All of these anions provide distinct features toward the entire conduct of the ionic liquid. If one of the types of the anion or cation of the mix is altered depending on the other, characteristics such as stability, conductivity, or viscosity can be strengthened.

This possibility of fine-tune ionic liquids allows a wide range of possibilities in different fields, including chemistry, materials science, and engineering. Different combinations can lead to applications in batteries, solvents, and catalysts. Due to the great potential of these compounds, their broad use in certain fields of industry and science, ionic liquids can be called an object of promising studies and developments.

Figure 1. Ionic liquids with cations of imidazole and pyridine.

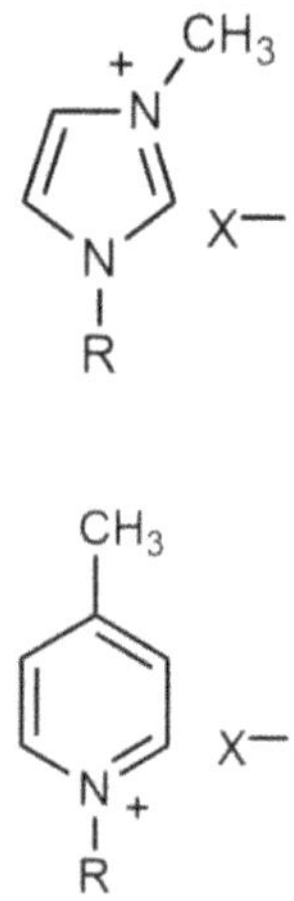

R: Methyl — X: CH_3OSO_3-
R: Ethyl — X: $BR^-,NO_3^-,BF_4^-,^-PF_6^-,CF_3SO_3^-$
R: Buthyl — X: $BR^-,Cl^-,BF_4^-,PF_6,CH_3OSO_3$
R: Hexyl — X: Cl^-,BF_4^-,PF_6
R: Octyl — X: Cl^-,BF_4^-

R: Butyl — X: Cl^-,BF_4^-,PF_6^-

PROPERTIES

Ionic liquids are rather special type of substances with several features which define their properties and potential uses. Important characteristics of the liquid's properties which would be defined include density of the liquid, that is the amount of mass for the given volume and the viscosity of the liquid, that is, the thickness of the liquid.

Furthermore, hydrophobicity deals with ability of the liquid to resist water, while solvation of the ionic liquid system concerns with the manner in which the ions of the system engage other molecules in the solution.

Depending on which ions form ionic liquids, the specific properties of an ionic liquid can vary widely. These properties depend on the choice of the cation which is positively charged ion and anion which is negatively charged ion. For the same charge, variations in the length and organisation of the alkyl chains bound to the cation can change the nature of the liquid. Such as longer alkyl chains result in an increase in viscosity and changes in the lipophilicity character. Hence, it is crucial to consider the choice of cations and anions in order to design the desired properties of ionic liquids for their application in certain areas of activity – chemistry, chemical engineering, environmental science and other ones. (Nazari et al., 2013; Greaves & Drummond, 2007; Ngo et al., 2000; Poole, 2004; Anthony et al., 2001).

Density

Ionic liquids have been recognized to have higher densities than water. Such liquids have densities falling between 1.1 and 1.6 grams per cubic centimeter (g/cm-3) of the substance. The specific density of an IL depends on the types of the cations and anions of the ILs regardless of their excess moles in the ILs. If the organic cation occupies more space or is generally larger than a corresponding n-alkyl cation, the density of the liquid typically drops.

Even though densities of ionic liquids are higher than those of molecular liquids, water can be always present in them in some traces. Nevertheless, the action of water is relatively unimportant for their properties, which are often deemed negligible for most programs in engineering.

It is possible to change the density of an ionic liquid by slight modifications to the cation being employed. This creates great opportunity to select the density of the liquid in compliance with the explicit needs of some procedure or another use.

A table of some ionic liquids and their densities will provide further insight into this topic.

Table 1. Densities of various ionic liquids

Ionic liquid	Density (g/cm^{-3})
[bmim]BF_4	1.320
[bmim]PF_6	1.510
[emim]Br	1.575
[emim]Br_6	1.426

Change of anion in ionic liquid brings more significant changes in their characteristics. For example, employing of the bromoaluminate melts, one obtains densities that are much higher than those in the normal organic solvents. This observation also brings out the fact that the density is very sensitive to the choice of anion (Sanders et al 1986).

In addition to this, there is a broader conclusion regarding the relationship between anions and cations in ionic liquids. While the specific density range is largely determined by the anion selected, careful selection of the cation can allow for fine-tuning of density within that range. This concept is supported by density measurements taken from ionic liquids that incorporate triflate or trifluoroacetate ions. These measurements show how different configurations can alter density, emphasizing the importance of both the anion and cation in determining the overall properties of ionic liquids (Keim et al 2000).

Viscosity

The viscosity of ionic liquids is not constant. It can change based on external factors. One important factor that influences their viscosity is the presence of water. When water is mixed with ionic liquids, the solution becomes diluted. This dilution results in a decrease in viscosity. As the ionic liquid becomes less thick, it can flow more freely. Understanding these changes is vital for their application in various fields, such as chemistry and materials science. Knowing how viscosity affects ionic liquids can help in choosing the right ones for specific tasks.

The viscosity of an ionic liquid is fundamentally determined by the strength of its Van der Waals forces and its ability to form hydrogen bonds (Davis J.H. et al 1999). These two factors work together to influence how thick or fluid the liquid feels. When looking at various hydrophobic ionic liquids containing [bmim]+ ions, it becomes clear how important these interactions are. The balance between hydrogen bonds and Vander Waals forces plays a key role in defining the overall viscosity of these liquids. This comparison sheds light on the characteristics of hydrophobic ionic liquids. A selection of viscosity measurements for certain ionic liquids is provided in Table 4, illustrating the differences and similarities among them. Each value helps to understand how these forces operate in these specific ionic liquids and the effect they have on liquid behavior.

Table 2. Viscosity measurements

Anion $[A]^-$	(cP) (Keim et al 2000)
$CF_3SO_3^-$	90
n-$C_4F_9SO_3^-$	373
$CF_3\ COO^-$	73
n-$C_3F_7\ COO^-$	182
$(CF_3\ (S0_2)_2N^-$	52

The analysis in Table 2 indicates a clear rise in viscosity when comparing two sets of ions. First, there is a notable increase in viscosity when moving from the triflate ion to the n-C4F9SO3- ion. A similar trend occurs when transitioning from the trifluoracetate ion to the n-$C_3F_7COO^-$ ion. This pattern implies that the larger viscosities are linked to stronger Vander- Waals forces present in the n-$C_4F_9SO_3$- and n-$C_3F_7COO^-$ ions. Even though the Vander-Waals interactions are stronger with the ionic liquids that contain (CF3(SO2)2N- ions, an interesting observation arises. The ionic liquid [bmim]CF_3SO_3- actually shows a lower viscosity compared to bmim2N-.

Viscosity does not solely depend on the strength of these interactions. The structure of the cation also plays a crucial role. When a cation, such as [emim]+, has a side chain that is flexible and has a low molar mass, it often results in lower viscosities. This flexibility allows the ionic liquid to flow more easily, contributing to reduced resistance to flow. Consequently, the combination of ion type and cation structure significantly influences the overall viscosity of ionic liquids.

Longer or fluorinated alkyl chains have stronger Van der Waals interactions, which leads to higher viscosities (Davis & Forrester, 1999), This means that liquids with these types of chains are thicker and flow less easily. The presence of these interactions enhances the internal friction of the liquid. On the other hand, when small amounts of organic co-solvents are added, the viscosity can decrease significantly. Co-solvents can help break down the stronger interactions between the molecules, making the liquid flow more freely. Additionally, simply increasing the temperature can also lower the viscosity of ionic liquids. Higher temperatures provide energy that helps the molecules move more freely. As a result, both the addition of co-solvents and an increase in temperature can make ionic liquids less viscous and easier to work with.

Hydrophobicity

Ionic liquids can be categorized as either hydrophilic or hydrophobic depending on the specific cations and anions selected. When the cations have long alkyl chains, the formed ionic liquid possess hydrophobic characteristics. The hydrophobic backbone excludes the ionic liquid from dissolving in water it cannot dissolve in water. Among such anions, [PF6]– and (CF3SO2)2N– prevent the ionic liquid from mixing with water through formation of an upper barrier. On the other hand, the ionic liquids that possess anions such as nitrates, acetates, trifluoroacetate exhibit good miscibility with water. These anions enhance a hydrophilic nature, which enable the ionic liquid to dissolve and misc in aqueous systems. For an ionic liquid to either reject water or dissolve in it depends on the selection of cations and anions. These characteristics are particularly important in different applications such as solvents and electrolytes.

CLASSICAL SYNTHETIC APPROACHES

Ionic liquids (ILs) are usually produced using conventional techniques such as quaternization and metathesis reactions. These traditional techniques serve as the basis for creating ILs. In a metathesis reaction, two chemicals exchange ions to create new ionic combinations. This reaction is necessary to produce several ILs with distinct properties (Zhao et al., 2005).

In contrast, quaternization is the process of forming a quaternary ammonium salt by adding a certain group to a molecule. Additionally, this mechanism is essential to the development of different ILs. When combined, these two techniques enable the production of a wide variety of ionic liquids, each with unique physical and chemical characteristics.

These methods' adaptability is crucial for creating ILs that work well for a variety of uses, such as electrolytes in batteries or solvents in chemical reactions. Further research into these techniques may result in even more customised ILs, increasing their efficacy across a range of domains. In order to enable a broad range of scientific and industrial applications, the established procedures continue to be essential to the continuous development of ionic liquid technology (Zhao et al., 2005).

Metathesis Reactions

In a chemical process known as a metathesis reaction, ions from two distinct salts exchange positions. A byproduct salt and an ionic liquid are produced as a result of this exchange. When a metal salt with the required anion, like silver tetrafluoroborate, combines with a halide salt that contains the desired cation, such imidazolium halide, this reaction is frequently seen.

The cation, which is frequently an imidazolium molecule, interacts with the halide anion in the original salt in this reaction. It also adds the desired anion to the mixture by exchanging the metal cation from the metal salt. The ionic liquid and a byproduct salt, like silver halide, are produced as a result. Filtration is typically used to eliminate the byproduct from the mixture, making purification easier.

The ability to produce ionic liquids with a wide variety of anions makes the metathesis approach effective and popular. For enterprises and researchers wishing to modify features for particular uses, this adaptability is a huge benefit. It is crucial to remember that this method frequently uses pricy or moisture-sensitive substances, such as silver salts. Because of the waste produced during synthesis, the usage of these components may raise production costs and raise environmental issues. When thinking about metathesis reactions for the creation of ionic liquids, it is essential to balance these factors. (Earle et al., 2006)

Quaternization Processes

Quaternization is a chemical process that creates an ionic liquid, a new kind of substance, from some chemicals that contain sulphur, phosphorus, or nitrogen. Using this approach, a cation is created by adding an alkyl group to the original chemical.

One well-known instance of quaternization is the creation of ionic liquids based on imidazolium. Imidazole and an alkyl halide react in this method, causing the halide's alkyl group to bind itself to the imidazole. This process produces a substance known as a quaternary ammonium salt.

Because of its ease of use and capacity to generate greater amounts of the intended output, this quaternization process is preferred in both labs and companies. But the creation of this novel compound is not the end of the process. Additional purification procedures are frequently needed after the ionic liquid has been synthesised. These procedures are essential because they aid in the removal of any leftover unreacted components and byproducts. It might be difficult to achieve high yields while maintaining the purity of the finished product, (Wasserscheid & Keim, 2000).

To improve the outcomes, the quaternization method must be optimised. Effectively eliminating contaminants that could degrade the product's quality is one of the main challenges. Since too much moisture might disrupt the process, controlling the amount of moisture in the surroundings is also very important. Furthermore, there is a growing need for safer and more affordable reagents for these procedures. By addressing these issues, production efficiency and ionic liquid quality can both be improved. In addition to helping research, this could result in improvements in the range of applications that use these ionic liquids.

SYNTHESIS OF IONIC LIQUIDS

Ionic liquids can be broadly classified into two groups. Simple salts are included in the first category. One kind of cation connected to one kind of anion makes up simple salts. A stable ionic structure with a variety of functions is the result of this combination. Ionic liquids that are more complicated fall into the second category. Due to the dynamic interactions between many ions, these liquids require a more complex force balance. They differ from simple salts due to the special qualities created by this interaction.

There are two necessary processes in the production of ionic liquids. Selecting which ions to mix is the initial stage. Because it directly affects the final ionic liquid's characteristics, this decision is very important. Features like conductivity, viscosity, and stability will depend on the particular cations and anions chosen. These ions are mixed in a controlled environment in the second stage. The ions will appropriately combine to generate the intended ionic liquid thanks to this meticulous mixing. In order to create ionic liquids with the precise properties required for various applications, both stages are essential. To get the desired outcomes, each step needs to be carried out properly, (Han et al 2019).

Formation of the Desired Cation

To achieve the desired cation, two main methods are commonly used in chemistry. The first method is the protonation of an amine. During this process, an acid donates a proton, which is a positively charged particle, to the amine. This donation changes the amine into a positively charged species. This transformation is significant because it increases the amine's ability to react with other chemical substances. The enhanced reactivity opens up new pathways for different chemical reactions, making the protonated amine a more useful compound in various applications. Understanding this process is essential for chemists looking to manipulate chemical reactions effectively.

Exchange of Anion

Halide salts can undergo treatment with Lewis acids, resulting in the creation of ionic liquids driven by Lewis acid chemistry. This treatment leads to the generation of novel chemical structures, paving the way for a variety of applications. In addition to this method, another way to modify halide salts is through a process called anion metathesis. In this method, one anion in a specific salt is swapped out for another anion taken from a different salt. This simple yet effective technique allows for the adjustment of the properties of the original compounds. Both of these methods—treatment with Lewis acids and anion metathesis—are useful in facilitating the exchange of anions. The outcomes of these processes are new ionic liquids that possess distinct and unique characteristics. The ability to create these ionic liquids opens up exciting possibilities for their use in diverse fields, including chemistry, engineering, and materials science. As researchers explore these approaches, they continue to find innovative ways to harness the potential of ionic liquids, making significant contributions to both academic and practical applications.

Table 3. List of commercially available ionic liquids

	NAME	STRUCTURE
1.	1-Allyl-3-methylimidazolium bis(trifluoromethylsulfonyl)imide	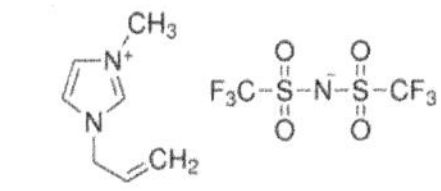
2.	1-Ethyl-3-methylimidazolium methanesulfonate	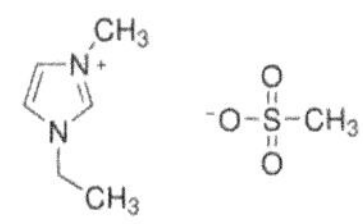
3.	1-Hexyl-3-methylimidazolium bis(trifluormethylsulfonyl)imide	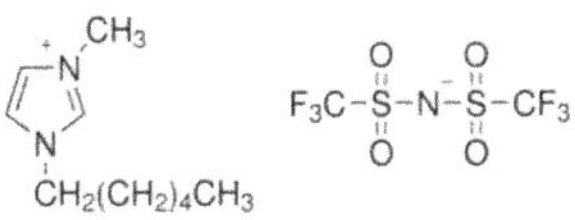
4.	1-Hexyl-3-methylimidazolium bis(trifluormethylsulfonyl)imide	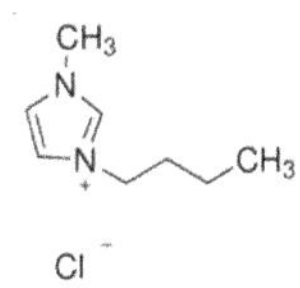
5.	1-Butyl-3-methylimidazolium hexafluorophosphate	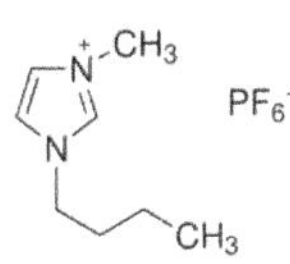
6.	1-Ethyl-3-methylimidazolium acetate	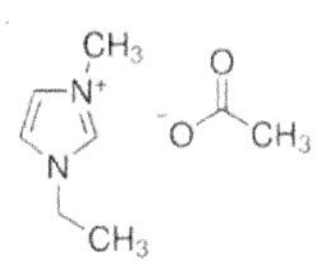
7.	1-Ethyl-1-methylpyrrolidinium bis(trifluoromethylsulfonyl)imide	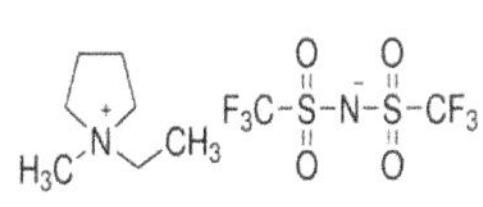
8.	5-(Trifluoromethyl) dibenzothiophenium trifluoromethanesulfonate	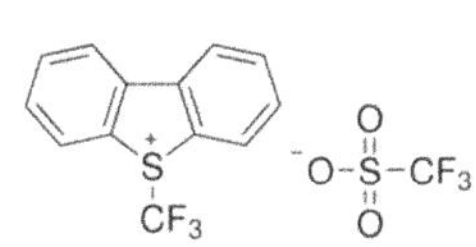

continued on following page

Table 3. Continued

9.	Tributylmethylammonium methyl sulfate	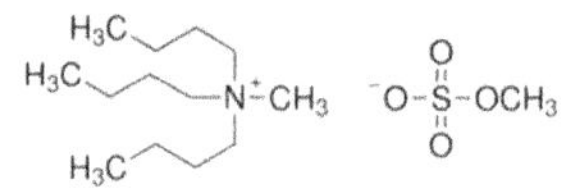
10.	1-Butyl-1-methylpyrrolidinium iodide	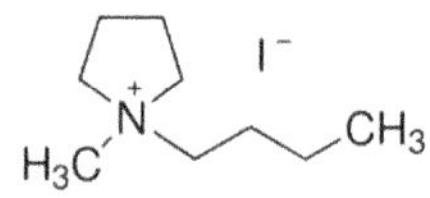
11.	Tributylmethylammonium bis(trifluoromethylsulfonyl)imide	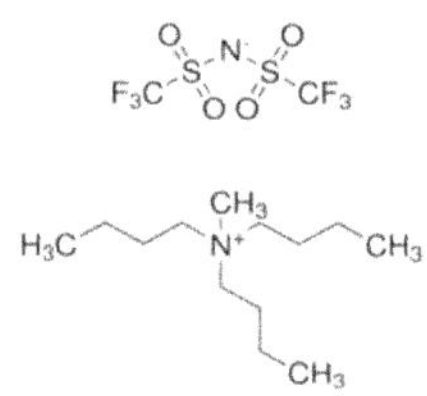
12.	4-(Triphenylphosphonio)butane-1-sulfonate	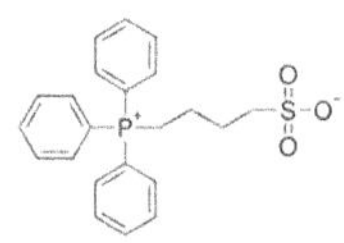
13.	(6-Bromo-1-oxohexyl)ferrocene	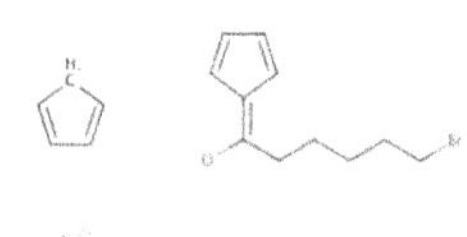
14.	Methyl-trioctylammonium bis(trifluoromethylsulfonyl)imide	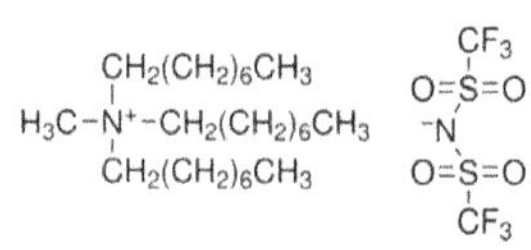
15.	1-Ethyl-3-methylimidazolium tetrachloroaluminate	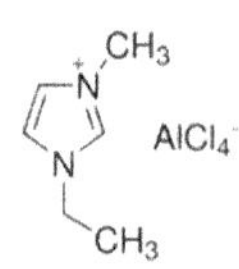

(Javvad Ali et al 2021)

VARIOUS SYNTHESIS METHODS OF IONIC LIQUIDS

Ionic liquids, often referred to as ILs, are specialized fluids made from various ions, including ammonium, phosphonium, and sulfonates, (Kärkkäinen, 2007). These unique substances have garnered interest due to their distinct properties and potential applications in numerous areas.

There are two main methods for producing ionic liquids: metathesis and acid-base neutralization. (Scheme 1). Metathesis is a process where ions are swapped between different compounds. This ion exchange results in the creation of new ionic liquids, each exhibiting unique characteristics based on the starting materials used.

On the other hand, acid-base neutralization works by combining an acid with a base. This reaction produces a salt, which can display the properties found in ionic liquids. This method is straightforward and effective, making it a popular choice for many researchers.

In addition to these primary methods, there is a third technique that focuses on creating imidazolium cations. This method was developed by researchers Earl and Seddon. It involves the reaction of strong bases known as imidazole carbenes. This approach is particularly effective for forming specific types of ionic liquids, allowing for tailored properties and functions (Earle & Richard, 2003)

Each of these methods showcases the versatility of ionic liquids and highlights their diverse range of applications. From industrial uses to their role in scientific research, ionic liquids offer unique benefits due to their low volatility and ability to dissolve various compounds. Their ongoing exploration continues to reveal potential uses in fields such as chemistry, materials science, and environmental technology.

Figure 2.

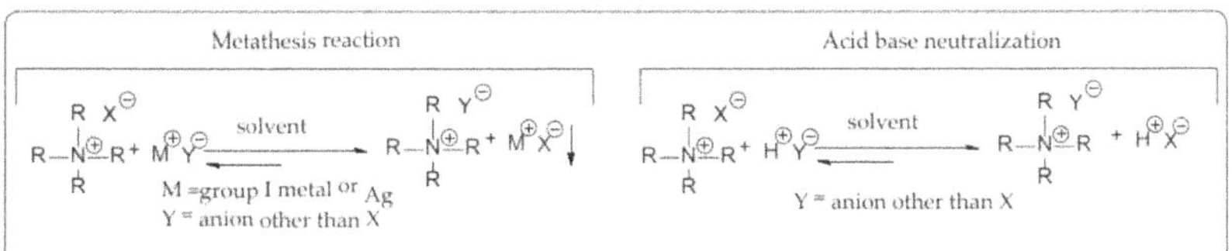

Scheme 1: General procedure for the metathesis reaction and acid base neutralization.

Different types of alkyl-ammonium halides are available on the market for commercial use. These compounds can be directly involved in metathesis reactions (Welton T 1999) In these reactions, components from different chemical compounds swap places. This process can facilitate the creation of new compounds with useful properties. However, there may be instances when alkyl-ammonium halides are not readily accessible. In such situations, alternative methods must be employed.

One effective method is known as quaternization. This is a chemical reaction where a base is alkylated using a haloalkane, resulting in the formation of an organic halide salt. This process involves adding an alkyl group to a halogen compound. The outcome is a new compound that possesses distinct characteristics compared to the original materials (Welton T 1999) Quaternization is versatile and not limited to just producing alkyl-ammonium halides. It can also help create other important halides, such as pyridinium and imidazolium halides (Scheme 2). These compounds have significant roles in numerous chemical reactions and applications in various fields. The steps involved in the quaternization process can be effectively illustrated through a reaction scheme, showcasing how the transformation occurs systematically.

Quaternization serves as a valuable tool in organic chemistry. It allows chemists to produce a wide range of compounds suited for different applications. By manipulating the components involved in this process, researchers can meet specific needs across various scientific disciplines, highlighting the importance of this method in chemical synthesis.

Figure 3.

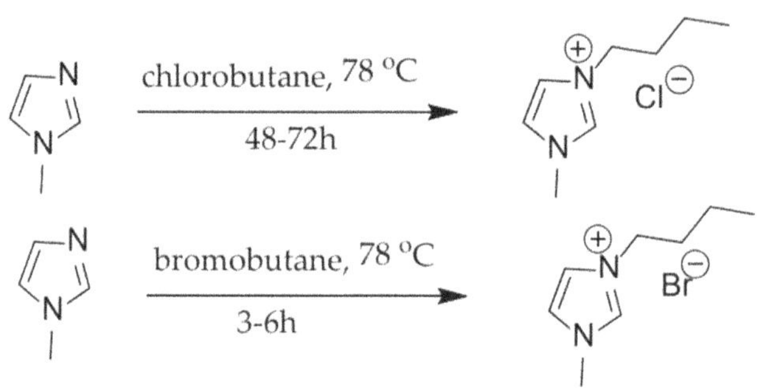

Scheme 2: Examples of the preparation of imidazolium halides using classic methods from Lévêque et. al [8].

Many researchers are now investigating various approaches to enhance chemical processes. Among these, nontraditional approaches such as microwave (Srour et al., 2013; Boros et al., 2008) and ultrasonic techniques (Srour et al., 2013; Namboodiri &Varma, 2002) are becoming more and more prevalent in quaternization processes, (Srour et al., 2013; Leveque et al., 2007). These approaches have a number of advantages over traditional ones, and they produce better outcomes in a variety of applications. High-frequency sound waves are used in ultrasound to speed up mixing and reaction times. Chemical operations may be completed more quickly as a result. Likewise, microwave energy offers quick heating, which can raise the reaction's efficiency. Both microwave and ultrasonic techniques have been demonstrated to produce more and better-quality items. Because of this, scientists are using these novel techniques more frequently in an effort to improve the results of their chemical interactions.

Metathesis is a chemical reaction that helps produce ionic liquids (ILs) that do not contain halides. An example of this process is the creation of the ionic liquid [EMIM][BF4]. This compound is made starting from a halide salt, specifically [EMIM]I. In addition to the halide salt, silver or a group 1 metal can be used in this reaction, with Ag[BF4] being one possible choice. The metathesis reaction effectively replaces the halide component with a more desirable anion, resulting in a halide-free ionic liquid. This method is significant because it allows for the synthesis of ionic liquids that are more environmentally friendly and have unique properties, making them useful in various applications (Welton T 1999). Other ILs have been prepared using metathesis reactions, such as thiocyanate, nonafluorobutanesulfonate, bis((trifluoromethyl)sulfonyl)imide, tris(trifluoromethyl)sulfonyl, methide, trifluoroacetate, and heptafluorobutanoate. Metathesis reactions are widely utilized in the synthesis of ILs due to their simplicity and ease of preparation. However, despite their advantages, these reactions do have some drawbacks. One of the inconveniences is the potential contamination with a small amount of halide ions, which can react with solute materials and affect the purity of the ILs. Additionally, the presence of water or silver can also be problematic in the metathesis reaction process. Careful attention must be paid to avoid these impurities and ensure the desired quality of the final IL product, (Welton, 1999; Srour et al., 2013)

Recently, Srour and colleagues (2019) reported a groundbreaking discovery where they successfully demonstrated a metathesis reaction that does not involve silver or water. This finding addresses a significant issue in the realm of metathesis reactions, which is the limited availability of commercially accessible metal salts. Furthermore, the challenge of coordinating transition metals with bioactive compounds such as amino acids adds another layer of complexity to this field of research (Fukumoto et al., 2005; Vidugiris et al., 1988).

One notable approach that circumvents issues related to contamination is the acid-base neutralization method (Ohno & Yoshizawa, 2002) as proposed by previous studies. By employing this method, researchers can simplify the process while ensuring the production of desired salts without the formation of unwanted by-products. This strategy has proven to be particularly effective when dealing with tertiary amines in combination with halide acids or certain organic acids. (Fukaya et al., 2007)

However, challenges arise when weaker acids, such as hydrohalic acids, are involved in the neutralization process, necessitating the use of hydroxide quaternary cations for effective neutralization, (Fukaya et al., 2007)

To address these challenges and explore alternative methodologies, recent advancements by Ohno and collaborators have introduced innovative ion exchange resin methods. These novel approaches offer promising solutions to improve the efficiency and versatility of metathesis reactions, ushering in a new era of exploration and discovery in this field, (Fukumoto, 2005; Ohno & Fukumoto, 2007).

The use of imidazole carbenes as a strong base (Earle & Richard, 2003) (Scheme 3) has gained attention due to its numerous advantages. One of the key benefits is that it allows for the creation of non-contaminated ionic liquids (ILs) that are free from halide ions or metal ions. This is significant as it ensures the purity and integrity of the ILs produced. Additionally, imidazole carbenes offer the unique advantage of being able to synthesize ILs or salts from various starting materials such as alcohols like methanol and propanol, carbonic acid, acetic acid, and alkyl sulphonic acid. This versatility opens up a wide range of possibilities for the production of ILs with different properties and applications. However, it is important to note that one limitation of using imidazole carbenes as a strong base is that it is only effective for the preparation of imidazolium-based ionic liquids.

Figure 4.

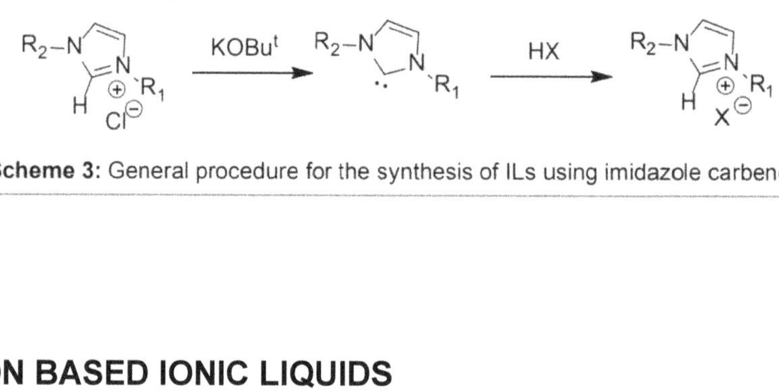

Scheme 3: General procedure for the synthesis of ILs using imidazole carbenes.

CATION BASED IONIC LIQUIDS

For any ionic liquid, a cation with a bulky structure and low symmetry is preferable. Ammonium, sulfonium, phosphonium, imidazolium, pyridinium, picolinium, pyrrolidinium, thiazolium, oxazolium, and pyrazolium cations provide the base for majority of ionic liquids that are presently being analyzed. As there is a change in size, symmetry, and alkyl chains connected to cations it has varying effects on the properties of ionic liquids, including density, viscosity, and melting temperature. Furthermore, by adding an appropriate function to the cation, other ionic liquids can be developed, forming third-generation ionic liquids—task-specific ionic liquids. Adjusting the product's stereochemistry is a crucial goal for researchers working on chemical transformations and total synthesis. Chiral ionic liquids have been specifically engineered to perform the task of asymmetric synthesis in this regard.

Synthesis of Cationic Based Ionic Liquids

The synthesis of cationic ionic liquids involves various methods, including quaternization reactions and direct synthesis using N-heterocyclic carbene intermediates. These methods allow for the design of specific cations tailored for particular applications, enhancing their functionality and effectiveness.

Various alkyl chains or anions, including nitrate, halogenides, phosphorium, tetrafluoroborate, trifluoromethanesulfonate, bistriflimide (Tf2N), hexafluorophosphate (PF6), and organic cations, including imidazolium, pyridinium, ammonium, alkyl ammonium, phophorium, and metal-based, are attached to the ionic liquids (Rabia Hassan et al 2022). Different combinations of these anions and cations give ILs some unique properties, such as miscibility and hydrophobicity/hydrophilicity. Also the length of alkyl chain can be changed to get the required hydrophobicity.

Synthesis of Imidazolium, Pyridiniun and Ammonium Based Ionic Liquids

The widely accepted method for synthesis of imidazolium, ammonium and pyridinium based cationic liquids are metathesis.

Chemical Required

Substituted imidazole or pyridine or ammonium (core structure), alkyl halide or alkyl sulfonate (alkylating agent), strong acid or suitable counter-ion salt (to exchange halide/sulfonate for other anions, if required), Organic solvents (e.g., acetonitrile, dichloromethane, or ethanol), Deionized water, Drying agents such as sodium sulfate or magnesium sulfate, Nitrogen gas (optional for inert atmosphere)

General Synthesis Steps

Formation of Imidazolium Salt or Quaternization

The imidazole and an alkyl halide (e.g., methyl iodide, ethyl bromide) was taken in round bottomed flask and mixed together in the molar ratio of 1:1 in a dry organic solvent such as acetonitrile or ethanol. The reaction mixture was heated at 50–100°C (depending on the reactivity of the alkyl halide) with constant stirring. The reaction mixture was refluxed from 6 to 48 hours, until the product is obtained. Nitrogen gas is passed through reaction container to avoid the loss of volatile alkyl halides during the process. The progress of reaction is monitored using technique like thin-layer chromatography (TLC). A rotary evaporator was used to extract the solvent under reduced pressure once the reaction was finished. To get rid of unreacted starting ingredients, the residue was cleaned using an organic solvent, like ethyl acetate. The resulting imidazolium halide was vacuum-dried.

Anion Exchange: The Ion Exchange Step is Optional

The imidazolium halide is dissolved in deionized water and aqueous solution of the desired anion source such as lithium bis(trifluoromethanesulfonyl)imide [$LiNTf_2$] or sodium tetrafluoroborate [$NaBF_4$]) is added into this. The mixture is thoroughly mixed at room temperature or slightly elevated temperatures (25–60°C) for ion exchange. Separate the organic phase containing imidazolium-based IL from the aqueous phase. Water is used to wash the organic phase so as to remove residual salts and dried over sodium sulfate or magnesium sulfate (drying agents). A reduced pressure is to be used to isolate the pure ionic liquid and the solvent. The product is purified by recrystallization or vacuum distillation. The product is characterized using different analytical techniques, (Hassan et al., 2022). Successfully synthesized three series of ionic liquids (ILs) using imidazolium, pyridinium, and ammonium cations with various anions, demonstrating that the unique combinations of cations and anions significantly influence the antibacterial activity of the ILs against pathogenic bacteria, including both Gram-negative and Gram-positive strains. The imidazolium cation combined with hydrogen sulphate (HSO_4–) and bis(trifluoromethanesulfonyl) imide (NTf2–) anions presented the strongest antibacterial activity among the studied ionic liquids. Therefore, they could represent a good alternative to conventional antibiotics and antiseptics, according to in vitro assays and in silico studies. Synthesis of three sets of ionic liquids were based on the usage of 1-methyl-3-butyl imidazolium, butyl pyridinium, and diethyldibutylammonium as cations. The anions used in this process are bromide (Br), methanesulfonate(CH_3SO_3), bis(trifluoromethanesulfonyl)imide (NTf_2), dichloroacetate($CHCl_2CO_2$), tetrafluoroborate (BF_4) and hydrogen sulfate(HSO4), (Hassan et al., 2022).

Synthesis of Bromide-Based Ionic Liquids

Figure 5. Synthesis of bromide-based ionic liquids (a) 1-butyl-3-methylimidazolium bromide (5); (b) butyl pyridinium bromide (6); (c) diethyl dibutyl ammonium bromide (7)

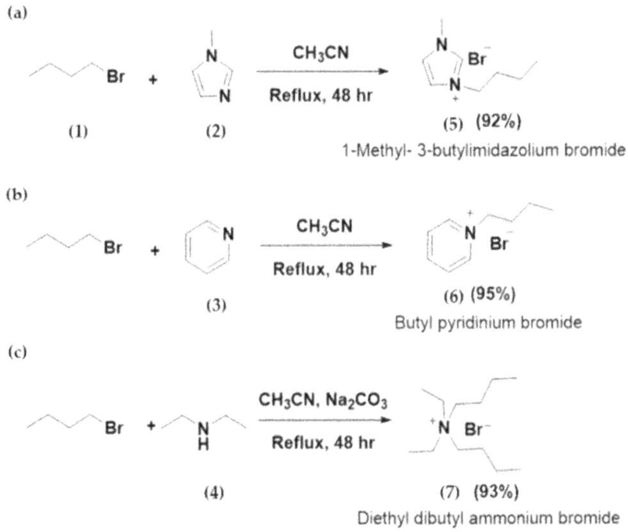

Figure 6. Synthesis of imidazolium-based ionic liquids

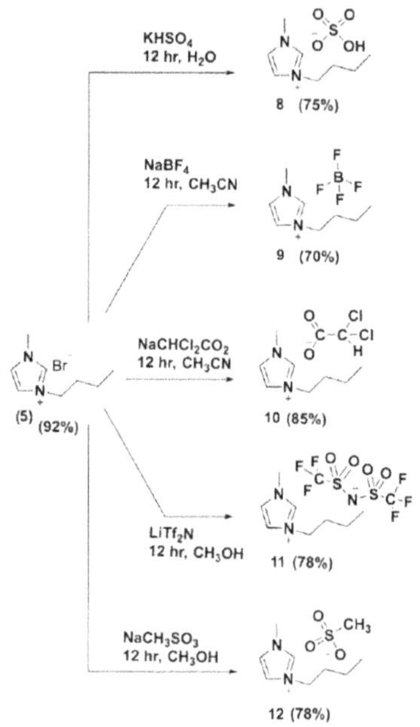

Figure 7. Synthesis of butyl pyridinium-based ionic liquids

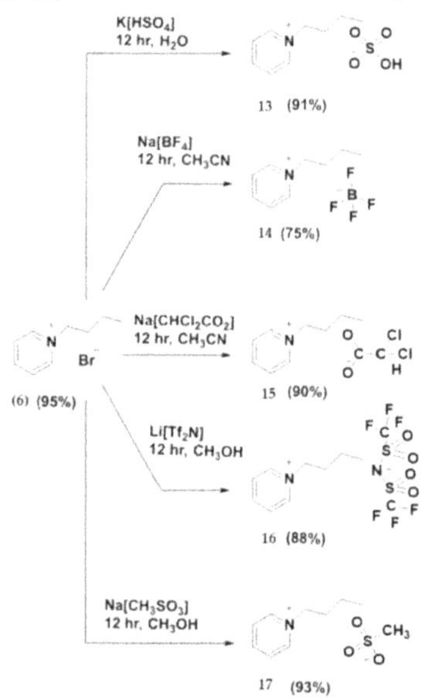

Figure 8. Synthesis of quaternary ammonium-based liquids

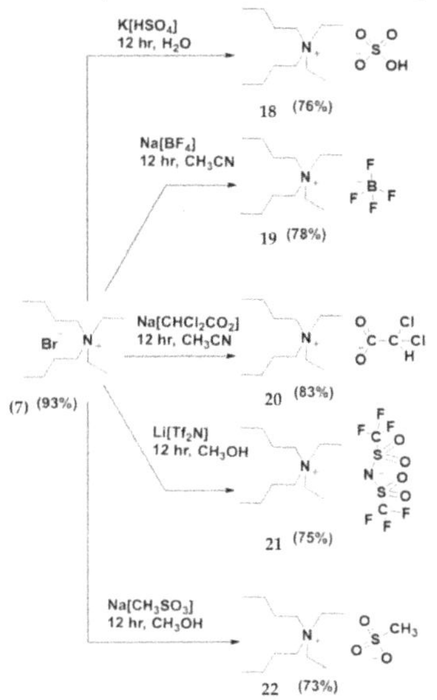

Metal-Based Ionic Liquids

Ionic liquids (ILs) is the base fluid for metal based ionanofluids-a subcategory of nanofluids that is composed of metal or metal oxide nanoparticles (NPs). These are used in a variety of applications as they combine adjustable properties of ionic liquids with unique thermal, electrical, and catalytic capabilities of metal nanoparticles. One feasible approach to synthesize metal-based ionanofluids is through a simple one-step route that enhances the thermal conductivity but maintains the stability of the fluid medium.

An effective way of achieving nanoscale dispersion homogeneity is represented by recent experiments that successfully carried out the reduction of copper sulfate by fructose and produced extremely stable copper nanofluids in a water-ethylene glycol mixture. This method shows the possibility of easy scale up with a low productions cost, which is very important because various sectors look for effective cooling solutions. Further, the exploration of various metal sources or the modification of the base fluid composition can lead to new formulations with certain properties, thus expanding the scope of applications for these advanced materials in thermal management systems. Such innovative approaches not only promise better performance but also open up avenues for sustainable material development processes, which align with international efforts to produce greener technology. (Kumar, S.

P.,2024). The ionanofluids are prepared using metal-based ionanofluids through various techniques which include thermal breakdown, sol-gel procedures, as well as chemical reduction. Each of these techniques has unique benefits and can be modified to produce ionanofluids with certain qualities, such as increased stability or thermal conductivity. Chemical reduction is the preferred method among the techniques listed above since it is capable of yielding uniformly sized and shaped nanoparticles, which are essential for maximizing the performance of ionanofluids in various applications.

However, the sol-gel method provides more control over the structure and composition of the ionanofluids, and metal ions can be added to further optimize their characteristics. Another important technique is thermal decomposition, which may create nanoparticles at high temperatures and yield materials with better mechanical qualities and crystallinity, which are advantageous for applications needing reliable performance. All of these synthesis methods demonstrate the flexibility and elasticity of ionanofluids, making them suitable for a wide range of applications in industries as advanced cooling systems, energy storage, and catalysis. In order to fully utilize ionanofluids in industrial applications, researchers are always looking for new ways to improve their stability and effectiveness, (Murshed et al, 2011).

Synthesis of Metal-Based Ionic Liquids

It can be done either by hydrothermal, or reduction of metal or dispersion method. The synthesis of metal-based ionic liquids often involves the use of metal salts and organic ligands in the presence of ionic liquids as solvents.

Thermal or Hydrothermal Method

Materials

Metal Salt: Commonly used metal salts include Copper(II) chloride (CuC12), Nickel(II) chloride (NiC12), or Zinc chloride (ZnC12).

Organic Ligand: Examples include 1,4-benzenedicarboxylic acid (BDC), 2-methylimidazole, or pyridine.

Ionic Liquid: An appropriate ionic liquid such as 1-butyl-3-methylimidazolium chloride (BMIMCl) or 1-ethyl-3-methylimidazolium tetrafluoroborate (EMIMBF4), a suitable solvent like acetonitrile or dichloromethane and catalyst like anhydrous potassium carbonate (K2CO3) to facilitate the reaction.

Procedure. In a reaction flask dissolve the Metal salt and potassium carbonate in the ionic liquid. Add the organic ligand to the reaction mixture while stirring and maintaining the temperature at around 60-70°C. Allow the reaction to proceed for several hours until the formation of the metal-based ionic liquid is complete. The monitoring can be done by thin-layer chromatography (TLC) or gas chromatography (GC). After the reaction is complete, cool the mixture and filter out the solid potassium carbonate.

General Methods by Reduction Process

In situ synthesis involves generating metal nanoparticles directly in the ionic liquid. First of all, Ionic Liquid Medium is prepared by adding the ionic liquid (IL) to a beaker or reaction vessel. Add suitable solvent such as water or ethanol to facilitate mixing. The metal precursor (e.g., $AgNO_3$, $HAuCl_4$) is dissolved in the IL or a small amount of a compatible solvent. A reducing agent (e.g., $NaBH_4$) is added dropwise under stirring at room temperature or slightly elevated temperatures (30–80°C). The color change of solution typically indicates nanoparticle formation (e.g., yellow for gold nanoparticles, brown for silver). After synthesis of nanoparticles stir the solution atleast for1–2 hours. Remove any by-products by centrifugation or filtration. If necessary, wash the product with ethanol or water and re-disperse in the IL.

Synthesis by Dispersion Method. Synthesis of metal based ionanofluides can be done by synthesizing nano particles followed by its dispersion into ionic liquids. For synthesis of Metal Nanoparticles a conventional nanoparticle synthesis method such as:

- Chemical reduction: Reduce metal salts using a reducing agent in the presence of a stabilizing agent.
- Thermal decomposition: Decompose metal complexes in a high-temperature furnace or hydrothermal reactor.
- Green synthesis: Employ plant extracts or biocompatible materials as reducing agents.
- The synthesizes nano particles are purified and separated by centrifugation step.

Dispersion in Ionic Liquid. Add the purified nanoparticles to the ionic liquid. Use ultrasonication (bath or probe sonicator) to disperse the nanoparticles homogeneously in the IL. If needed, add stabilizing agents to improve dispersion stability.

ANION BASED IONIC LIQUIDS

Various anions can be added to generate ionic liquids with a variety of characteristics. There are two types of ionic liquid anions: (1) fluorous anions like PF6-, BF4-, CF3SO3-, and (CF3SO3)2N-, and (2) non-fluorous anions like AlCl4-. The anions as such as chloride, nitrate, acetate, hexafluorophosphate, and tetrafluoroborate are the most frequently found in ionic liquids. However, due to their unique features, fluorous anions are frequently used in the construction of ionic liquids. Ionic liquids containing 1-n-butyl-3-methylimidazolium cation and PF6-anion, for example, are water-immiscible, but Ionic liquids containing the same cation and BF4-anion is water soluble, as was previously mentioned. This is an example of "designer solvent" property of the ionic liquids', which states that the density, hydrophobicity, viscosity, and solvation characteristics of the system can be changed by altering the anion.

The two anion types that are used in the majority of ionic liquid applications are PF_6- and BF_4-, although they have a significant disadvantage. When heated in the presence of water, these anions decompose and release HF. After this phenomenon was discovered, the use of fluorous anions with a C-F bond that is resistant to hydrolysis began. As a result, fluorine-carbon linked ionic liquids containing CF_3SO_3- and (CF_3SO_3)2N anions have been created. Fluorinated anions, however, are typically costly and environmentally hazardous. Because of their non-toxic and biodegradable structures, alkylsulfate anions made from cheap bulk chemicals have been identified as the most well-known non-fluorous anions

Characterization of Ionic Liquid

To understand the characteristics of ionic liquids properly, it is important to characterize their chemical properties. These characteristics enables the identification of distinct features of ionic liquids when compared to those of regular solvents. Viscosity and density make it possible to study the system's bulk characteristics, but a number of additional internal characteristics can also be investigated. It has already been explained how an ionic liquid's constituent cations and anion affect its density, viscosity, surface tension, melting point, and solvent characteristics. These systems must, however, undergo a thorough investigation that includes conductivity tests, thermal analysis, decomposition temperature measurement, structural analysis, and toxicity investigations. The conductivity (and associated impedance) and diffusion coefficients of Ionic liquids can be useful in determining its solvent-solute interactions and transport characteristics. Ionic liquids based on imidazolium have been examined and found to have low electrical resistance values. This is predicted owing to the charges of the ions and higher concentration of these charges in a pure ionic liquid, (Clare et al., 2009).

Phase equilibrium studies and other thermochemical studies of ionic liquids have thrown light on the solubility and stability of ionic liquids as well as how they interact with their environment. As ionic liquids are made up of charges, any lattice formation or other comparable architecture that may exist inside them could have a direct impact on properties like melting point. The practical application of ionic liquids depends on the temperature at which they maintain the liquid state, making melting temperature a crucial parameter, (Nanda et al., 2012). The relationship between ionic liquid structure and physical properties can be confirmed by using differential scanning calorimetry (DSC) and can help in determining the melting, crystallization, and glass transition (Tg) points of ionic liquids. The melting point of ionic liquids can be significantly altered by increasing ion size, which results in weaker coulombic interactions in the crystal lattice. Additionally, when side chain length increases, the system's stronger hydrogen bonds may be diminished by the consequent weak van der Waals interactions. When combined with DSC data, thermogravimetric analysis (TGA) can be used to examine decomposition temperatures, determine useful practical limits, and assess the general stability of ionic liquids. Research has shown that in some circumstances, per fluorinating anions with stable C-F bonds, like those in Nuclear Transport Factor 2, can also improve the heat stability of ionic liquids, which is essential for their use in batteries. The structure of ionic liquids has been studied using spectroscopic analytical techniques. Diffusional and structural examination of ionic liquids is now possible by techniques like infrared (IR) and nuclear magnetic resonance (NMR) spectroscopy, (Zhu et al., 2021). This aids in developing an understanding of solvent-solvent interactions and how they affect the transport characteristics of liquids. More thorough analysis of breakdown fragments is being made possible by mass spectrometry. Physical characteristics like the enthalpy of vaporization, which is related to the heat stability, can be determined by studying pattern of these events. Furthermore, immobilization in HPLC stationary phases makes it easier to examine the ionic liquids' ion exchange characteristics as well as more specialized interactions like those with aromatic molecules, (Nanda et al., 2012). Ionic liquids have shown significant levels of toxicity, although very few studies have been conducted on this subject. According to reports, phosphonium-based ILs can be identified as both corrosive and irritating. Recent research has also revealed that the alkyl chain length, especially the imidazolium cations, can have a significant impact on the toxic effects of ionic liquids. Nevertheless, biocompatibility of inert ions can be a good option for the current ionic liquids, which are highly toxic. Tao et al. has synthesized ionic liquids based on amino acids that are biodegradable. Furthermore, other biocompatible molecules, such as the sugar-based anions, succinate, and lactate, can be utilized to create ionic liquids with significantly lower toxicity, (Nanda et al., 2012).

APPLICATIONS OF IONIC LIQUIDS IN ORGANIC SYNTHESIS

In the last decade, researchers have shown significant interest in the use of ionic liquids for synthetic processes, (Wilkes et al., 1992, Chauvin et al., 1995; Bonhote et al., 1996; Davis et al., 1999). This growing attention has led to numerous studies examining various applications of these substances. The two key areas of focus are their role as solvents in traditional organic reactions and their use in biphasic catalysis systems.

Among the different types of ionic liquids, imidazole-based ionic liquids and chloroaluminates are especially noteworthy. This examination will primarily concentrate on chloroaluminates because they were the first ionic liquids to be discovered. These melts have a historical importance, as they were utilized in the initial experiments that revealed the potential of ionic liquids in chemical reactions.

In this discussion, chloroaluminates will be presented first, highlighting their unique properties and advantages. Following this, attention will shift to imidazole-based ionic liquids, comparing their features and applications to those of chloroaluminates. This structured approach aims to provide a clear understanding of the roles that different types of ionic liquids play in synthetic chemistry.

Electrophilic Aromatic Substitution reactions in Chloroaluminate-based Salts

Electrophilic substitution reactions were the first kind of reaction studied in room temperature chloroaluminate(III) ionic liquids, (Boon et al., 1986). Using their roles as a solvent and a catalyst, simple arenes are perfect for these reactions due to their high concentration of chloroaluminate (UI) species and their solubility in acidic ionic liquids. Only nitrobenzene failed to react when tested with 1-chloropropane out of all the arenes. Polyalkylation was seen, as in the traditional reactions with bench solvents. For instance, when too much chloroethane was reacted with benzene, a mixture of mono (12%), di(9%), tri (33%), tetra (24%), penta (17%), and hexa (2%), substituted products were produced, (Boon et al., 1986). Using extra arene in the process can minimise this polyalkylation. It is important to remember that basic ionic liquids lack the necessary catalytic activity to support alkylation reactions.

Among the first investigations to use ionic liquids as Lewis acid catalysts, Friedel-Crafts alkylations and acylations are highly significant from a commercial standpoint (Boon et al 1986). $AICI_3$ is the catalyst typically employed in Friedel-Crafts processes, which leads to issues with disposal and byproducts. Boon and colleagues performed the Friedel-Crafts acylation for the first time in the most popular chloroaluminate(III) ionic liquid [emim]Cl-$AlCl_3$. Wilkes and his team used simple benzene derivatives and primary and secondary halides to perform Friedel-Crafts alkylations.

The primary catalyst for the process was Lewis's acid equilibrium of the ionic liquid:

$$2\ AICI_4^- \Longleftrightarrow Al_2Cl_7^- + Cl^-$$

where $Al_2Cl_7^-$ was the Lewis acid and Cl^- the Lewis base.

They demonstrated that the Lewis acid catalyst in the reaction was the $Al_2Cl_7^-$ species found in the acidic mixture. Thus, it was demonstrated that the ionic liquid served as both the solvent and the catalyst for the reaction.

In 1996, it was reported that ferrocene and acetic anhydride underwent a Friedel-Crafts acylation process in [emim] $C1^-$ $A1C1_3$ as the ionic liquid, (Surette et al., 1996). The concentration of $AICI_3$ in the melt regulated the rate of the reaction and dictated whether a mono or bis substitution reaction occurred. The acylation is shown in Scheme 3 below.

Figure 9. Ferrocene acylation in ionic liquid

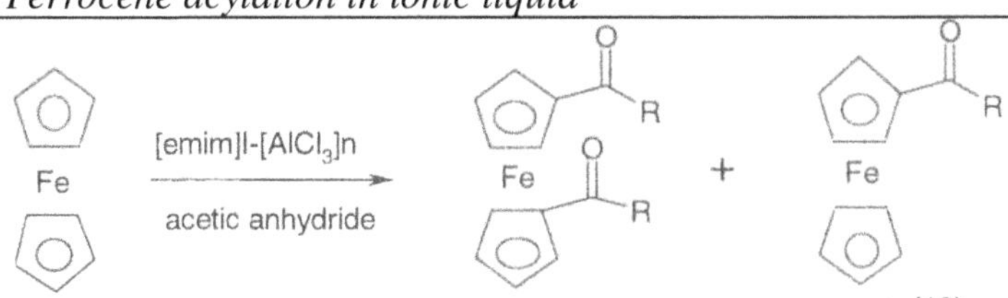

(10) (11) (12)

The acylative cleavage of cyclic and acyclic ethers is the last synthesis that makes use of the Lewis acidity of chloroaluminate(III) species. There are numerous ways to cleave ethers, but no technique to cleave ether connections specifically has been documented, (Green et al., 2000). For the acylative cleavage of a number of cylic and acyclic ethers, including THF, diethyl ether, and diisopropyl ether, the ionic liquids obtained from [emim] I/$AlCl_3$ have been employed as solvents and Lewis acid catalysts. These chloroaluminate (HI) melts were found to have another special application by Green and colleagues (Green et al 2000), whose characteristics enabled superior synthetic chemistry outcomes.

Figure 10. Acylative cleavage of ethers

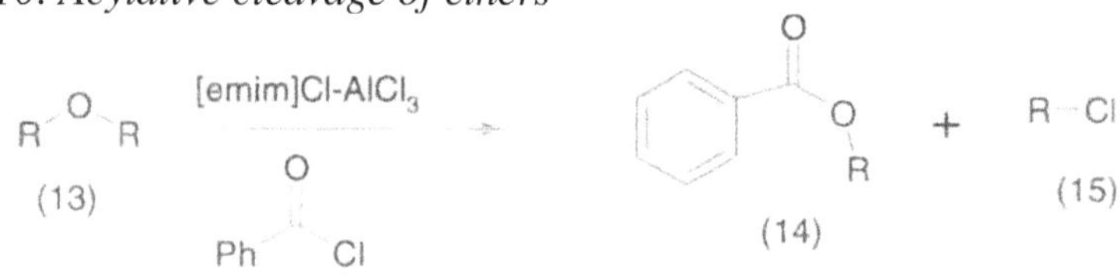

Imidazole-based or Neutral Ionic Liquids

These ionic liquids are currently favoured because of their characteristics, simplicity of handling, and convenience of preparation (Wilkes et al., 1992; Chauvin et al., 1995). These ionic liquids have been used for a variety of processes. The Diels-Alder reaction (Jaeger, 1989) will be discussed because it had great results and paved the way for further synthetic procedures to be investigated in the moisture stable room temperature ionic liquids, but only the more recent reactions will be covered in this literature survey (Keim et al., 2000).

Diels-Alder Reaction

As previously established, cyclopentadiene and methylacrylate react to produce exo and endo compounds. Numerous molecular solvents have been used to study the reaction, and the effects of these solvents on the reaction's endo/exo selectivity are well known. The more polar (endo) activated complex is believed to be caused by the solvent's "polarity." Solvophobic interactions, which create an internal pressure and encourage the reagents to associate in the solvent cavity during the activation process, have recently been ascribed to the effect.

Ionic liquids were believed to have the ability to significantly alter Diels-Alder processes because of their highly organised hydrogen bonding systems. One of the most practical C-C bond forming processes in organic chemistry is still the Diels-Alder reaction. For the Diels-Alder process, water has been employed as a solvent. The ionic liquid ethylammonium nitrate (EAN), a low melting fused salt with a melting point of 12°C, was employed by Jaeger and colleagues . EAN and water have a lot in common, including a high cohesive energy and characteristics related to its hydrophobic impact.

In order to produce endo (18) and exo (19) cycloaddition products, cyclopentadiene (16) was reacted with methyl acrylate (17a) and methyl vinyl ketone (17b). The Diels-Alder cycloaddition is seen in Scheme 5.

Figure 12. Diels-Alder reaction in ionic liquid

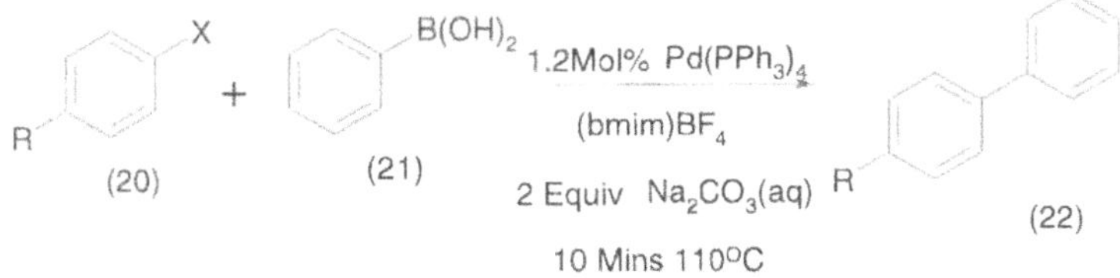

X = I, Br, Cl
R = H, Cl, $COCH_3$, CH_3, OCH_3

In comparison to non-polar solvents, EAN provided significant selectivity improvements for the reaction of (16) with (17a) and (17b), as well as a rate enhancement for the former (17a) (Jaeger, 1989). Although EAN was a better reaction medium in one way, its rate and endo selectivity were not as good as those of water in these reactions. Neutral organic molecules generally dissolve better in EAN than in water. This essentially resulted in the employment of EAN in water-sensitive Diels-Alder processes. Because of its potential for explosion in specific situations, EAN is not currently in use, (Hammond, 1983).

Suzuki Cross-Coupling Reaction

The synthesis of biaryls is the most effective use of the Suzuki cross-coupling reaction, which is a very flexible technique for creating new C-C bonds (Stranforth et al., 1998). Despite its utility, the reaction has some drawbacks, including poor reagent solubilities, catalyst breakdown, and catalyst loss into the result. In an attempt to address these issues, Stranforth and colleagues selected [bmim]BF4 as the ionic liquid for the reaction because of its wide liquid range (Mathews et al., 2000). 4-The original Suzuki conditions were used to react bromotoluene with phenylboric acid in [bmim]BF_4 (Suzuki et al., 1981).

After six hours, this produced 4-methylbiphenyl in a 30% yield using a catalyst that decomposed 1.2Mol% Pd(PPh3)4. By altering the process, higher reaction rates were attained and catalyst breakdown was avoided. The catalyst with the aryl halide in [bmim]BF_4 was heated to 110°C while being vigorously stirred as part of the novel process (see Scheme 6). After cooling the catalytic solution to room temperature, aryl boronic acid and a sodium carbonate aqueous solution were added to initiate the reaction. The results obtained for the different areas used are displayed in Table 4 overleaf.

Figure 13. Suzuki-cross coupling in ionic liquid

$$Ph_3P{=}CH\text{-}Y \quad + \quad PhCHO \quad \Rightarrow \quad PhCH{=}CH\text{-}Y \quad + \quad Ph_3PO$$

(25) (26) (27) (28)

a : Y = COMe b : Y = CO_2Me c : Y = CN

Additionally, the concentration of the catalyst was decreased, leading to higher yields. Overall, there were several benefits to the Suzuki cross-coupling reaction being successfully conducted in the ionic liquid $[bmim]BF_4$ at room temperature;

- Reaction showed a significant increase in reactivity at reduced catalyst concentration.
- Homo-coupled products can be eliminated, affording isolated products in relatively high purity.
- The reactions can be performed under air without loss of yield or catalyst decomposition.
- Catalyst can be recycled without loss of catalytic activity.

Table 4. Results of Suzuki cross-coupling reactions in [bmim]BF

X	R	% Yield
I	H	86
Br	H	93
Cl	H	1
Br	Cl	17
Br	$COCH_n$	67
Br	OCH_n	81

Nucleophilic Displacement Reactions

Phase transfer catalysis (PTC), as shown in scheme 8, is frequently used to perform nucleophilic displacements in order to promote the reaction between the organic reactants and the inorganic salts that supply the nucleophile (Chapman and Hall 1994). Between a polar phase containing the salt and a non-polar phase containing the organic reactant, the phase transfer catalyst typically a tetraalkylammonium salt serves as a shuttle for the reactant anion. Wheeler and colleagues investigated the use of ionic liquids in these processes (Wheeler et al 2001). Using [bmim]PF_6 as the ionic liquid, they decided to use potassium cyanide to displace the chloride of the benzyl chloride (23) and produce phenylacetonitrile (24) instead.

Figure 14. Nucleophilic displacement reaction in ionic liquid

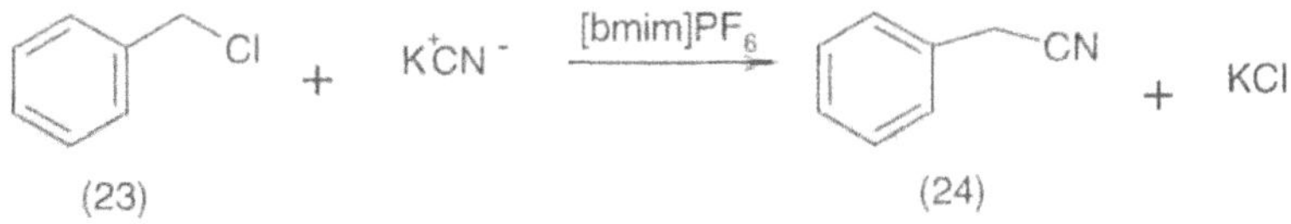

Ionic liquids appeared to be a good fit for reactants where PTC works because of their large organic cations (Wheeler et al., 2001). Due to the ionic liquid's decreased viscosity and the KCN salt's increased solubility, the ionic liquid-based reaction was conducted at a variety of temperatures, with the maximum rate occurring at 80°C. All things considered, it is evident that using ionic liquids in these PTC reactions appears to be advantageous, as the products were readily separated using safe techniques. The fact that the ionic liquid could be easily cleaned with water once the reaction was finished, removing the inorganic salts, and then reused in another reaction was a key component of this procedure.

This shows that ionic liquids are viable as solvents for interactions between organic chemicals and inorganic salts, opening the door to environmentally benign manufacturing.

Wittig Reaction

One of the most widely used techniques for creating alkenes is still the Wittig reaction, which typically provides good stereocontrol. Crystallisation and chromatography are typically used to solve the classical challenge of separating the alkene from the by-product (Ph_3PO). As demonstrated by Boulaire and colleagues in 2000, [bmim]BF_4 can be employed as a medium for Wittig reactions utilising stabilised ylides (see to scheme 9). This facilitates the recycling of the ionic liquid and the simpler separation of alkenes from Ph_3PO, (Boulaire et al., 2000).

Biginelli Reaction Using Ionic Liquids as Catalyst

Because of their pharmacological and therapeutic qualities, 3,4-dihydropyrimidin-2(lH)-ones(32) (Biginelli compounds) and their derivatives have recently attracted more attention (Ma et al., 2000). The dihydropyrimidine nucleus is synthesised using one-pot or multi-step methods. Using 3-dicarbonyl compounds with aliphatic and aromatic aldehydes and urea or thiourea in an ethanol solution with catalytic quantities of acid, the traditional Biginelli synthesis (scheme 10) is a one-pot condensation (Kappe, 1993) This method's drawbacks include poor yields, severe reaction conditions, and lengthy reaction periods. Lewis's acids, transition metals, and an appropriate proton source were found to be excellent catalysts for the process by Hu and colleagues (Hu et al., 1998).

Peng and Deng (2001) developed a novel method that used room temperature ionic liquids [bmim]BF4 and [bmim]PF6 as solvent-free catalysts for the Biginelli condensation.

Figure 15. Biginelli condensation reaction in ionic liquid

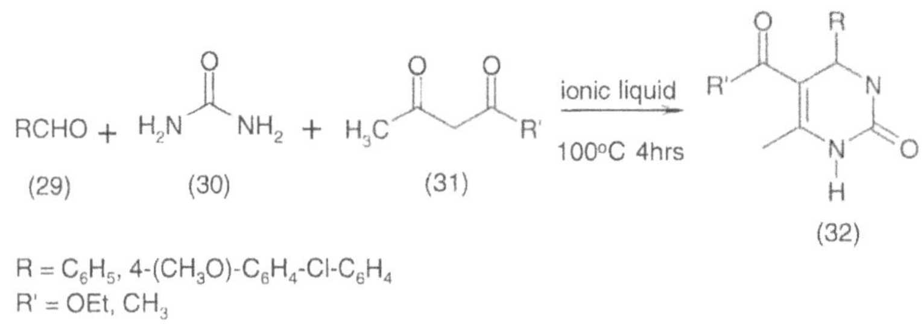

It was demonstrated that carrying out the reaction with only the reactants present and no ionic liquid produced no product, demonstrating the need for a catalyst. The ionic liquids [bmim]BF_4 and [bmim]PF_6 were used as catalysts in a range of processes. For the first time, the reaction was executed successfully and produced high yields, up to 95% in certain instances.

The following were the new reaction conditions' primary benefits:

- A basic catalyst system was employed.
- Quicker responses
- Easy synthetic process

The ionic liquid's cation/anion connection and other physical characteristics enable it to function as a catalyst in this reaction even if it isn't being employed as a solvent.

Nitration of Aromatic Compounds

An important reaction for industry is the nitration of aromatic compounds, which yields important organic intermediates. Concerns about the environment, disposal issues, and the regeneration of spent acids—all of which are produced via electrophilic aromatic nitration using common solvents—remain. Problems with product recovery arise because the $ArNO_2$ compounds generated are typically soluble in concentrated nitric acid. Additionally, the water byproduct—H_2SO_4 and HNO_3—that is produced during mixed acid nitration reduces the acidity and, consequently, the efficiency for continuous operations. Using potassium nitrate, Boon and colleagues (Boon et al 1987) conducted the first investigation of catalysis of electrophilic aromatic substitution processes in acidic [emim] chloroaluminates.

Using a variety of nitrating systems, [emim]X compounds, and aromatic substrates, Laali and colleagues (Laali & Gettwert 2000) investigated the use of ionic liquids as solvents for aromatic nitration.

Figure 16. Nitration with nitronium tetrafluorborate

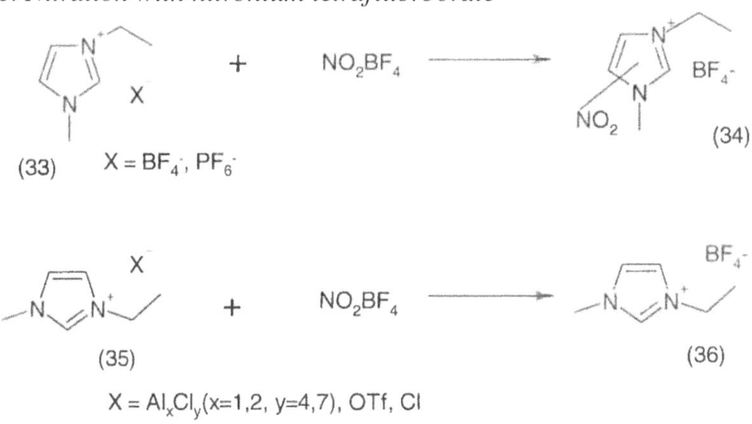

ROLE OF IONIC LIQUIDS IN GREEN SOLVENT CHEMISTRY

The development of ecologically safe and sustainable methods for the synthesis and design of ionic liquids is becoming more and more important as their use in a variety of applications grows. This section discusses the idea of "green synthesis" for ionic liquids and how important it is to lessen the environmental impact of these multipurpose solvents.

From Fossil Sources to Renewables

Ionic liquids were traditionally made from raw materials sourced from fossil fuels. Nonetheless, the use of sustainable and renewable feedstocks is a major trend in current research (Chen et al., 2019). The carbon footprint related to the manufacture of ionic liquids can be considerably decreased by substituting biomass-derived or other environmentally acceptable sources for petroleum-based starting materials. Green chemistry and environmental responsibility are in line with this shift. In addition to lessening the impact on the environment, using renewable feedstocks helps the chemical sector become more sustainable and conserve resources.

Designing Low-Toxicity Ionic Liquids

A key component of green synthesis is the creation of ionic liquids with reduced toxicity profiles. The selection of cations and anions utilised in the ionic liquid structure can affect toxicity. Combinations that preserve the intended qualities of ionic liquids while reducing their possible negative impacts on the environment and human health are being actively investigated by researchers (Zakrewsky et al., 2014). This method encourages the use of ionic liquids in a variety of applications in a safe and responsible manner. The creation of ionic liquids with low toxicity makes them more appealing as eco-friendly solvents and promotes their usage in a greater variety of applications, including those with strict safety and health regulations.

Energy-Efficient Production

Another area where sustainable design is being focused is on energy consumption during the synthesis of ionic liquids. The environmental advantages of ionic liquids may be compromised by high-energy operations. To lessen the energy footprint of production, researchers are creating more energy-efficient techniques such ionic liquid recycling and microwave-assisted synthesis (Suresh et al., 2020). Cost-effectiveness and environmental impact are both decreased by energy-efficient production methods. Ionic liquids' green credentials can be further supported by industries lowering their energy consumption and related greenhouse gas emissions through the optimisation of production processes.

Green Solvents in Ionic Liquid Production

Researchers are investigating the use of ionic liquids as solvents for their own industrial processes in an effort to further improve the green credentials of ionic liquids (Dupont et al., 2002). By using this method, fewer conventional volatile organic solvents—which can be bad for the environment—are used. A closed-loop system that reduces waste and its impact on the environment can be developed by utilising the solvation properties of ionic liquids. Green chemistry, which minimises waste and emissions throughout the production process, is in line with this closed-loop method, which lessens the environmental impact of producing ionic liquids.

Life Cycle Assessment

One useful method for assessing how ionic liquids affect the environment is a thorough life cycle assessment (LCA). LCAs take into account these chemicals' complete life cycle, from the extraction of raw materials to their synthesis, use, and disposal. Researchers and businesses can identify areas for improvement to lessen environmental effect and advance sustainability by performing life cycle assessments (LCAs) (Zhang et al., 2019). Making educated decisions on the usage of ionic liquids is made possible by this comprehensive approach. By ensuring that environmental factors are incorporated into the manufacturing and use of ionic liquids, life cycle assessments (LCAs) promote a more sustainable and accountable methodology.

REGULATORY CONSIDERATIONS AND ENVIRONMENTAL POLICIES

Examining the regulatory environment and environmental laws that impact the acceptance of ionic liquids is essential as their use grows. With an emphasis on their effects on sustainability and responsible use, we examine the environmental regulations and laws pertaining to ionic liquids in this area.

Green chemistry programs and ideas have been adopted by many nations and areas as a basis for chemical innovation and manufacturing. These programs promote the employment of sustainable and eco-friendly substitutes, such as ionic liquids, in chemical reactions (Anastas & Warner, 2000). The larger objectives of lowering the chemical industry's environmental impact and promoting sustainable practices are in line with support for green chemistry. Businesses and researchers show their dedication to environmentally friendly and responsible chemical processes by actively taking part in green chemistry projects.

Global Harmonization

The harmonization of regulations concerning chemical substances, such as ionic liquids, is an important global initiative that is currently underway. This process aims to create consistent regulations across different countries and regions. By doing so, it ensures a unified approach to managing the environmental impacts of ionic liquids on an international level. When regulations are consistent, it leads to better environmental protection and enhances sustainable practices. This consistency also supports industries and researchers working with ionic liquids, providing them with a fair and equal framework within which to operate.

A globally synchronized strategy for regulating ionic liquids plays a key role in reinforcing the commitment to environmental stewardship and sustainability across borders. It helps establish a shared understanding of what responsible use looks like, fostering cooperation among various stakeholders. This is especially important as the use of ionic liquids grows, given their potential benefits in many applications.

The relationship between regulations and environmental policies is crucial in guiding the responsible management of ionic liquids. Industries and researchers must comply with these regulations while also engaging actively with relevant environmental policies. Such engagement is vital for those who aim to enhance the sustainability of their projects. By observing these guidelines, the chemical community can assure that ionic liquids are used responsibly and in ways that meet environmental objectives. This approach not only protects the environment but also promotes innovation and safe use in various applications.

Traditional volatile organic solvents have been shown to be less harmful to the environment than ionic liquids. They improve safety in chemical processes, lower waste, and improve product quality. In line with the tenets of green chemistry, their application represents a real and useful transition towards more environmentally friendly industrial processes (Anastas & Warner, 2000). This transforming role is not without difficulties, though. Their wider implementation is hampered by high production costs, toxicity worries, and biodegradability problems.

The potential of ionic liquids to bring about change is significant. However, challenges remain that must be addressed to unlock their full capabilities. It is important to tackle these issues through new and creative methods of synthesis. Designing safer ionic liquids and creating biodegradable alternatives also play a crucial role in this process. These strategies are essential to ensure that ionic liquids can serve as effective eco-friendly and responsible solvents.

Looking ahead, the future of ionic liquids appears bright. Their applications are expanding beyond just green chemistry into various industries. This includes energy storage, where ionic liquids can improve battery performance. In pharmaceuticals, they show promise in drug delivery and formulation. They also have potential in environmental cleanup, helping to remove pollutants from water and soil.

Numerous real-world case studies illustrate how ionic liquids can help solve urgent global problems. These examples highlight their ability to promote a more sustainable and efficient approach in various sectors. The combination of innovative uses and practical applications positions ionic liquids as valuable tools for industry.

The connection between rules and environmental policies is vital for using ionic liquids responsibly. Industries and researchers need to comply with regulations to ensure they operate sustainably. Engaging actively with environmental guidelines is essential for maximizing the benefits of ionic liquids. A consistent set of global regulations is necessary. This ensures a unified approach to managing the environmental effects of ionic liquids worldwide.

The commitment of scientists and industries to these goals suggests a healthier and more eco-friendly future for the chemical industry. By addressing current limitations and aligning with regulatory frameworks, ionic liquids can play a key role in creating an environmentally responsible industrial landscape.

CONCLUSION

Ionic liquids (ILs) have become increasingly relevant in commercial processes, not only for their environmentally friendly properties but also for other beneficial characteristics. Their low volatility and low flammability add to their appeal in various applications. Because ILs consist entirely of ions, they offer unique benefits in situations that require high levels of electrochemical and thermal stability and conductivity, which are hard to find in traditional pure substances.

Though ILs typically come with a higher price tag compared to standard solvents, this initial cost can be justified. Their ability to be recycled effectively can lead to savings in the long run. Enhanced catalyst recovery, increased reaction rates, better selectivity, and improved separation of products are some of the advantages that can offset the upfront expenses. Numerous processes have demonstrated these benefits in real-world applications.

ILs are not limited to basic uses as solvents or catalysts. They are also finding roles in niche areas like performance additives and analytical materials. Their strong ability to dissolve various substances, along with their adjustable properties, creates significant competitive advantages over conventional solvents. However, simply swapping ILs into an existing process requires careful evaluation to ensure compatibility with that process.

While ILs may not be classified as "inherently green," they can enhance the sustainability of processes. They contribute to environmental and economic improvements through better solvent recyclability and more effective catalyst recovery. Adjusting the hydrophilicity of an IL can aid in separating products effectively. Nevertheless, it is essential to consider how changes in the structure of an IL can influence its toxicity.

Looking ahead, it is likely that the use of ILs in commercial settings will continue to grow over the next decade. As more industries adopt these versatile solvents, the range of applications will expand. ILs are expected to gain recognition as standard solvents worth considering in experimental setups for various reactions.

REFERENCES

Anastas, P. T., & Warner, J. C. (2000). Green Chemistry. *Green Chemistry*. Advance online publication. DOI: 10.1093/oso/9780198506980.001.0001

Anthony, J. L., Maginn, E. J., & Brennecke, J. F. (2001). Solution thermodynamics of imidazolium-based Ionic liquids and water. *The Journal of Physical Chemistry B*, *105*(44), 10942–10949. DOI: 10.1021/jp0112368

Bonhôte, P., Dias, A., Papageorgiou, N., Kalyanasundaram, K., & Grätzel, M. (1996). Hydrophobic, highly conductive ambient-temperature molten salts. *Inorganic Chemistry*, *35*(5), 1168–1178. DOI: 10.1021/ic951325x PMID: 11666305

Boon, J. A., Levisky, J. A., Pflug, J. L., & Wilkes, J. S. (1986). Friedel-crafts reactions in ambient-temperature molten salts. *The Journal of Organic Chemistry*, *51*(4), 480–483. DOI: 10.1021/jo00354a013

Brennecke, J. F., & Maginn, E. J. (2001). Ionic liquids: Innovative fluids for chemical processing. *AIChE Journal. American Institute of Chemical Engineers*, *47*(11), 2384–2389. DOI: 10.1002/aic.690471102

Chauvin, Y., Mußmann, L., & Olivier, H. (1995). Flüssige 1,3-Dialkylimidazoliumsalze als Lösungsmittel fur die Katalyse in Zweiphasensystemen: Durch Rhodiumkomplexe katalysierte Hydrierung, Isomerisierung und Hydroformylierung von Alkenen. *Angewandte Chemie*, *107*(23-24), 2941–2943. DOI: 10.1002/ange.19951072341

Davis, J. H.Jr, & Forrester, K. J. (1999). Thiazolium-ion based organic Ionic liquids (OILs).1,2 novel OILs which promote the benzoin condensation. *Tetrahedron Letters*, *40*(9), 1621–1622. DOI: 10.1016/S0040-4039(99)00025-8

Dupont, J., Fonseca, G. S., Umpierre, A. P., Fichtner, P. F., & Teixeira, S. R. (2002). Transition-metal nanoparticles in Imidazolium Ionic Liquids: Recycable catalysts for biphasic hydrogenation reactions. *Journal of the American Chemical Society*, *124*(16), 4228–4229. DOI: 10.1021/ja025818u PMID: 11960449

Earle, M. J., Esperança, J., Gilea, M. A., Canongia Lopes, J. N., Rebelo, L. P., Magee, J. W., Seddon, K. R., & Widegren, J. A. (2006). The distillation and volatility of Ionic liquids. *Nature*, *439*(7078), 831–834. DOI: 10.1038/nature04451 PMID: 16482154

Freemantle, M., Welton, T., & Rogers, R. D. (2009). An introduction to Ionic liquids. DOI: 10.1039/9781839168604

Greaves, T. L., & Drummond, C. J. (2007). Protic Ionic Liquids: Properties and applications. *Chemical Reviews*, *108*(1), 206–237. DOI: 10.1021/cr068040u PMID: 18095716

Green, L., Hemeon, I., & Singer, R. D. (2000). 1-Ethyl-3-methylimidazolium halogenoaluminate Ionic liquids as reaction media for the acylative cleavage of ethers. *Tetrahedron Letters*, *41*(9), 1343–1346. DOI: 10.1016/S0040-4039(99)02289-3

Han, H., Geng, X., Zhang, B., Meng, J., Liu, X., He, X., Liu, Z., Gao, Y., Liu, D., & Hu, X. (2019). Synthesis of novel functional Ionic liquids and their application in biomass. *RSC Advances*, *9*(51), 29652–29658. DOI: 10.1039/C9RA06243B PMID: 35531511

Hassan, R., Nazir, F., Roosh, M., Qaisar, A., Habib, U., Sajini, A. A., & Iqbal, M. (2022). Synthesis, characterization, biological evaluation, and in Silico studies of imidazolium-, pyridinium-, and ammonium-based Ionic liquids containing N-butyl side chains. *Molecules (Basel, Switzerland)*, *27*(19), 6650. DOI: 10.3390/molecules27196650 PMID: 36235187

Hu, E. H., Sidler, D. R., & Dolling, U. (1998). Unprecedented catalytic three component one-pot condensation Reaction: An efficient synthesis of 5-Alkoxycarbonyl- 4-aryl-3,4-dihydropyrimidin-2(1H)-ones. *The Journal of Organic Chemistry*, *63*(10), 3454–3457. DOI: 10.1021/jo970846u

Jaeger, D. A., & Tucker, C. E. (1989). Diels-Alder reactions in ethylammonium nitrate, a low-melting fused salt. *Tetrahedron Letters*, *30*(14), 1785–1788. DOI: 10.1016/S0040-4039(00)99579-0

Md. Javvad Ali, Praveen M., Ghouse A. and Fatima F. (2021) A brief review of ionic liquids: synthesis and applications. International J. of Biomed Research. 1(2); DOI: . (2021). *Application of Ionic Liquids in Drug Delivery*, 165-186. https://doi .org/DOI: 10.31579/IJBR-2021/011

Karadas, F., Atilhan, M., & Aparicio, S. (2010). Review on the use of Ionic liquids (ILs) as alternative fluids for CO_2Capture and natural gas sweetening. *Energy & Fuels*, *24*(11), 5817–5828. DOI: 10.1021/ef1011337

Laali, K. K., & Gettwert, V. J. (2000). Electrophilic nitration of aromatics in Ionic liquid solvents. *The Journal of Organic Chemistry*, *66*(1), 35–40. DOI: 10.1021/jo000523p PMID: 11429927

Ma, Y., Qian, C., Wang, L., & Yang, M. (2000). Lanthanide Triflate catalyzed Biginelli reaction. one-pot synthesis of Dihydropyrimidinones under solvent-free conditions. *The Journal of Organic Chemistry*, *65*(12), 3864–3868. DOI: 10.1021/jo9919052 PMID: 10864778

Mathews, C. J., Smith, P. J., & Welton, T. (2000). Palladium catalysed Suzuki cross-coupling reactions in ambient temperature Ionic liquids. *Chemical Communications (Cambridge)*, (14), 1249–1250. DOI: 10.1039/b002755n

Mutelet, F., Butet, V., & Jaubert, J. (2005). Application of inverse gas chromatography and regular solution theory for characterization of Ionic liquids. *Industrial & Engineering Chemistry Research*, *44*(11), 4120–4127. DOI: 10.1021/ie048806l

Nazari, S., Cameron, S., Johnson, M. B., & Ghandi, K. (2013). Physicochemical properties of imidazo-pyridine protic Ionic liquids. *Journal of Materials Chemistry. A, Materials for Energy and Sustainability*, *1*(38), 11570. DOI: 10.1039/c3ta12022h

Ngo, H. L., LeCompte, K., Hargens, L., & McEwen, A. B. (2000). Thermal properties of imidazolium Ionic liquids. *Thermochimica Acta*, *357-358*, 97–102. DOI: 10.1016/S0040-6031(00)00373-7

Oliver Kappe, C. (1993). 100 years of the biginelli dihydropyrimidine synthesis. *Tetrahedron*, *49*(32), 6937–6963. DOI: 10.1016/S0040-4020(01)87971-0

Peng, J., & Deng, Y. (2001). Ionic liquids catalyzed Biginelli reaction under solvent-free conditions. *Tetrahedron Letters*, *42*(34), 5917–5919. DOI: 10.1016/S0040-4039(01)01139-X

Plechkova, N. V., & Seddon, K. R. (2008). Applications of Ionic liquids in the chemical industry. *Chemical Society Reviews*, *37*(1), 123–150. DOI: 10.1039/B006677J PMID: 18197338

Poole, C. F. (2004). Chromatographic and spectroscopic methods for the determination of solvent properties of room temperature Ionic liquids. *Journal of Chromatography. A*, *1037*(1-2), 49–82. DOI: 10.1016/j.chroma.2003.10.127 PMID: 15214660

Sanders, J. R., Ward, E. H., & Hussey, C. L. (1986). Erratum: "Aluminum bromide-1-Methyl-3-Ethylimidazolium bromide Ionic liquids. I. Densities, viscosities, electrical conductivities, and phase transitions" [J. Electrochem. Soc., 133, 325 (1986)]. *Journal of the Electrochemical Society*, *133*(7), 1526–1526. DOI: 10.1149/1.2152158

Seddon, K. R., Stark, A., & Torres, M. (2000). Influence of chloride, water, and organic solvents on the physical properties of Ionic liquids. *Pure and Applied Chemistry*, *72*(12), 2275–2287. DOI: 10.1351/pac200072122275

Surette, J. K., Green, L., & Singer, R. D. (1996). 1-Ethyl-3-methylimidazolium halogenoaluminate melts as reaction media for the Friedel–crafts acylation of ferrocene. *Chemical Communications (Cambridge)*, (24), 2753–2754. DOI: 10.1039/CC9960002753

Wasserscheid, P., & Keim, W. (2000). Ionic liquids—New "Solutions" for transition metal catalysis. *Angewandte Chemie*, *39*(21), 3772–3789. DOI: 10.1002/1521-3773(20001103)39:21<3772::AID-ANIE3772>3.0.CO;2-5 PMID: 11091453

Welton, T. (1999). Room-temperature Ionic liquids. Solvents for synthesis and catalysis. *Chemical Reviews*, *99*(8), 2071–2084. DOI: 10.1021/cr980032t PMID: 11849019

Wheeler, C., West, K. N., Eckert, C. A., & Liotta, C. L. (2001). Ionic liquids as catalytic green solvents for nucleophilic displacement reactions. *Chemical Communications (Cambridge)*, (10), 887–888. DOI: 10.1039/b101202a

Wilkes, J. S. (2002). A short history of Ionic liquids—From molten salts to neoteric solvents. *Green Chemistry*, *4*(2), 73–80. DOI: 10.1039/b110838g

Wilkes, J. S., & Zaworotko, M. J. (1992). Air and water sTable 1-ethyl-3-methylimidazolium based Ionic liquids. *Journal of the Chemical Society. Chemical Communications*, (13), 965. DOI: 10.1039/c39920000965

Zakrewsky, M., Lovejoy, K. S., Kern, T. L., Miller, T. E., Le, V., Nagy, A., Goumas, A. M., Iyer, R. S., Del Sesto, R. E., Koppisch, A. T., Fox, D. T., & Mitragotri, S. (2014). Ionic liquids as a class of materials for transdermal delivery and pathogen neutralization. *Proceedings of the National Academy of Sciences of the United States of America*, *111*(37), 13313–13318. DOI: 10.1073/pnas.1403995111 PMID: 25157174

Zhu, H., & O'Dell, L. A. (2021). Nuclear magnetic resonance characterisation of Ionic liquids and organic Ionic plastic crystals: Common approaches and recent advances. *Chemical Communications*, *57*(46), 5609–5625. DOI: 10.1039/D1CC02151F PMID: 34048516

Chapter 4 Characterization Techniques of Single and Hybrid Nanoparticles

Pallavi Singh Chouhan
https://orcid.org/0000-0001-8755-3422
School of Pharmacy, Indira University, Pune, India

Amruta Prasann Walvekar
https://orcid.org/0000-0001-7262-8380
School of Pharmacy, Indira University, Pune, India

Shraddha Prakash Devarshi
https://orcid.org/0000-0002-6725-8917
School of Pharmacy, Indira University, Pune, India

Tejal Nikhil Bele
https://orcid.org/0000-0002-4170-6232
School of Pharmacy, Indira University, Pune, India

Rutuja Vijay Kamble
https://orcid.org/0000-0002-7919-7197
School of Pharmacy, Indira University, Pune, India

ABSTRACT

Nanostructures, including nanoparticles, are gaining significant interest due to their wide range of applications. Their size, elemental composition, crystal structure, and physical properties—such as reactivity, mechanical strength, conductivity, and optical behavior—are crucial for determining their performance. Smaller particles have greater surface area-to-volume ratios, enhancing reactivity, while elemental

DOI: 10.4018/979-8-3373-1245-3.ch004

2026, IGI Global Scientific Publishing. Copying or distributing in print or electronic forms without written permission of IGI Global Scientific Publishing is prohibited. Use of this chapter to ative artificial intelligence (AI) technologies is expressly prohibited. The publisher reserves all rights to license its use for generative AI training and machine learning model development.

composition affects stability and biological compatibility. Crystal structure influences hardness and conductivity, and other factors like shape, surface charge, and modification determine interactions. Various characterization techniques like XRD, SEM, TEM, FTIR, and VSM help analyze these properties. Each technique has its advantages, and combining methods often provides a more comprehensive understanding. Additional techniques, including EDX, UV-Vis, DLS, TGA, and spICP-MS, further enhance analysis by revealing elemental composition, optical properties, stability, and particle size distribution, aiding the development of nanoparticles for advanced applications.

INTRODUCTION

In modern medical field, the use of nanoparticles has become a popular trend. Research about particles ranging from 1 to 100 nm has gained increasing focus because of their diverse uses spanning drug delivery systems and cancer diagnosis and therapy routines, (Brigger et al., 2012; De Jong & Borm, 2008).

As nanoparticles' size, shape, and morphology differ from those of bulk materials, their catalytic qualities get improved. Many types of metallic nanoparticles, such as gold, silver, alloys, magnetic, etc., have wide applications in a variety of industries. (Figure 1). In the fields of biotechnology and medicine, the use of nanoparticles helps in transforming processes, making them more straightforward, secure, affordable, and portable. In current years, nanoparticles have gained interest with their unique qualities, such as their antimicrobial activity and tolerance to extreme heat and oxidation (Hasan, 2015).

Figure 1. Applications

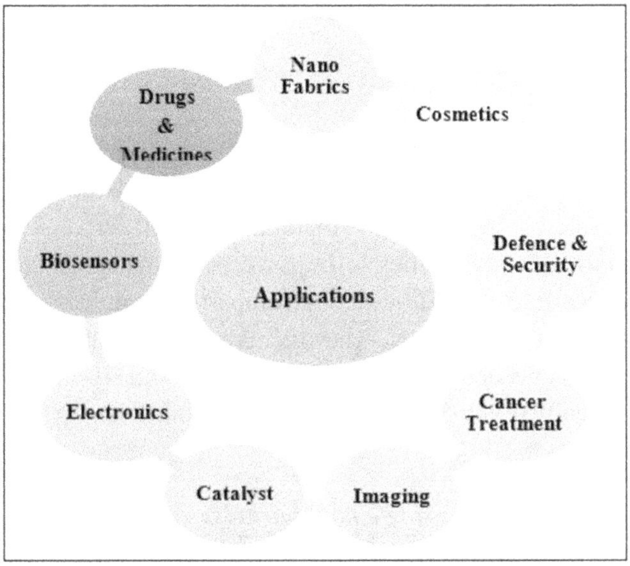

There are two general approaches for the synthesis of nanoparticles: top-down and bottom-up. Chemical procedures include adding a reducing agent which carry out reduction of metal ions into metal nanoparticles, while physical processes include a ball milling technique and other methods (Sin Cheow et al., 2013). Biological methods can also be used to synthesize nanoparticles. Microorganism-based (Jayaseelan et al., 2013) and green synthesis-related processes (Kumar et al., 2014; Elango et al., 2015) are the primary focus of biological approaches. Synthesis based on microorganisms is tedious and needs special attention. The plant extract's secondary metabolites aid in the process by serving as the reducing agent. It has been found that practically every section of the plant yields good extract preparation results.

Additional techniques for synthesis include reduction method, sol-gel, laser irradiation, microwave irradiation, radiolytic co-reduction, sonochemical co-reduction, electrochemical method and micelle synthesis, (Roopan et al., 2014).

Single Nanoparticles

These are discrete particles made up of a single substance or composition. Drug delivery, imaging, sensing, and catalysis are few of the uses for these nanoparticles. Metals (like gold and silver), metal oxides (like titanium dioxide), and polymers are examples of common materials used for them. These are preferred when simplicity and uniformity in material properties are required. They are cost-effective as compared to Hybrid particles.

Hybrid Nanoparticles

The nanoscale integrations of two or more distinct components make up hybrid nanoparticles. These could be various inorganic materials or combination of inorganic and organic components. The main aim of synthesizing them is to utilize each component's special qualities to improve performance. E.g., for improving stability or drug loading capabilities, gold nanoparticles could be mixed with silica or polymers. They are preferred over the other type when enhanced performance or multifunctionality is needed, (Rahman et al., 2022; Vargas-Bernal, 2020). Table 1 gives a comparison between Single & Hybrid Nanoparticles

Table 1. Comparison between single nanoparticles & hybrid nanoparticles

Feature	Single Nanoparticles	Hybrid Nanoparticles
Composition	Made from a single material (metal, polymer, etc.)	Made from two or more different materials
Properties	Depends on a single material's characteristics	Exhibit combined properties from multiple materials
Applications	Drug delivery, Sensors, catalysis, Imaging	Drug delivery, biosensing, Energy, Environmental applications
Stability	Often stable, but may be susceptible to aggregation	Improved stability, especially when combining organic/inorganic components
Customization	Limited to properties of a single material	High customization potential, tailoring properties for specific applications

(Altammar, 2023)

CHARACTERIZATION OF NANOPARTICLES

To ensure that the produced particles are at the nanoscale, characterization is very essential. "Characterization" in material science refers to the broad and general procedures used to study the structure and properties of the material. This basic process is crucial for a scientific understanding of the material. Characterization includes techniques needed to investigate microscopic structures and material qualities, i.e., any material analysis process which involves density calculation, mechanical testing, and thermal analysis. Characterisation approaches that have been used for years always find newer and sophisticated methods joining them. A few techniques are quantitative, others are qualitative. The many facets of nanoparticle characterisation are explored in this chapter, with an emphasis on popular and efficient methods for it. (Figure 2). The characterization is crucial for understanding the properties of nanoparticles. The unique properties of nanomaterials, arising from their size and composition, necessitate a diverse arsenal of characterization methods to fully elucidate their structure, morphology, and functionality.

Figure 2. Nanoparticle characterization

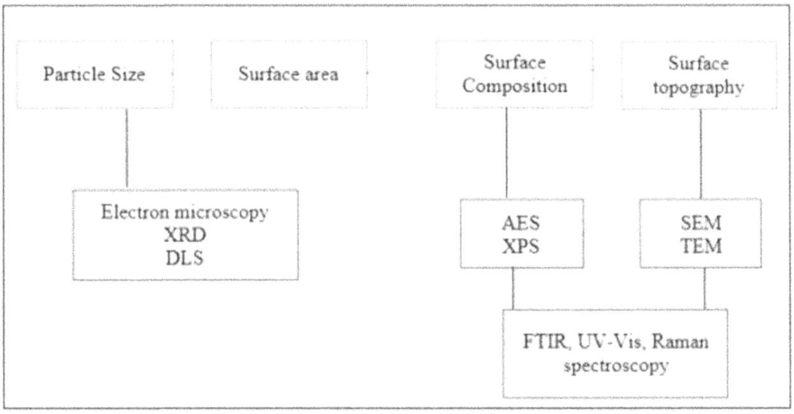

The chapter will discuss following key techniques-

1. X-ray Diffraction (XRD): Deciphering Crystalline Structure and Size
2. Scanning Electron Microscopy (SEM): Visualizing Morphology and Surface Features
3. Transmission Electron Microscopy (TEM): High-Resolution Imaging and Structural Analysis
4. Fourier Transform Infrared (FTIR) spectroscopy: Identifying Functional Groups and Surface Chemistry
5. Vibrating Sample Magnetometry (VSM): Measuring Magnetic Properties

The discussion will help in providing a detailed overview of their principles, experimental procedures, data analysis, and applications in the context of single and hybrid nanoparticles. Further, we will examine other complementary techniques to provide a holistic understanding of nanoparticle characterization.

X-RAY DIFFRACTION (XRD) ANALYSIS

The XRD method determines the crystalline nature of the examined material. Identification serves as the main purpose for XRD analysis but the technique also delivers extra details. The XRD pattern serves to calculate element ratios when samples consist of mixed components. The obtained data enables researchers to determine both crystallinity levels and deviations from pristine compositions of particular elements while revealing the sample's structural nature.

Principle

X-ray beam interaction with atomic planes leads to beam partial transmission while absorption and refraction and scattering and diffraction take place in the sample. X-rays reflect from elements differently based on atomic distribution and atom type.

The wavelength of electromagnetic X-ray energy occurs at shorter frequencies. Sufficiently energetic electrically charged particles will create X-rays after their deceleration process. High voltage connections between electrodes produce electron attraction that draws them toward the metal target. X-ray generation produces emissions from the impact area that spread out across all directions. The produced X-rays undergo collimation before they are directed toward a single finely ground sample. The detector picks up X-ray signals so the computer with a microprocessor or electronic system processes this information.

The X-ray scan procedure depends on changing the alignment between the X-ray source and sample and detection system. A crystal lattice scatters X-ray photons upon impact. Self-interference of light particles eliminates maximum scattering because this phenomenon is identified as destructive interference. Diffraction takes place when scattering waves match up with scatterings from different planes. New wave fronts develop through constructive interference at this point from the reflected signals.

Crystalline materials possess individual atomic arrangements that generate their exclusive X-ray diffraction pattern. Bragg's equation (Bragg, 1913) serves to determine the angle of diffraction which is given by

$$2d\ n\ \sin\theta\ \lambda$$

where d denotes the spacing between the planes, θ denotes the angle of incidence, n denotes an integer, and λ denotes the beam wavelength (Figure 3) (Taha Sochi, 2010)

Figure 3. Angle of refraction for crystalline materials

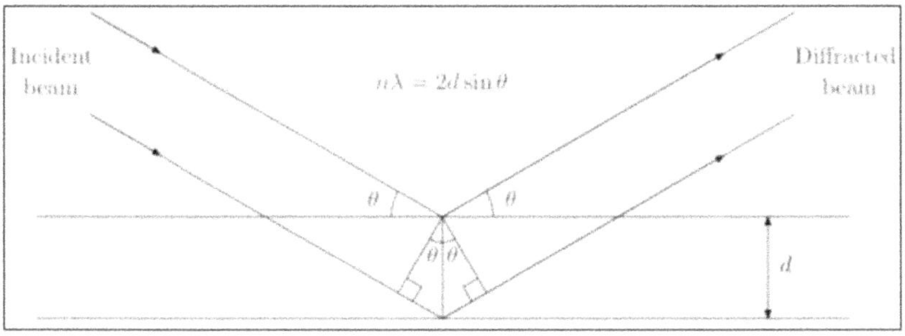

EXPERIMENTAL SETUP AND DATA ACQUISITION

A typical XRD instrument comprises an X-ray source (usually a sealed tube or rotating anode generating monochromatic X-rays), a sample holder, and a detector (e.g., scintillation counter, position-sensitive detector) Figure 4 (Omori et al., 2023). The sample, typically a finely powdered material, is placed in the holder, and the X-rays are diffracted at various angles. The detector measures the intensity of the diffracted X-rays at each angle, generating the diffraction pattern. Sample preparation is critical for accurate XRD analysis of nanoparticles. Careful dispersion of the nanoparticles in a suitable medium is essential to minimize preferred orientation effects. Two divergent slits exist both before the sample and X-ray source and also before the detector and sample. Divergent slits placed between the specimen and the X-ray source while simultaneously reducing background noise and constrained scattering radiation lead to the radiation collimation. The goniometer joins the sample holder and detector units to perform aligned specimen rotation compared to detector rotation at a 2:1 ratio. For some applications, such as the characterization of thin films, grazing incidence XRD may be employed (Kumar et al., 2023). This technique provides enhanced surface sensitivity, allowing for the study of thin layers or surface coatings on nanoparticles.

Figure 4. Scan of the sample's surface

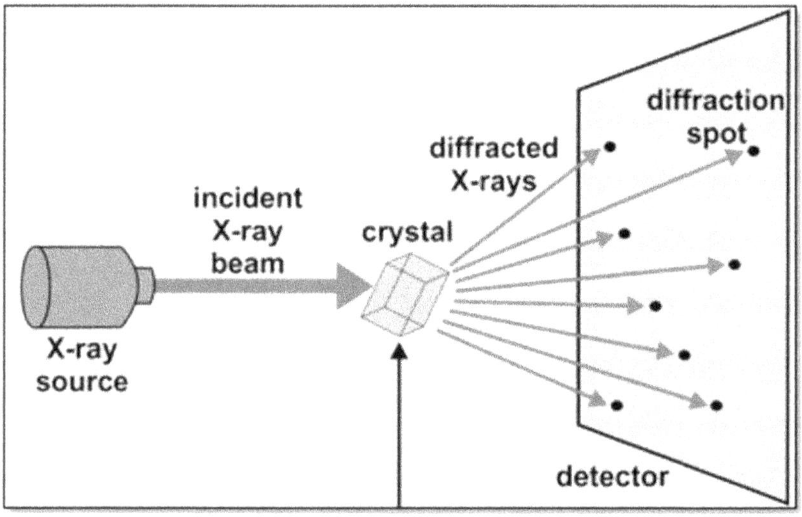

DATA ANALYSIS AND INTERPRETATION

Analysing XRD data involves several steps. Firstly, peak identification involves assigning observed diffraction peaks to specific crystallographic planes using established databases. This process allows the determination of the crystal structure and phase composition of the nanoparticles. Secondly, crystallite size calculation is performed using the Scherrer equation, as described above. However, peak broadening can also be caused by factors other than crystallite size, such as micro strain within the crystallites. The Williamson-Hall plot serves as a method which distinguishes between crystallite size effects and strain effects that broaden peaks. (Iyengar et al., 2014). This plot involves plotting $\beta \cos \theta$ against $\sin \theta$, where the slope and intercept provide information on the strain and crystallite size, respectively (Sagayaraj et al., 2017). The analysis of XRD patterns can provide valuable information about the crystalline nature, phase purity, and crystallite size of nanoparticles. Nevertheless, it's important to understand that crystalline areas are where XRD becomes particularly sensitive. Nanoparticles that are amorphous or weakly crystalline could not show distinct, sharp diffraction peaks, which might result in challenges in accurately determining size.

APPLICATIONS IN NANOPARTICLE CHARACTERIZATION:

XRD has been extensively used to characterise numerous single and hybrid nanoparticle types. Ullah (2024), for example, used the aqueous seed extract of Myristica fragrans to synthesise and characterise CuO, NiO, alongside hybrid nanoparticles comprising both CuO and NiO. The XRD results revealed the crystalline nature of the prepared nanoparticles and their estimated average sizes. Studies on the synthesis of silver nanoparticles (Sharma et al., 2022) as well as studies on the characterisation of hybrid polymer-metal-carbon nanomaterials synthesized (Ozkan et al., 2018), XRD results have been presented to ascertain whether the phases responsible for the end product were properly formed or if the material synthesis was of an amorphous nature. Based on a detailed study of magnetite/maghemite based hybrid nanoparticles associated with gadolinium hydroxide (Schneider et al., 2023) the hybrid formation has been revealed by XRD. Sagayaraj et al., 2017 has characterized PVP coated $ZnFe2O4$ nanoparticles. The XRD pattern confirmed the spinel structure and phase purity In fact, XRD has been widely used in several studies to characterize the dimension and crystalline structure of many other nanoparticles such as CuO (Amin, 2021), CuO & ZnO (Patel et al., 2022), $Fe3O4$ (Elmobarak et al., 2021) etc. The examples clearly reflect the versatility and widespread applicability of XRD in nanoparticle characterization.

Scanning Electron Microscopy (SEM) Analysis

Materials may be observed with a decent level of detail at the micron level thanks to optical microscopy. It is challenging to obtain higher resolution in optical microscopes due to aberrations and wavelength limitations. Consequently, other imaging methods like as AFM (Binnig et al., 1986), SEM, TEM, and STM (Pasquini et al., 2005) have been developed technologically to identify submicron-sized compounds. Although the underlying principles of these methods differ, they always result in a picture that is greatly enlarged. These imaging methods allow for the observation of nanoparticles.

`Scanning electron microscopy (SEM) employs an electron beam to view the object, in contrast to an optical microscope, which uses a light beam (Iyengar et al., 2014). At the top of the device is an electron cannon that blasts a highly concentrated electron stream. The two primary kinds of electron guns are thermionic guns, which heat the filament until the electrons stream away, and field emission guns, which create a powerful electric field that rips electrons from the atom.

High-energy electron beams are used by the SEM to scan the sample's surface (Sagayaraj et al., 2017). Because SEM uses light waves to generate a magnified picture, it is different from traditional light microscopes. The electron beam in a

SEM interacts with the specimen surface as it hits it. X-rays and three different kinds of electrons—primary, secondary, and Auger—are released when the object is struck by the incoming electron beam. The primary, or backscattered, and secondary electrons are used in SEM. SEM uses secondary electrons to create high-resolution pictures that expose details of around 1–5 nm. EDX is a technology that uses distinctive X-rays to detect elemental compositions. This method likewise uses the backscattered electrons to create the picture. Figure 5 (Sharma et al., 2022) displays a schematic picture showing how SEM operates. The electron beam is directed to the final lens by the scanning coils in the electron column. In order to accomplish raster scanning throughout the rectangular region of the surface, this deflects the beam in both vertical and horizontal directions. Electronic equipment are used to detect and amplify signals, which are then shown as pictures on a cathode ray tube. The microscope and raster scanning are in sync. The graphic displays a dispersion map of the signal strength originating from the scanned area of the specimen.

Figure 5. Working of SEM

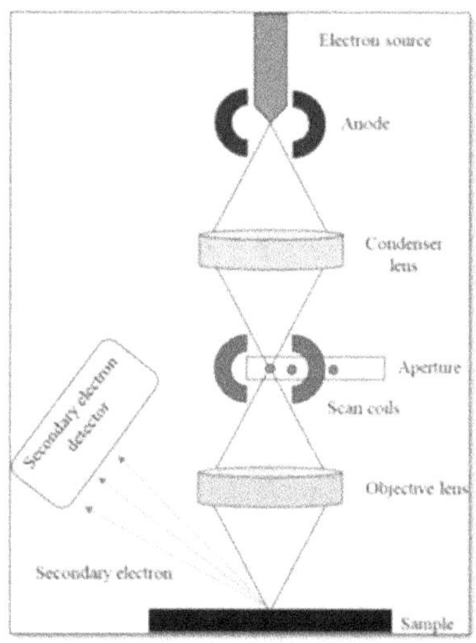

SEM Sample Preparation

Proper sample preparation is essential for SEM analysis to function. Metals don't need to be prepared because they already conduct electricity when exposed to electron bombardment. However, a sputter coater is required to prepare the non-metals (Nixon, 1971). A thin coating of conducting substance, often gold, is applied to the specimen. Gold is produced by using an electric field and argon gas. Positively charged ions are created when the electric field extracts an electron from the argon. These positively charged ions are drawn to the negatively charged gold foil. Gold atoms are released by the argon ions and land on the object, creating a thin layer of conductivity. Environmental SEM or SEM in low voltage mode can be used to photograph non-conductive materials (Hortolà, 2005). Since water molecules would vaporise in a vacuum, obstructing the electron beams and reducing picture clarity, water removal is also crucial in conventional SEM. Modern SEMS can now function without a complete vacuum. Although the pictures produced by these devices are of lesser resolution, they make it possible to examine a wide variety of samples that were previously unexplored, which is crucial for companies and other fields.

Data Acquisition and Image Analysis

SEM images are acquired by scanning an electron beam travelling across the surface of the sample and detecting this emitted signals. These signals are then converted into an image, providing a visual representation of the sample's surface topography and composition. Image analysis techniques are used to quantify the nanoparticle size, shape, morphology, and surface features (Ullah et al., 2024), (Dada et al., 2018). Software packages are commonly used for automated particle size analysis, providing statistical data on size distribution, average particle size, and other relevant parameters (Ozkan et al., 2018). Careful image analysis is crucial for obtaining reliable and reproducible results.

APPLICATIONS IN NANOPARTICLE CHARACTERIZATION

Single and hybrid nanoparticle characterisation has made extensive use of SEM.In the Myristica fragrans-mediated synthesis of CuO, NiO, along with CuO-NiO hybrid nanoparticles, SEM images revealed the form, shape and dimension of the resulting nanoparticles, showing differences in shape and dispersion between CuO, NiO, and the hybrid nanoparticles. Amin et al, 2021 has synthesized copper oxide nanoparticles using Aerva javanica extract from leaves was utilised to produce copper oxide nanoparticles, and SEM was used for characterisation. The obtained images demonstrated that the particles were spherical and ranged in size from 15 to 23 nm. Since the geometry of metal nanoparticles primarily affects their optical and electrical properties, (Dalir et al. 2020) used SEM to discover that the synthesised metal particles were spherically shaped and widely distributed (Moradi et al., 2021) has synthesized nickel magnetic nanoparticles to create magnetic graphene oxide, which is then used as a new support to create a copper catalyst. This catalyst has a nanosheet structure, according to the SEM data. Understanding the aggregation state, shape distribution, and surface characteristics of nanoparticles—all of which are essential for comprehending their characteristics and their uses—can be gained from SEM pictures.

However, it is important to note that SEM images can be affected by sample preparation artifacts, such as charging or aggregation. To get accurate and repeatable findings, careful picture analysis is essential.

Transmission Electron Microscopy (TEM) Analysis

TEM offers the highest resolution imaging capabilities among microscopy techniques,

That makes it possible to visualize individual nanoparticles and their inner structure. A thin sample is exposed to a high-energy electron beam during TEM. The picture produced by the electrons' interaction with the material shows its morphology, crystal structure, and defects. Figure 6 shows the basic components in TEM. Different TEM modes High-resolution TEM (HRTEM), dark-field, as well as bright-field imaging provide different types of information. Bright-field imaging provides information on the sample's morphology and contrast, while dark-field imaging highlights specific crystallographic planes or features. HRTEM provides atomic-resolution images, allowing for the direct visualization of crystal lattices and defects (Dada et al., 2018).

Figure 6. Basic components of TEM

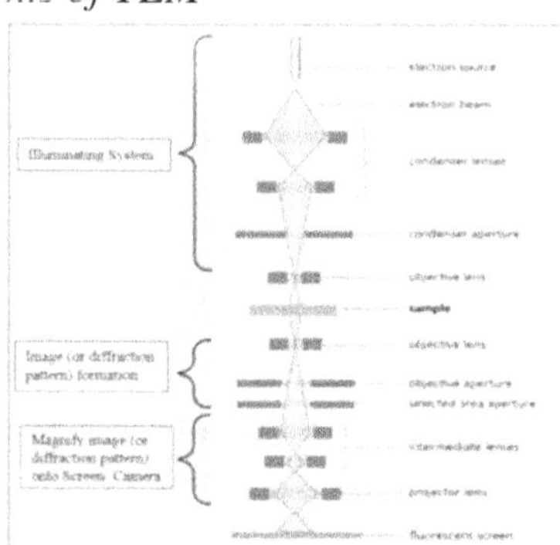

Principle

TEM images have a different contrast than light microscopic imagesWhen the electron beam interacts with the material, diffraction occurs rather than absorption.

The diffraction strength varies according on the plane's orientation with regard to the electron beam.

At certain angles, the electron beam is transmitted, whereas at other angles, it is considerably diffracted off the axis. The specimen may be tilted to achieve a certain diffraction condition thanks to holders that are fitted for it. The term “light field” refers to the process of creating a contrast image by allowing the unscattered electrons to pass through an aperture that blocks the deflected electrons. A dark field picture is another image that may be created using the deflected electrons.

Sample Preparation for TEM

Sample preparation for TEM analysis of nanoparticles is typically more complex than for SEM. The sample must be extremely thin (typically less than 100 nm) to allow for electron transmission. Various methods are employed for sample preparation, including drop-casting, embedding in resin, and ion milling (Kumar et al., 2023). The choice of method depends on the nature of the nanoparticles and the desired information. A diluted sample containing the specimen is deposited onto support grids to create materials with lower dimensions. Under cryogenic temperatures, an ultramichrotome equipped with a diamond knife is used to cut certain substances into thin films. Extremely thin samples that can tolerate the high vacuum inside the device are most preferred. For some applications, staining techniques can be used to enhance contrast (Cheville & Stasko, 2013).

Data Acquisition and Image Analysis

TEM images are acquired by focusing the electron beam on the sample and detecting the transmitted electrons. The images provide a high-resolution representation of the sample's structure and morphology. HRTEM images can be analysed to determine the crystal structure and orientation of the nanoparticles. Software packages are available for automated particle size analysis, providing statistical data on size distribution, average particle size, and other relevant parameters.

APPLICATIONS IN NANOPARTICLE CHARACTERIZATION

TEM has been widely used to characterize single and hybrid nanoparticles, providing high-resolution images that reveal details not accessible by other techniques. In order to compile information on the morphological characteristics of the nanomaterials based on TEM images, Lee et al, 2020 used a mass-throughput image analysis technique to analyze a large number of nanoparticles. This can be a useful tool for assessing the dependability and building large data sets of nanoparticles, in addition to the numerous applications where the morphological properties of the particles are crucial, as in catalysis and optics.

The study by Elmobarak, 2021 has highlighted the preparation of Fe_3O_4–SiO_2 nanoparticles as a magnetic demulsifier (MD) using a modified Stober process at room temperature. The analytical techniques including TEM supported the formation of grafted magnetite Fe_3O_4 nanoparticles covered with a silica surface.

Zhang et al. (2023) et al has proposed an analysis method that combines the use of the commonly used software, *ImageJ*, with *Origin* to analyze TEM images. With only a few clicks, this approach has been found to provide an accurate particle size distribution using SiO2 nanoparticles as a template. Time was saved and the possibility of human error in manual measurement was decreased with this semi-automated method. Additionally, this approach was said to have the ability to transform the analysis of nanoparticle size and produce more precise and effective outcomes.

Similarly, TEM has been instrumental in characterizing a wide range of other nanoparticles as silver nanoparticles (Sharma et al., 2022), polymer-metal-carbon hybrid nanomaterials (Ozkan et al., 2018), and as ZnO and CuO (Patel et al., 2022).

However, it is important to consider that electron beam damage can occur, especially for sensitive materials. Additionally, the sample preparation process can introduce artifacts. Therefore, careful sample preparation and image analysis are essential for obtaining reliable results.

Fourier Transform Infrared (FTIR) Spectroscopy

Fourier transform infrared (FTIR) spectroscopy is an important method that has proved effective for identification of functional groups and surface chemistry of organic, inorganic, and polymeric materials. It easily reflects alterations in the characteristic pattern of absorption bands with regard to a change in the composition of material. A schematic diagram of FTIR is represented in Figure 7 (Nuha Al-Harbi & Abd-Elrahman, 2024). The standard FTIR spectrometer comprises of a source, sample cell, detector, amplifier, A/D convertor, and a computer. The interferometer is used in the passage of radiation from the sources to the detector. The signal is converted to a digital signal through the A/D convertor and amplifier. The computer then makes the signal transfer where Fourier transform is carried out. The method is based on the principle that molecules absorb infrared radiation at specific frequencies that correspond to their vibrational modes. The sample transmits between 10,000 and 100 cm−1 of infrared light, of which some are absorbed and others flow through. The sample transforms the absorbed radiation into rotational or vibrational energy. The detector's signal, which typically ranges from 4000 to 400 cm−1, is a representation of the molecular fingerprint of the material. Since each molecule has a distinct fingerprint, FTIR is a crucial technique for identifying chemicals. The FTIR spectrum is a plot of absorbance or transmittance versus wavenumber (cm^{-1}), which provides a fingerprint of the molecular structure and functional groups present in the sample. This information is particularly valuable for characterizing the surface chemistry of nanoparticles, including the identification of capping agents, ligands, and surface modifications (Taha et al., 2013).

Figure 7. Schematic diagram of FITR

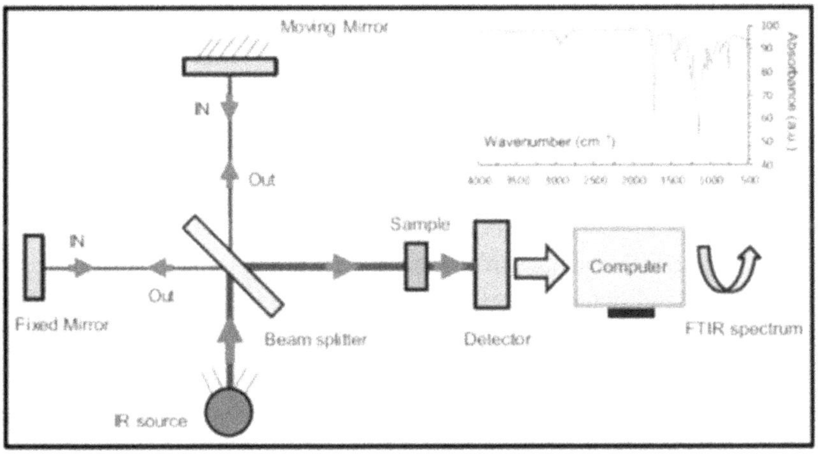

Sample Preparation for FTIR

Sample preparation for FTIR analysis of nanoparticles is a function of the physical state of the sample. For example, solid samples can be analyzed through techniques such as KBr pellet preparation where nanoparticles are mixed with KBr and pressed into a pellet. On the other hand, attenuated total reflectance (ATR) spectroscopy can be employed, which requires minimal sample preparation and is specifically suitable for direct analysis of nanoparticles. A liquid cell can directly analyze liquid samples. The selection of sample preparation method depends on the nature of the nanoparticles and the information that is desired.

Data Acquisition and Spectral Interpretation

The FTIR spectra are compared with known spectral databases for the interpretation of the observed absorption bands, which identifies the specific functional groups and facilitates determination of the surface chemistry. Particular absorption bands evidenced the presence of surface modification or specific ligands, which are necessary to be understood in the respect of properties, stability, and application of the nanoparticles. Overlapping absorption bands can sometimes make the spectral interpretation challenging.

APPLICATIONS IN NANOPARTICLE CHARACTERIZATION

FTIR spectroscopy is a useful tool for characterizing the surface chemistry of nanoparticles, especially in hybrid nanoparticles where the interaction between different components is of interest. In the Myristica fragrans-mediated synthesis of CuO, NiO, and CuO-NiO hybrid nanoparticles (Ullah et al., 2024), FTIR analysis confirmed the presence of functional groups in the plant extract that acted as reducing and capping agents during the synthesis process. Hussain Ali et al., 2016 has synthesized Iron Oxide NPs through a green synthesis route from leaves of Ficus Palmata in which FTIR analysis showed that several groups were involved in reduction and stabilization and helped in confirming the capping of Iron Oxide NPs by the phytochemicals present in the plant extract. Similar studies on silver nanoparticles (Sharma et al., 2022) and other nanoparticles have used FTIR to identify capping agents and surface modifications. results revealed the surface molecular interactions crucial for nanoparticle functionalities. (Dulog et al., 2024) used FTIR to analyze the generated AuNP/TiO_2 hybrid nanoparticles and discovered that they had a bigger pore size than TiO_2 NPs, which suggests that they have a better adsorption capacity. Furthermore, TiO_2 NPs are an excellent candidate for photocatalysis due to their distinct band gap and electron-hole pair production.

Vibrating Sample Magnetometry (VSM) Techniques

It is a technique used to measure the magnetic properties of materials, including nanoparticles. In VSM, a sample is vibrated in a uniform magnetic field, and the induced voltage in a near pickup coil is measured. The voltage is proportional to the magnetization of the sample, thus allowing the determination of magnetic properties such as saturation magnetization (Ms), remanence (Mr), coercivity (Hc), and magnetic susceptibility. Coercivity (Hc)" is the strength of the magnetic field needed to completely demagnetize a material, essentially measuring its resistance to demagnetization; "magnetization (Ms)" is the maximum magnetic moment per unit volume that a material can achieve; and "remanence (Mr)" is the residual magnetization that remains in a material even after the removal of applied magnetic field. All of these properties are commonly represented on a hysteresis loop. These properties are vital for the comprehension of magnetic nanoparticle behavior and, consequently, for its application in the areas of MRI, magnetic hyperthermia, and magnetic drug delivery (Wang et al.,2016).

Sample Preparation for VSM

Sample preparation consists of collecting nanoparticles in powdered form. For the case of liquid samples, the solvent must be completely removed so that it does not interfere with the measurement. The sample amount needed to take VSM measurements depends on the sensitivity and the magnetic property of the nanoparticle. Thus, careful sample preparation is necessary for one to obtain accurate and reproducible results (Kumar et al., 2023). Inconsistent packing density can lead to errors in the magnetization measurements.

Data Acquisition and Magnetization Curve Analysis

VSM measurements involve applying a magnetic field to the sample and measuring the resulting magnetization. A typical VSM measurement generates a magnetization curve (hysteresis loop), which shows the relationship between the applied magnetic field and the sample's magnetization. It provides crucial understanding about the magnetic behaviour of the nanoparticles and their suitability for various applications.

APPLICATIONS IN NANOPARTICLE CHARACTERIZATION:

VSM is widely used to characterize the magnetic properties of magnetic nanoparticles, particularly those containing iron oxides, cobalt ferrites, or other magnetic materials. Studying the magnetic properties of nanoparticles is important because they exhibit unique behaviours that differ from bulk materials, such as superparamagnetism. These properties make them valuable for various applications, including biomedical uses (e.g., drug delivery and cancer treatment), MRI contrast agents, data storage, sensors, and environmental clean-up. Understanding their magnetic characteristics helps improve technologies in computing, medicine, and energy, while also advancing our knowledge of magnetism at the nanoscale.

Nasirpouri et al. (2023) have used a technique of direct reduction–precipitation with tetramethylammonium hydroxide to create size-controlled magnetite nanoparticles. Zero remanence and zero coercivity on the magnetization loop demonstrated that all of the samples under study were superparamagnetic. The NP size has a linear relationship with the saturation magnetization. (Andrade et al., 2012).

First order reversal curves are an assembly of partial hysteresis loops that originate from the primary loop, as explained by Kumari et al. In a magnetic system, first order reversal curves (FORC) are effective for determining the domain size, composition, and interaction. It is an important technique for determining the effective magnetic particle size in a semi-quantitative manner. In some circumstances, FORC could make it easier to detect secondary or minor magnetic properties. VSM magneto metry was used to obtain such FORC measurements, (Kumari et al.,2014)

Nasirpouri et al (2023) has synthesized γ-Fe_2O_3 nanoparticles. Vibrating sample magnetometry (VSM) measurements of hysteresis loops revealed that mesoporous γ-Fe2O3 nanoparticles were superparamagnetic. When compared to non-calcined samples, the first-order reversal curve (FORC) diagram showed that calcined materials create a mesoporous structure that decreases coercive distribution (Hc) and magnetostatic interaction (Hu). To remove Pb(II) heavy metal ions from an aqueous solution, they used mesoporous γ-Fe2O3 nanoparticles as an adsorbent.

However, it is important to note that VSM measurements can be affected by sample preparation artifacts, such as aggregation or preferred orientation. Therefore, careful sample preparation and data analysis are essential for obtaining reliable results.

OTHER RELEVANT CHARACTERIZATION TECHNIQUES: A BROADER PERSPECTIVE

Beyond the five primary techniques discussed above, several other complementary methods are frequently employed for a comprehensive characterization of nanoparticles. These techniques often provide information that is not readily accessible through the primary methods.

Dynamic Light Scattering (DLS)

It is one of the method for figuring out the hydrodynamic size and size distribution of nanoparticles in solution (Loeschner et al., 2023). Using this non-destructive spectroscopic examination method, one can measure the size of particles suspended in a liquid between 1 and 1000 nm.

The particles suspended in a liquid experience a random movement termed as Brownian motion. The Doppler effect, which says that the frequency of light scattered by a particle is shifted depending on the particle velocity, serves as the basis for the DLS measurement. A quicker fluctuating intensity signal is produced because little particles go through the medium more quickly than large ones. Information on the physical characteristics of the sample can be obtained by detecting dispersed light from light-matter interactions. It also offers information on the state of nanoparticle aggregation. While the unaggregated solution will have a diameter that is comparable to or greater than TEM size, the aggregated solution will have hydrodynamic diameters that are larger than TEM size. The size and form of the macromolecules affect the hydrodynamic diameters (Jia et al., 2023)

Energy Dispersive X-ray Spectroscopy (EDX)

Energy dispersive X-ray spectroscopy (EDX), often integrated with SEM or TEM, is a technique used to determine the elemental composition of materials. The energy of the X-rays is specific to each element, allowing the identification and quantification of the elements present in the sample. The conducting sample emits X-rays when an electron beam with an energy of 10–20 KeV falls on its surface; the energy of the X-rays depends on the substance being studied. Since the X-rays are generated in an area that is roughly two microns deep, EDX does not fall under a surface science technique. It is possible to create an image of every element in the sample by guiding the electron beam across the substance. Due to the low X-ray intensity, obtaining the photographs typically takes many hours.

The composition or quantity of nanoparticles near and at the surface can be estimated using the EDX if there are any heavy particles present. For instance, nanoparticles like silver, gold, and palladium on the surface can be easily identified using EDX. Elements of low atomic number are difficult to detect by EDX.

UV-Vis Spectroscopy

The components of this technique include a monochromator, a detector, sample and reference beams, and a deuterium or tungsten lamp for the ultraviolet and visible area wavelengths, respectively. When the sample is exposed to UV radiation, the UV spectrum is produced.

Principle and Working

A portion of the light beam may be absorbed when it passes through the solution, while the remainder will be transmitted. Transmittance is defined as the ratio of light entering the sample to light leaving the sample at a specific wavelength. Absorbance is the negative logarithm of transmittance. The Beer-Lambert law$A = \varepsilon cl$, where A is the absorbance, ε is the absorptivity, c is the concentration, and l is the route length, indicates that the absorbance of a sample at a given wavelength is directly related to thc conccntration of the absorbing material and the path length. The absorbing material should be uniform and free of radiation scattering, and the incident radiation should be monochromatic.

Nanoparticles often exhibit unique optical properties due to their size and shape, which can be characterized by UV-Vis spectroscopy. For example, metallic nanoparticles exhibit surface plasmon resonance (SPR), a phenomenon where the electrons in the nanoparticle collectively oscillate in response to incident light. The SPR peak is highly sensitive to the size, shape, and dielectric environment of the nanoparticles, providing valuable information about their optical properties. Silver particles, for example, exhibit a plasmon resonance peak with an absorption at approximately 420–500 nm. In the case of silver nanoparticles, the shape and location of the plasmon absorption are influenced by the dielectric medium and particle size. (Mehrin et al.,2024)

Thermogravimetric Analysis (TGA)

The amount of surface coverage and the mass to mass ratio of NP to stabiliser are not revealed by FTIR, despite the fact that it provides information about the NP–stabiliser interaction and confirms the type of stabiliser. These factors are crucial for normalising saturation magnetisation values to purely metallic content, for example.

Information about the mass and composition of the stabilisers is provided by TGA. By heating a sample of nanomaterials, components with varying degradation temperatures decompose and evaporate, and a change in mass is noted. The TGA instrument records the temperature and mass loss, and it determines the kind and quantity of NP organic ligands based on the initial sample mass.

The same thermal breakdown mechanism as TGA is used by a technique called microthermogravimetric analysis (μ-TGA), however the mass of the sample under investigation is in the order of 1 μg, having the ability to detect mass changes below 1 nanogram. The detection limits of traditional TGA can be greatly enhanced in this manner. Mansfield et al. confirmed the occurrence of PEG coating on silica NPs and determined the amount and presence of surface-bound ligand coverage on Au NPs using μ-TGA.140 Their findings showed that this method is a reliable way to quantify NP coatings, and that it is occasionally possible to obtain data on the NPs' purity and composition. The authors emphasized the benefits of TGA, a straightforward method that requires no specific sample preparation other than the sample being dry. (Mansfield et al.,2014)

Single Particle Inductively Coupled Plasma Mass Spectrometry (spICP-MS)

A cutting-edge method that has transformed nanoparticle characterisation is Single Particle Inductively Coupled Plasma Mass Spectrometry (SP-ICP-MS), which allows the analysis of individual nanoparticles at extremely low concentrations. Because SP-ICP-MS offers real-time single-particle identification, it is an effective method for determining the concentration, elemental makeup, and size distribution of nanoparticles in complicated samples.

Grasso et al. (2021) used SP-ICP-MS to analyze engineered silver nanoparticles (AgNPs) in food matrices. This research focused on assessing the particle size distribution and concentration of AgNPs in various food products, revealing how these nanoparticles might behave in the human digestive system and their potential to bioaccumulate. SP-ICP-MS provided detailed information about nanoparticle dispersion, size, and elemental composition, which could inform regulatory decisions on the use of nanomaterials in food.

Heilgeist et al. (2021) used SP-ICP-MS to investigate the fate and transport of titanium dioxide (TiO_2) nanoparticles in aquatic environments. The study successfully tracked the size distribution and concentration of TiO_2 nanoparticles in water samples, providing valuable insights into the potential environmental risks posed by these nanoparticles. SP-ICP-MS was crucial in identifying TiO_2 nanoparticles at environmentally relevant concentrations and understanding their distribution in various aquatic matrices.

COMPARATIVE ANALYSIS OF CHARACTERIZATION TECHNIQUES

The nature of the nanoparticles and the particular attributes of interest play a major role in the choice of characterisation approach or techniques. A variety of methodologies is typically needed to acquire a comprehensive knowledge, as no single methodology can provide a complete picture. Table 2 showcase a comparison between features of different characterization techniques.

Table 2. Comparative analysis of characterization techniques

Technique	Strengths	Limitation	Information Provided
XRD	Provides information on crystal structure, phase purity, and crystallite size. Relatively inexpensive and widely available.	Limited information on morphology and surface chemistry. Not suitable for amorphous materials.	Crystal structure, phase composition, crystallite size.
SEM	Provides high-resolution images of surface morphology and topography. Can be coupled with EDX for elemental analysis.	Lower resolution than TEM. Sample preparation can introduce artifacts. Charging effects can be problematic for non-conductive samples.	Surface morphology, topography, elemental composition (with EDX)
TEM	Provides highest resolution images of morphology, crystal structure, and lattice spacing. Can provide atomic-resolution images (HRTEM).	Requires thin samples. Electron beam damage can occur. Sample preparation can introduce artifacts. Expensive and requires specialized expertise.	High-resolution morphology, crystal structure, lattice spacing, elemental composition (with EDX)

continued on following page

Table 2. Continued

Technique	Strengths	Limitation	Information Provided
FTIR	Provides information on functional groups and surface chemistry. Relatively simple and inexpensive.	Spectral interpretation can be challenging due to overlapping bands. Not suitable for all types of nanoparticles.	Functional groups, surface chemistry, capping agents
VSM	Provides information on magnetic properties such as saturation magnetization, remanence, and coercivity. Relatively straightforward technique.	Requires sufficient sample mass. Sample preparation can introduce artifacts. Not suitable for non-magnetic materials.	Magnetic properties (Ms, Mr, Hc, magnetic anisotropy)
DLS	Provides information on hydrodynamic size and size distribution in solution. Relatively simple and rapid.	Measures hydrodynamic size, which can differ from core size. Sensitive to aggregation and polydispersity.	Hydrodynamic size, size distribution
EDX	Provides elemental composition. Often coupled with SEM or TEM.	Lower sensitivity than other techniques for some elements. Requires high vacuum.	Elemental composition
UV-Vis	Provides information on optical properties, including band gap and plasmon resonance. Relatively simple and inexpensive.	Not suitable for all types of nanoparticles.	Optical properties (band gap, plasmon resonance)
TGA	Provides information on thermal stability and decomposition behavior. Relatively simple and straightforward.	Can be destructive.	Thermal stability, decomposition behavior
spICP-MS	High-throughput nanoparticle characterization, size distribution, number concentration, elemental composition. Excellent for dilute suspensions.	Requires specialized equipment and expertise.	Size distribution, number concentration, elemental composition

Complementary Use of Techniques

The synergistic application of several approaches is what gives nanoparticle characterization its power. For example, TEM can show high-resolution morphology and crystallinity, whereas XRD can provide crystal size and structure. This can be enhanced using FTIR, which can detect ligands and surface functional groups. DLS measures the hydrodynamic size in solution, whereas VSM provides the magnetic characteristics. UV-Vis spectroscopy exposes optical qualities, while EDX offers elemental composition. When these methods are combined, a significantly more thorough understanding is obtained than when they are used separately. This complementary approach is demonstrated by a large number of the studies covered in this chapter. For instance, (Ullah et al., 2024) efficiently combined XRD, SEM, TEM, FTIR, and EDX to comprehensively characterize synthesized nanoparticles. Similar combined approaches are evident in many other studies.

Applications of Ionanofluids

Ionanofluids, which are formed by distributing nanoparticles in ionic liquids, are a new class of materials with tunable characteristics that can be used for a variety of applications.

I. Heat Transfer

Ionanofluids have better heat transfer qualities than conventional fluids thus are ideal for applications requiring efficient thermal management. Adding nanoparticles to ionic liquids improves their critical boiling flux, convective heat transfer coefficients, and thermal conductivity. (Minea et al., 2021)

1. Solar Collectors

Ionanofluids enhance solar energy conversion, especially in flat plate and hybrid photovoltaic-thermal (PVT) collectors, by improving temperature rise and energy efficiency. Graphene-based ionanofluids, due to their lower specific heat capacity, boost energy efficiency, outperforming traditional fluids like water and pure ionic liquids. (Moulefera et al,.2025)

2. Heat Exchangers

Ionanofluids, with superior thermal conductivity, enhance heat exchanger performance. In plate heat exchangers, they boost energy efficiency and reduce environmental impact. A study by (Zahan et al,. 2025) found that a counterflow setup using graphene-based ionanofluids achieved a 76.23% thermal enhancement over a parallel flow configuration.

3. Microchannel Heat Sinks (MCHS)

Due to the excellent stability, ionic conductivity, non-flammability, and minimal volatility, ionanofluids present a promising alternative for thermal management in MCHSs and are therefore proven to be an ideal coolant for electronic devices.

4. Lubricant in Industrial Machinery

Ionanofluids are suitable for lubricant applications due to their special qualities, which include high thermal stability and adjustable viscosity. It is possible to produce lubricants with increased wear resistance, decreased friction, and improved load-bearing capacity by dispersing nanoparticles in ionic liquids. (Chereches et al,.2019). They are used to lubricate engines, compressors, and gears which leads to reducing energy consumption, extending equipment life, and enhancing performance.

II. Chemical and Engineering Applications

Ionic liquids, the base of ionanofluids, are effective in catalytic processes, supporting applications in organic synthesis, polymerization, and biomass conversion.

1. Green Solvents

Ionanofluids due to low vapor pressure and tuneable properties, act as green solvents. Thus, they can replace traditional organic solvents to minimize environmental impact and enhance safety, especially in VOC sensitive applications.

2. Extraction Processes

The unique solvent properties of ionic liquids make ionanofluids suitable for extraction processes. They can be used to extract valuable compounds from various sources, including biomass, natural products, and waste streams.

III. Electrical Applications

Electric Conductivity

The electrical conductivity of ionanofluids can be customized for different applications by adjusting the type and concentration of nanoparticles. For example, incorporating MXene nanoparticles into a diethylene glycol-based ionic liquid significantly enhanced conductivity, making it a promising candidate for cell applications. (Bakthavatchalam et al,. 2020)

IV. Emerging Applications

1. Biomedical Applications

Although still in the early stages of research, the biocompatibility and adjustable properties of certain ionic liquids and nanoparticles open up exciting possibilities in the biomedical field. Ionanofluids could be designed for use in drug delivery, bioimaging, and therapeutic applications. However, their safety and long-term effects on biological systems must be carefully evaluated. (Musia et al,. 2021)

2. Advanced Materials

Ionanofluids can be utilized as building blocks to create sophisticated materials with unique characteristics. Nanocomposites, thin films, and coatings with specific functions can be produced by adjusting the ionanofluid's composition and structure. Applications for these materials are numerous and include energy storage, electronics, and catalysis.

3. Environmental Remediation

Certain ionanofluids can be designed for environmental remediation applications, such as the removal of pollutants from water and air. The tuneable properties of ionic liquids and nanoparticles allow for the selective capture and degradation of various contaminants. This approach offers a sustainable and cost-effective solution for addressing environmental challenges. (Lourenço et al,.2021)

CONCLUSION

A comprehensive overview of the characterization methods frequently employed to examine single and hybrid nanoparticles has been discussed in this chapter. Along with the main techniques as XRD, SEM, TEM, FTIR, and VSM, a number of supplementary techniques like DLS, EDX, UV-Vis, TGA, and spICP-MS were explored. Each technique has its own set of benefits and drawbacks, providing different aspects of nanoparticle properties. The key takeaway is that a multi-technique approach is necessary for a thorough knowledge of nanoparticle properties, combining data from multiple sources to create a complete understanding. The examples taken from the literature demonstrate the effectiveness of this integrated approach by showing how integrating several methodologies results in a more thorough and refined understanding of the behaviour and usefulness of nanoparticles.

REFERENCES

Al-Harbi, N. & Abd-Elrahman, N. K. (2024). Physical methods for preparation of nanomaterials, their characterization and applications: A review. *Journal of Umm Al-Qura University for Applied Sciences*, *209*. Advance online publication. DOI: 10.1007/s43994-024-00165-7

Ali, A., Zafar, H., Zia, M., ul Haq, I., Phull, A. R., Ali, J. S., & Hussain, A. (2016). Synthesis, characterization, applications, and challenges of iron oxide nanoparticles. *Nanotechnology, Science and Applications*, *9*, 49–67. DOI: 10.2147/NSA.S99986 PMID: 27578966

Altammar, K. A. (2023). A review on nanoparticles: Characteristics, synthesis, applications, and challenges. *Frontiers in Microbiology*, *14*, 1155622. DOI: 10.3389/fmicb.2023.1155622 PMID: 37180257

Amin, F., Fozia, , Khattak, B., Alotaibi, A., Qasim, M., Ahmad, I., Ullah, R., Bourhia, M., Gul, A., Zahoor, S., & Ahmad, R. (2021). Fozia, Khattak, B., Alotaibi, A., Qasim, M., Ahmad, I., Ullah, R., Bourhia, M., Gul, A., Zahoor, S., & Ahmad, R. (2021). Green Synthesis of Copper Oxide Nanoparticles Using Aerva javanica Leaf Extract and Their Characterization and Investigation of In Vitro Antimicrobial Potential and Cytotoxic Activities. *Evidence-Based Complementary and Alternative Medicine : eCAM*, *2021*(1), 1–12. DOI: 10.1155/2021/5589703 PMID: 34239581

Andrade, Â. L., Valente, M. A., Ferreira, J. M., & Fabris, J. D. (2012). Preparation of size-controlled nanoparticles of magnetite. *Journal of Magnetism and Magnetic Materials*, *324*(10), 1753–1757. DOI: 10.1016/j.jmmm.2011.12.033

Binnig, G., Quate, C. F., & Gerber, Ch. (1986). Atomic Force Microscope. *Physical Review Letters*, *56*(9), 930–933. DOI: 10.1103/PhysRevLett.56.930 PMID: 10033323

Bragg, W. L. (1913). The Structure of Some Crystals as Indicated by Their Diffraction of X-rays. *Proceedings of the Royal Society of London. Series A, Containing Papers of a Mathematical and Physical Character*, *89*(610), 248–277. DOI: 10.1098/rspa.1913.0083

Brigger, I., Dubernet, C., & Couvreur, P. (2012). Nanoparticles in cancer therapy and diagnosis. *Advanced Drug Delivery Reviews*, *64*, 24–36. DOI: 10.1016/j.addr.2012.09.006 PMID: 12204596

Cheville, N. F., & Stasko, J. (2013). Techniques in Electron Microscopy of Animal Tissue. *Veterinary Pathology*, *51*(1), 28–41. DOI: 10.1177/0300985813505114 PMID: 24114311

Dada, A. O., Adekola, F. A., Adeyemi, O. S., Bello, O. M., Oluwaseun, A. C., Awakan, O. J., & Grace, F.-A. A. (2018). *Exploring the Effect of Operational Factors and Characterization Imperative to the Synthesis of Silver Nanoparticles.* Silver Nanoparticles - Fabrication, Characterization and Applications., DOI: 10.5772/intechopen.76947

De Jong, W. H., & Borm, P. J. (2008). Drug delivery and nanoparticles: Applications and hazards. *International Journal of Nanomedicine*, *3*(2), 133–149. DOI: 10.2147/IJN.S596 PMID: 18686775

Dulog, J. C., Unabia, R., Delicana, J. D. P., Lalem, A., Sayson, N. L. B., Capangpangan, R. Y., Lubguban, A. A., & Alguno, A. C. (2024). Synthesis and Characterization of AuNP/TiO_2 Hybrid Nanoparticles for Possible Photocatalytic Application. *Materials Science Forum, 1113*, 21–26. DOI: 10.4028/p-ogEu80

Elango, G., Kumaran, S. M., Kumar, S. S., Muthuraja, S., & Roopan, S. M. (2015). Green synthesis of SnO2 nanoparticles and its photocatalytic activity of phenolsulfonphthalein dye. *Spectrochimica Acta. Part A: Molecular and Biomolecular Spectroscopy*, *145*, 176–180. DOI: 10.1016/j.saa.2015.03.033 PMID: 25770967

Elmobarak, W. F., & Almomani, F. (2021). Application of Fe3O4 magnetite nanoparticles grafted in silica (SiO2) for oil recovery from oil in water emulsions. *Chemosphere*, *265*, 129054. DOI: 10.1016/j.chemosphere.2020.129054 PMID: 33280845

Grasso, A., Ferrante, M., Arena, G., Salemi, R., Zuccarello, P., Fiore, M., & Copat, C. (2021). Chemical characterization and quantification of silver nanoparticles (Ag-NPs) and dissolved ag in seafood by single particle icp-ms: Assessment of dietary exposure. *International Journal of Environmental Research and Public Health*, *18*(8), 4076. DOI: 10.3390/ijerph18084076 PMID: 33924319

Hasan, S. (2015). A review on nanoparticles: Their synthesis and types. *Research Journal of Recent Sciences*, *2277*, 2502.

Heilgeist, S., Sekine, R., Sahin, O., & Stewart, R. A. (2021). Finding nano: Challenges involved in monitoring the presence and fate of engineered titanium dioxide nanoparticles in aquatic environments. *Water (Basel)*, *13*(5), 734. DOI: 10.3390/w13050734

Hortolà, P. (2005). SEM examination of human erythrocytes in uncoated bloodstains on stone: Use of conventional as environmental-like SEM in a soft biological tissue (and hard inorganic material). *Journal of Microscopy*, *218*(2), 94–103. DOI: 10.1111/j.1365-2818.2005.01477.x PMID: 15857371

Iyengar, S. J., Joy, M., Ghosh, C. K., Dey, S., Kotnala, R. K., & Ghosh, S. (2014). Magnetic, X-ray and Mössbauer studies on magnetite/maghemite core–shell nanostructures fabricated through an aqueous route. *RSC Advances*, *4*(110), 64919–64929. DOI: 10.1039/C4RA11283K

Javan bakht Dalir, S., Djahaniani, H., Nabati, F., & Hekmati, M. (2020). Characterization and the evaluation of antimicrobial activities of silver nanoparticles biosynthesized from Carya illinoinensis leaf extract. *Heliyon, 6*(3), e03624. DOI: 10.1016/j.heliyon.2020.e03624

Jayaseelan, C., Rahuman, A. A., Roopan, S. M., Kirthi, A. V., Venkatesan, J., Kim, S. K., Siva, C., & Iyappan, M. (2013). Biological approach to synthesize TiO2 nanoparticles using Aeromonas hydrophila and its antibacterial activity. *Spectrochimica Acta. Part A: Molecular and Biomolecular Spectroscopy*, *107*, 82–89. DOI: 10.1016/j.saa.2012.12.083 PMID: 23416912

Jia, Z., Li, J., Gao, L., Yang, D., & Kanaev, A. (2023). Dynamic light scattering: A powerful tool for in situ nanoparticle sizing. *Colloids and Interfaces*, *7*(1), 15. DOI: 10.3390/colloids7010015

Kumar, D. A., Palanichamy, V., & Roopan, S. M. (2014). Green synthesis of silver nanoparticles using Alternanthera dentata leaf extract at room temperature and their antimicrobial activity. *Spectrochimica Acta. Part A: Molecular and Biomolecular Spectroscopy*, *127*, 168–171. DOI: 10.1016/j.saa.2014.02.058 PMID: 24632169

Kumar, S., Yadav, D. V., & Nehra, D. A. (2023). A Review Of Characterization Tools And Techniques For Nanomaterials And Nanocomposites. *International Journal of Psychosocial Rehabilitation*, *27*(2), 42–53. DOI: 10.61841/rpe30a86

Kumari, M., Widdrat, M., Tompa, É., Uebe, R., Schüler, D., Pósfai, M., & Hirt, A. M. (2014). Distinguishing magnetic particle size of iron oxide nanoparticles with first-order reversal curves. *Journal of Applied Physics*, *116*(12), 124304. DOI: 10.1063/1.4896481

Lee, B., Yoon, S., Lee, J. W., Kim, Y., Chang, J., Yun, J., Ro, J. C., Lee, J.-S., & Lee, J. H. (2020). Statistical Characterization of the Morphologies of Nanoparticles through Machine Learning Based Electron Microscopy Image Analysis. *ACS Nano*, *14*(12), 17125–17133. DOI: 10.1021/acsnano.0c06809 PMID: 33231065

Loeschner, K., Johnson, M. E., & Montoro Bustos, A. R. (2023). Application of single particle ICP-MS for the determination of inorganic nanoparticles in food additives and food: A short review. *Nanomaterials (Basel, Switzerland)*, *13*(18), 2547. DOI: 10.3390/nano13182547 PMID: 37764576

Mansfield, E., Tyner, K. M., Poling, C. M., & Blacklock, J. L. (2014). Determination of nanoparticle surface coatings and nanoparticle purity using microscale thermogravimetric analysis. *Analytical Chemistry*, *86*(3), 1478–1484. DOI: 10.1021/ac402888v PMID: 24400715

Mehrin, S., Tanisa, N. Y., Awal, R., Alam Khan, M. K., Shaqur, A., Ahmed, S., Islam, S., & Mia, A. (2024). Efficient Bioelectrochemical Cell Generation and Green Synthesis of Silver Nanoparticles Using Pomegranate and Pineapple Peel Extracts: A Comprehensive Characterization Study. *Advances in Materials Science and Engineering*, *2024*(1), 6681559. DOI: 10.1155/2024/6681559

Moradi, P., & Hajjami, M. (2021). Magnetization of graphene oxide nanosheets using nickel magnetic nanoparticles as a novel support for the fabrication of copper as a practical, selective, and reusable nanocatalyst in C–C and C–O coupling reactions. *RSC Advances*, *11*(42), 25867–25879. DOI: 10.1039/D1RA03578A PMID: 35479448

Nasirpouri, F., Fallah, S., Ahmadpour, G., Moslehifard, E., Samardak, A. Y., Samardak, V. Y., & Samardak, A. S. (2023). Microstructure, ion adsorption and magnetic behavior of mesoporous γ-Fe 2 O 3 ferrite nanoparticles. *RSC Advances*, *13*(36), 25140–25158. DOI: 10.1039/D3RA01663C PMID: 37622013

Nixon, W. C. (1971). The general principles of scanning electron microscopy. *Philosophical Transactions of the Royal Society of London. Series B, Biological Sciences*, *261*(837), 45–50. DOI: 10.1098/rstb.1971.0035

Omori, N., Bobitan, A. D., Vamvakeros, A., Beale, A. M., & Jacques, S. D. M. (2023). Recent developments in X-ray diffraction/scattering computed tomography for materials science. *Philosophical Transactions. Series A, Mathematical, Physical, and Engineering Sciences*, *381*(2259), 20220350. Advance online publication. DOI: 10.1098/rsta.2022.0350 PMID: 37691470

Ozkan, S. Z., & Karpacheva, G. P. (2018). Hybrid Materials Based on Poly-3-amine-7-methylamine-2-methylphenazine and Magnetite Nanoparticles Immobilized on Single-Walled Carbon Nanotubes. *Polymers*, *10*(5), 544–544. DOI: 10.3390/polym10050544 PMID: 30966578

Pasquini, A., Picotto, G. B., & Pisani, M. (2005). STM carbon nanotube tips fabrication for critical dimension measurements. *Sensors and Actuators. A, Physical*, *123-124*, 655–659. DOI: 10.1016/j.sna.2005.02.036

Patel, M., Mishra, S., Verma, R., & Deep Shikha. (2022). *Synthesis of ZnO and CuO nanoparticles via Sol gel method and Its Characterization by using XRD and FT-IR Analysis*. DOI: 10.21203/rs.3.rs-1234162/v1

Rahman, A., Chowdhury, M. A., & Hossain, N. (2022). Green synthesis of hybrid nanoparticles for biomedical applications: A review. *Applied Surface Science Advances, 11*, 100296. DOI: 10.1016/j.apsadv.2022.100296

Roopan, S. M., Surendra, T. V., Elango, G., & Kumar, S. H. S. (2014). Biosynthetic trends and future aspects of bimetallic nanoparticles and its medicinal applications. *Applied Microbiology and Biotechnology, 98*(12), 5289–5300. DOI: 10.1007/s00253-014-5736-1 PMID: 24728718

Sagayaraj, R., Aravazhi, S., Praveen, P., & Chandrasekaran, G. (2017). Structural, morphological and magnetic characters of PVP coated ZnFe2O4 nanoparticles. *Journal of Materials Science Materials in Electronics, 29*(3), 2151–2158. DOI: 10.1007/s10854-017-8127-4

Sharma, N. K., Vishwakarma, J., Rai, S., Alomar, T. S., AlMasoud, N., & Bhattarai, A. (2022). Green Route Synthesis and Characterization Techniques of Silver Nanoparticles and Their Biological Adeptness. *ACS Omega, 7*(31), 27004–27020. DOI: 10.1021/acsomega.2c01400 PMID: 35967040

Sin Cheow, W., Xu, R., & Hadinoto, K. (2013). Towards sustainability: New approaches to nano-drug preparation. *Current Pharmaceutical Design, 19*(35), 6229–6245. DOI: 10.2174/13816128113193500002 PMID: 23469999

Sochi, T. (2010, December 20). *High Throughput Software for Powder Diffraction and its Application to Heterogeneous Catalysis*. ResearchGate. https://www.researchgate.net/publication/48172215_High_Throughput_Software_for_Powder_Diffraction_and_its_Application_toHeterogeneous_Catalysis

Taha, M., Hassan, M., Essa, S., & Tartor, Y. (2013). Use of Fourier transform infrared spectroscopy (FTIR) spectroscopy for rapid and accurate identification of Yeasts isolated from human and animals. *International Journal of Veterinary Science and Medicine, 1*(1), 15–20. DOI: 10.1016/j.ijvsm.2013.03.001

Transmission Electron Microscopy. (2024). Nanoscience Instruments. https://www.nanoscience.com/techniques/transmission-electron-microscopy/

Ullah, A., Rehman, U. U., Ahmad, R., & Rahman, F. (2024). Bioinspired green synthesis of copper, nickel, and hybrid nanoparticles using Myristica Fragrans seeds: Biomedical applications and beyond. *Materials Research Express, 11*(8), 085401. DOI: 10.1088/2053-1591/ad68cf

Vargas-Bernal, R. (2020). Introductory Chapter: Hybrid Nanomaterials. *IntechOpen EBooks*. DOI: 10.5772/intechopen.92012

Wang, Y., Zhang, W., Luo, C., Wu, X., & Yan, G. (2016). Superparamagnetic FeCo@ SnO2 nanoparticles on graphene-polyaniline: Synthesis and enhanced electromagnetic wave absorption properties. *Ceramics International*, *42*(10), 12496–12502. DOI: 10.1016/j.ceramint.2016.05.038

Zhang, S., & Wang, C. (2023). Precise Analysis of Nanoparticle Size Distribution in TEM Image. *Methods and Protocols*, *6*(4), 63. DOI: 10.3390/mps6040063 PMID: 37489430

ABBREVIATIONS

AES: Atomic emission spectroscopy
AFM: Atomic Force Microscopy
BSE: Backscattered Electrons
DLS: Dynamic light scattering
FORC: First order reversal curves
FTIR: Fourier Transform Infrared Spectroscopy
HRTEM: High-Resolution Transmission Electron Microscopy
NPs: Nanoparticles
SE: Secondary Electrons
SEM: Scanning Electron Microscopy
STM: Scanning Tunnelling Microscopy
TEMl: Transmission Electron Microscopy
Uv-Vis: Ultraviolet-Visible
XPS: X-ray Photoelectron Spectroscopy
XRD: X-ray diffraction

Chapter 5
Ionic Liquids as Catalysts, Solvents, and Sustainable Manner for Biodiesel Preparation

M Amin Mir
https://orcid.org/0000-0003-4689-0507
Prince Mohammad Bin Fahd University, Saudi Arabia

Eskilla Venkata Ramana
University of Aveiro, Portugal

Syed M. Hasnain
https://orcid.org/0000-0002-5102-3101
Prince Mohammad Bin Fahd University, Saudi Arabia

Maythem Mahmoud
https://orcid.org/0000-0002-4534-3213
Prince Mohammad Bin Fahd University, Saudi Arabia

Kim Andrews
https://orcid.org/0000-0002-0304-9630
Prince Mohammad Bin Fahd University, Saudi Arabia

ABSTRACT

Biodiesel is a promising alternative biofuel that is a subject of much research because of growing demand for sustainable energy sources. In the synthesis of biodiesel, ionic liquids (ILs), a class of organic salts with adjustable physicochemical characteristics, have drawn a lot of interest as effective and eco-friendly substitutes for

DOI: 10.4018/979-8-3373-1245-3.ch005

2026, IGI Global Scientific Publishing. Copying or distributing in print or electronic forms without written permission of IGI Global Scientific Publishing is prohibited. Use of this chapter to ative artificial intelligence (AI) technologies is expressly prohibited. The publisher reserves all rights to license its use for generative AI training and machine learning model development.

traditional solvents and catalysts. This chapter examines the various functions of ILs in the production of biodiesel, including as their application as extraction solvents, reaction media, and homogeneous and heterogeneous catalysts. ILs are very useful in transesterification and esterification reactions because of their strong catalytic activity, recyclability, and improved enzyme stability. Particular attention is paid to basic and acidic ILs, their impact on reaction efficiency, and their potential to take the role of conventional catalysts while lowering environmental issues associated with the process.

INTRODUCTION

Ionic liquids (ILs) are a category of organic mixtures composed of organic cations paired with either Melting temperatures of organic or inorganic anions are less than 100 °C. Typically, their synthesis entails the exchange of anion after cation production. The investigation of ILs began in 1914 when Walden (1914) studied the physical characteristics of ethyl ammonium nitrate, thay is created when concentrated nitric acid reacts with ethylamine. However, the discovery did not acquire much traction because of its explosive character. Until the 1990s, when their potential in chemical synthesis was acknowledged, ILs were mostly ignored. This resulted in the creation of ILs with unique physicochemical properties, (Seddon, 1997).

Because of their adjustable characteristics and structural versatility, ILs have become a potential class of green solvents in recent years. Because their properties may be modified by choosing the right cations and anions to fit particular reactions, this 2nd generation of ILs has produced a lot of confirmation, (Vancov et al., 2012). This flexibility has been used by researchers to create ILs that are tailored for a range of uses, improving characteristics like volatility, viscosity, melting points, and thermal and electrochemical stability, (Wasserscheid & Welton, 2002). High-volatility solvent replacement, biomass pre-selection, purification, medium of reaction, catalysis, cellulose dissolving, and heavy metal removal are just a few of the processes in which ILs have been employed, (Tadesse & Luque, 2011).

As seen in Figure 1 (Farra et al., 2015), research is currently moving toward the third generation of ILs, which are designed to have desired biological characteristics such enzyme stability and activation. Different ILs have been used to study a variety of enzymatic processes, (Tang, Baker, & Zhao, 2012). Furthermore, because of their low vapor pressure, which lessens negative impacts on the environment and human health, ILs have drawn attention as environmentally benign solvents, (Passos, Freire, & Coutinho, 2014). Recent research indicates that certain mixed-strain consortia may be able to biodegrade ILs (Troter et al., 2016), as seen in Table

1, despite previous findings raising concerns about the poor biodegradability and possible toxicity of some ILs.

Due to environment frindly benign and adjustable, ILs are seen as a good substitute for conventional solvents. Their application in the synthesis of biodiesel has been the subject of numerous investigations; in this process, ILs provide benefits by taking the place of traditional solvents during the reaction and purification phases. Compared to conventional catalysts, ILs reduce wastewater formation and make product separation and purification easier by functioning as co-solvents, catalysts, extracting solvents, or enzyme supports, (Yoo, Pu, & Ragauskas, 2017). High prices, high viscosities, toxicity issues, low solid loading capacity, limited water tolerance, and intricate recycling procedures are some of the barriers should get overcome for widespread commercialization, (Niedermeyer et al., 2013). The following sections explore the uses of IL, current advancements, related difficulties, and potential future breakthroughs in the manufacture of biodiesel.

Table 1. Ionic liquid and deep eutectic solvents application in biodiesel production

Feedstock	Reactor Type & Volume (cm^3) / Agitator Type / Agitation Intensity (rpm)	Alcohol	Alcohol:Oil Molar Ratio (mol/mol)	Catalyst / Loading (wt%)	Temperature (°C)	Optimal Reaction Condition	Conversion (C) or Yield (Y) % / Time (h)
Cotton Seed Oil	Three-neck flask, 100 / Magnetic Stirrer / 550	Methanol	4.5:1 – 16.5:1	IMC2OH / 0.15 - 1.75 wt%	36 - 66	12.5:1; 0.5 wt%: 55.5°C	99.0 (Y) / 4.5
Glycerol Trioleate	Batch Reactor, 200 / Mechanical Stirrer / 600	Methanol	3.5:1 - 9.5:1	(BMIm)(OH)	21 - 121	9.5:1; 121.5°C	88.0 (Y) / 8.5
Rapeseed Oil	Three-neck flask, 500 / Mechanical Stirrer / 650	Methanol	3.5:1 - 8.5:1	(BMIm)(Im) / 1.5 - 10.5 wt%	31 - 71	6.5:1; 4.5 wt%: 61.5°C	≥ 96.0 (Y) / 1.5
Soybean Oil	Flask / Magnetic Stirrer / 500	Methanol	6.5:1 - 18.5:1	CHOH / 2.5 - 10.5 wt%	31 - 66	9.5:1; 6.5 wt%: 61.5°C	≥ 96.0 (Y) / 1.5

continued on following page

Table 1. Continued

Feedstock	Reactor Type & Volume (cm^3) / Agitator Type / Agitation Intensity (rpm)	Alcohol	Alcohol:Oil Molar Ratio (mol/mol)	Catalyst / Loading (wt%)	Temperature (°C)	Optimal Reaction Condition	Conversion (C) or Yield (Y) % / Time (h)
Sunflower Oil	Flask / Magnetic Stirrer / 600	Methanol	6.5:1 - 8.5:1	CHOMe / 4.5 wt%	31 - 66	9.5:1; 6.5 wt%: 61.5°C	96.0 ±1.8 (Y) / 2.5
Jatropha Oil	Round-bottom flask / Mechanical Stirrer / 650	Methanol	3.5:1 - 18.5:1	CHOH / 1.5 - 5.5 wt%	31 - 71	9.5:1; 4.5 wt%: 61.5°C	89.0 ±1.8 (Y) / 2.5

Figure 1. Chemical compositions of ionic liquids that are often utilized

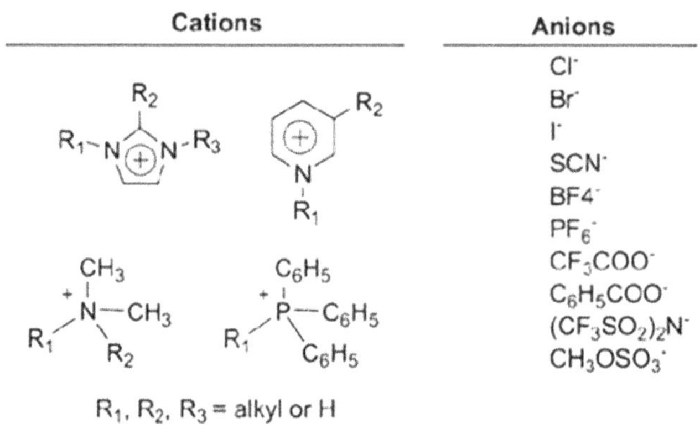

(*Mir & Banik, 2025*)

IONIC LIQUIDS AND THEIR APPLICATION IN BIODIESEL PRODUCTION

Ionic liquids (ILs), which are in line with green chemistry principles and have a great deal of promise for industrial applications, are interesting materials for the manufacture of biodiesel due to their special features. ILs can be used as supports for the enzymatic preparation of biodiesel or as independent catalysts, solvents, or reaction media. Because they are in the same phase as the reactants, they are usually categorized as homogeneous catalysts. However, the advantages of heterogeneous catalysis also include their capacity to be recycled following reactions and to make product separation simpler. Consequently, ILs successfully combine the benefits of heterogeneous and homogeneous catalytic systems.

Additionally, ILs have been investigated for the modification of materials like ceramics and for the confinement of metal complexes and traditional catalysts in the generation of biodiesel. Because they reduce enzyme deactivation and increase yields, they play a particularly important role in the enzymatic biodiesel synthesis process. The several uses of ILs in the production of biodiesel will be covered in detail in the sections that follow.

Figure 2. Deep eutectic solvents that are commonly used

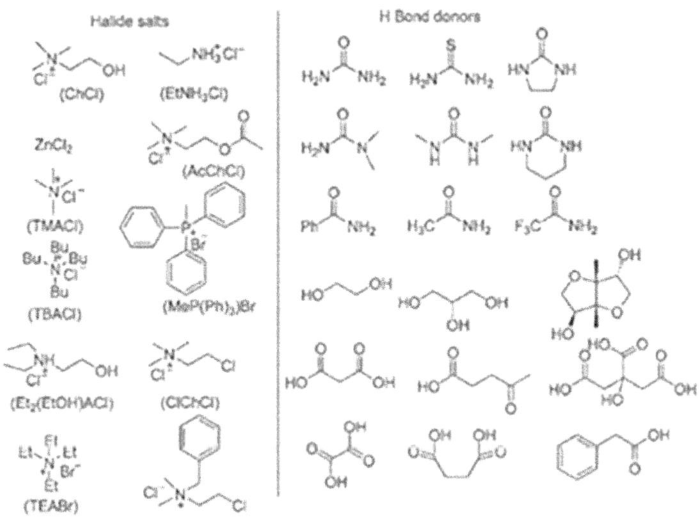

(Mir & Banik, 2025)

ACIDIC IL AS CATALYSTS

Many different types of acidic ionic liquids (ILs) are investigated as possible catalysts for the generation of biodiesel, especially Brønsted acidic ILs. Both the esterification of oils and free fatty acids (FFAs) have been successfully accomplished by these catalysts, which usually have sulfonic acid parts in their cations. Wu et al.

(2007b) investigated the use of ILs as enzymes in cottonseed oil transesterification with CH_3OH in one of their groundbreaking experiments. The study used the HSO4 ion in conjunction with three distinct cations of differing acidity, such as trimethylamine and N-methylimidazole.

Under ideal circumstances (oil/methanol/IL = 2:21:1.125 molar ratio at 169°C for 4.9 hours), 1-(4-sulfonic acid) butyl pyridinium hydrogen sulfate produced the greatest fatty acid methyl ester (FAME) yield of 91.9% among the ILs tested. For instance, when tested under the same reaction circumstances, concentrated sulfuric acid (H_2SO_4) produced 81% FAME after three hours, with an H2SO4 concentration of 86%. According to the study's findings, the IL catalysts produced yields that were on par with those seen in prior investigations. In a different study, jatropha oil was transesterified utilizing (BSPy)(CF3SO3), yielding a FAME yield of up to 91.9% and stable catalytic property during seven cycles in a row, (Li et al., 2010).

Similar to this, Guo et al. (2011) studied the combination of (BMIm)(CH_3SO_3) and $FeCl_3$ for the formation of biodiesel from jatropha oil, where metal ions functioned as Lewis acidic points, and under ideal circumstances, achieved an outstanding FAME yield of 99.7%. Single- and multi-SO_3H functional Brønsted acidic ILs are further ILs of interest in the manufacture of biodiesel, (Han et al., 2009). A multi-SO3H IL with high ionicity and acidic nature was assessed by Liang et al. (2010) as a catalyst for rapeseed oil transesterification, and under ideal circumstances, a 98% biodiesel yield was obtained after 7 hours. Additionally, the catalyst showed outstanding resistance to water; when the water content rose from 0.3% to 2.0%, the yield only dropped by 3%.

Propyl sulfonic acid groups on the cations of dicationic Brønsted acidic polar liquids (DAILs) have demonstrated strong catalytic activity and reusability in the preparation of biodiesel, (Zhou et al., 2011). The FFAs esterification to biodiesel utilizing small alcohols catalyzed by both monocationic and dicationic ILs was investigated by Fang, Yang, & Jiao (2011). According to the study, DAILs produced greater conversion rates (93–96%) than monocationic ILs (85-87%). Additionally, DAILs showed excellent catalytic stability, sustaining activity with little loss over six cycles, (Liu et al., 2015). Table 2, which demonstrates the use of ILs in the production of biofuel, demonstrates the paucity of research on the esterification and transesterification of animal fats and vegetable oils using basic ILs, (Liu et al., 2015).

Table 2. Ionic liquid use in biofuel production

IL	Catalyst	Source of Oil	Conditions of Reaction	(%) Yield of Biodiesel
[BMIm]PF_6	Sn (3-hydroxy-2-methyl-4-pyrone)$_2$ $(H_2O)_2$	Soy	62°C, 1.5 h	57
[BMIm]$InCl_4$	Sn (3-hydroxy-2-methyl-4-pyrone)$_2$ $(H_2O)_2$	Soy	81.5°C, 4.5 h	85
[Et_3NH]Cl	AlC13	Soy	71.5°C, 4 h	80
[BMIm]NTf_2	K2CO3	Soy	71.5°C, 2.5 h	96
[BMIm]NTf_2	H2SO4	Soy	71.5°C, 24.5 h	88
[EMIm]TfO	Lip	Soy	51.5°C, 13.5 h	80
[BMIm]NTf_2	Lip	Soy	26.5°C, 30 h	96
[BMIm]PF_6	Lip	Corn	41.5°C, 25 h	71
[BMIm]PF_6	Lip	Tri	49.5°C, 6.5 h	80
[C16MIM] NTf_2	Lip	Tri	61.5°C, 4 h	92
[C16MIM] NTf_2	Lip	Tri	61.5°C, 6 h	98
[BSPy]CF_3SO_3	[BSPy]CF_3SO_3	Jatropha	99.5°C, 5 h	92
[BMIm]HSO_4	[BMIm]HSO_4	FA	121.5°C, 3 h	93
[NMP]CH_3SO_3	[NMP]CH_3SO_3	FA	71.5°C, 9 h	95

Lipase= Lip, Soybean= Soy, Triolein= Tri, Fatty acid = FA

BASIC IONIC LIQUID CATALYSTS

Although less common, basic ionic liquids (ILs) as those based on imidazolium and choline have also been investigated for biodiesel synthesis in addition to acidic ILs, (Luo et al., 2013). In order to prevent soap production, the majority of basic ILs require substrates with low free fatty acid (FFA) content. Five fundamental depictional functional ILs with imidazolium structures were examined as catalysts in cottonseed oil transesterification in a noteworthy study by Liang et al. (2010). The length of the cations' carbon chains affected these ILs' catalytic activity. Under ideal circumstances (oil/methanol = 1:12 molar ratio at 55 °C for 4 hours), the FAME fraction and selectivity achieved 98.5% and 99.9%, respectively, by making use of bis-(3-methyl-1-imidazolium)-ethylene di-hydroxide catalyst with 0.4% amount. This illustrated how basic ILs might be used as catalysts to produce biodiesel from cottonseed oil. Furthermore, a number of imidazolium hydroxides have been investigated as recoverable catalysts for the methanol-glycerol trioleate transesterification process. By using 1-butyl-3-methylimidazolium hydroxide ([BMIm]OH), Zhou et al. (2012) were able to achieve an 87% methyl ester yield and show that it could be reused for up to six cycles without experiencing a substantial yield loss. Figure 3 provides an overall function of ILs in biodiesel processes.

Figure 3. Mechanism of ionic liquid-catalyzed biodiesel production

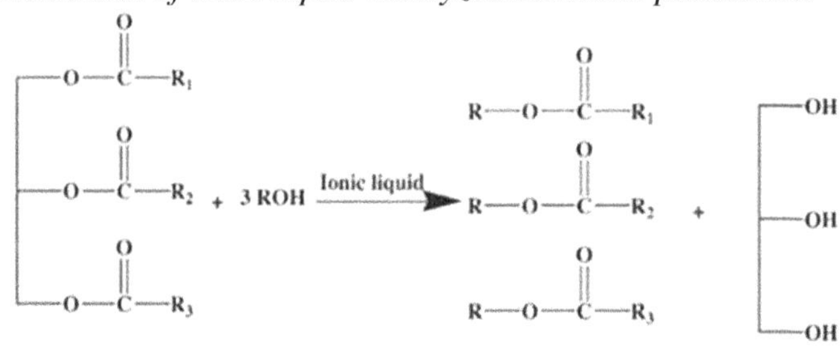

(Liu et al., 2015)

SOLVENT AND CO-SOLVENT USE OF IONIC LIQUIDS

Enzymes and organic solvents like, CH_3OH, was historically employed together as catalysts to produce biodiesel. The oils and fats as with poor miscibility inactivation of enzymes, and disturbance of intra-protein hydrophobic reactions are some of the difficulties that come with using them, (Akoh et al., 2007). With beneficial qualities including hydrophobicity, hydrogen bonding, and polarity that aid in stabilizing enzymes while reducing environmental issues, ionic liquids (ILs) present a possible substitute. Furthermore, ILs require little energy inputs and permit mild reaction conditions, which makes them reusable. The use of co-solvents in the preparation of biodiesel was investigated by a number of research groups (Moniruzzaman et al., 2010), e.g. Ha et al. (2007) used immobilized Candida antarctica lipase as the catalyst to test 23 distinct ILs in the methanolysis of soybean oil. After 11.5 hours at 50°C, the hydrophilic IL (EMIm)(TfO) outperformed solvent-free alternatives and other largely utilized solvents to obtain the greatest FAME production of 80%. In a different work, Sunitha et al. (2007) used hydrophobic ILs, such as 1-butyl-3-methylimidazolium (BMIm)(PF6) and 1-ethyl-3-methylimidazolium (EMIm)(PF6), with Novozym as the catalyst to methanolyze sunflower oil for 10 hours, yielding FAME yields of up to 98%. Remarkably, hydrophilic ILs based on BF4 produced significantly worse outcomes in comparable circumstances.

EXTRACTION OF SOLVENTS IN THE SYNTHESIS OF BIODIESEL BY IONIC LIQUIDS

Using ionic liquids (ILs) as extraction solvents at different phases of the biodiesel synthesis process is a promising usage for ILs. Even though ILs have a number of desirable qualities, such as synthetic versatility, thermal stability, non-flammability, non-volatility, and recyclability, as still a number of barriers should be overcomes before they can be extensively used for large-scale applications.

EXTRACTION OF LIPIDS

A crucial stage in the synthesis of biodiesel is the extraction of lipids from biomass, particularly when employing novel and promising oil feedstocks. Because of their strong lipid parts, efficient in photosynthetic activity, rapid preparation rate, and widespread availability, microalgae have garnered a lot of interest as a source of biofuel, (Mir et al., 2025). After harvesting and dewatering the biomass, lipid extraction from microalgae can be done. This procedure can be carried out immediately from wet concentrated biomass (wet method) or with the dried biomass powder (dry approach), (Xu et al., 2011). Solvents are essential in both methods because they facilitate cell rupture and increase the biomass's mass transfer characteristics, which improves the efficiency of lipid extraction.

Using solvents such as Benzine, n-C_6H_6, C_2H_5OH mixes, and n-C_6H_6-C_2H_5OH combinations, are the main extraction ways include Soxhlet extraction and the Bligh and Dyer method, (Bligh & Dyer, 1959). These solvents do, however, have a number of disadvantages, including low miscibility with water (apart from ethanol), flammability, toxicity, and volatility, (Mata, Martins, & Caetano, 2010).

Due to their capacity to either auto-partition lipids or increase the accessibility of co-solvents to intracellular lipids, ionic liquids (ILs) have attracted interest as a possible medium for enhancing the efficiency of lipid extraction. Furthermore, ILs have benefits over traditional solvents, such as synthetic flexibility, thermal stability, and non-volatility. Therefore, ILs in conjunction with organic solvents such CH_3OH, CH_3COOH, CH_3Cl, CH_3CO, Benzine, n-C_6H_6, C_2H_5OH have been the subject of several investigations.

For instance, Young et al. (2010) looked into the lipid extraction from microalgae using 1-ethyl-3-methylimidazolium methyl sulfate (EMIm) and CH_3OH (mass ratio 2.4:2). The extraction efficiency was increased by the co-solvent's facilitation of the lipids' auto-partitioning into a distinct immiscible phase. With yields from other biomass types, such as Chlorella and Dunaliella species, falling within the limits documented in the literature, the study obtained a lipid yield of 49.9% from jatropha oil seed. In a related work, Kim et al. (2012) investigated a number of IL-methanol co-solvents for Chlorella vulgaris lipid extraction, evaluating their effectiveness against the conventional Bligh and Dyer method.

The Bligh and Dyer method produced a lipid yield of 10.6%, while the IL co-solvents (BMIm)(MeSO$_4$), (EMIm)(MeSO$_4$), and (BMIm)(CF$_3$SO$_3$) produced lipid yields of 11.84%, 11.88%, and 12.54%, respectively. These findings showed that the IL negative part structure and the hydrophobic or hydrophilic characteristics of the ILs affected the effectiveness of lipid extraction. In their comparison of 12 distinct ILs for extraction of lipid from Chlorella vulgaris, Choi et al. (2014) discovered that ILs in association with organic chemicals typically produced larger lipid yields due to the synergistic effects of various anions. Although ILs are demonstrated potential for efficient lipid extraction, more development is necessary to satisfy the environmental and financial requirements for industrial-scale use, (Harris et al., 2018). (Harris and others, 2018).

Even though ILs have many benefits, there are still issues with their practical use, including the difficulty of their synthesis, the use of volatile or possibly poisonous solvents, and production-related environmental issues. The goal of ongoing research is to overcome these obstacles by creating more effective synthesis techniques, investigating less expensive cations (such pyridinium, ammonium, and phosphonium) for the creation of IL, and employing deep eutectic solvents (DESs) to increase productivity and lower expenses, (Yang et al., 2017).

FREE FATTY ACID EXTRACTION

Vegetable oil biodiesel has a lot of promise, due to presence of free fatty acids (FFAs) frequently prevents the conversion process from proceeding as planned. Transesterification with an alcohol is the usual conversion technique, although excessive FFA concentrations can drastically lower this process' effectiveness. This happens as a result of FFAs reacting with the basic natured catalyst to produce soap, that lowers the ester yield. Glycerolysis, hydrolysis, and caustic stripping are traditional techniques for eliminating FFAs. Despite their effectiveness, these techniques have significant disadvantages, including the need for vacuum and high temperatures, as well as the complexity of additional stages. Ionic liquids (ILs) are being investigated by researchers as a potential solution to these problems.

Manic et al. (2011) looked at the extraction of FFAs (more especially, linoleic acid) from soybean oil using ILs and poly(ethylene glycol) solvents by various molar masses. The ILs studied included 1-butyl-3-methylimidazolium dicyanamide (BMImDCA) and AMMOENG100, both of that are miscible with linoleic acid. AMMOENG100 had the highest dispersion coefficient for linoleic acid among the

ILs that were examined. These results imply that ILs have a great deal of promise for enhancing the deacidification step of the biodiesel production process.

UNSATURATED FATTY METHYL ACID ESTERS EXTRACTION

Unsaturated fatty acid methyl esters (uFAMEs) must frequently be eliminated throughout the biodiesel synthesis process in order to lessen the fuel's susceptibility to oxidation. Additionally, the procedure is compulsory to meet European biodiesel standards like EN 14214 and EN 14213, which set limits of 119.5 and 129.9 g iodine/100 g, respectively, for the total amount of unsaturated biodiesel (European Committee for Standardization, 2003). Urea insertion in complexation (Mir, Chang, & Hefni, 2024), small-temperature fractional crystallization (Mudhaffar & Salimon, 2010), molecular distillation, supercritical fluid extraction, and lipase concentration are some of the conventional techniques for eliminating uFAMEs. Despite their effectiveness, many techniques have drawbacks that impact yields and real-world implementation. Ionic liquids have drawn interest as a potential remedy for these issues.

The utilization of ILs as extractants in the segregation of uFAMEs during the preparation of biodiesel were investigated in a number of researches. For example, Li, Pham, & Pittman, 2009a) used new π-complexing sorbents to extract FAMEs from soybean biodiesel, like, methyl linoleate (17:1.9) and methyl linolenate (17:2.99). ILs were covalently immobilized onto silica to generate these sorbents, and silver salts were subsequently applied to the silica-supported ILs. Higher extraction capacity and improved reusability were demonstrated by the AgBF4/SBA-15-IL-PF4 sorbent, which continued to function after eight consecutive cycles. Building on this, Li, Pham, and Wang (2009b) addressed issues with silver salts, including their greater cost, increased viscosity, and decreased activity over time, by using an IL-co-solvent mixture. The study effectively separated PUFA FAMEs utilizing ILs immobilized on mesoporous SBA-15, increasing extraction capabilities and reusability.

Pham et al. (2009) achieved excellent selectivity in a related work by using mesoporous silica-based adsorbents to separate polyunsaturated triacylglycerols. Additionally, using reversible π-π complexation, Cheong et al. (2011) investigated ILs with cyclic rings for the extraction and enrichment of n-3 polyunsaturated fatty acids (PUFAs) and ethyl esters. Their research showed that ILs may attain greater purity and improved extraction capabilities, indicating a strong relationship between ILs' aromatic/delocalized cation structure and better n-3 PUFA extraction.

NEW GENERATION OF IONIC LIQUIDS-DEEP EUTECTIC SOLVENTS

Deep eutectic solvents (DESs), the newest generation of ionic liquids (ILs), have drawn a lot of interest lately because they may eventually take the place of traditional ILs in the manufacturing of biodiesel. DESs are usually made up of a complexing agent and a solid organic salt mixture that stays liquid at temperatures below 99.9°C, (Zhao, Zhang, & Crittle, 2013). By interacting with the anion in these solvents, the complex forming substance increases the anion's effective size and decreases its reaction with the cation. The freezing point of the mixture drops as a result of this interaction, (Zhao, Zhang, & Crittle, 2013). The first demonstration of such solvents was made by Abbot et al. (2007), who reported that a sequence of DESs created by combining a complexing agent with a solid organic salt could turn liquids at temperature lower than 99.9°C.

The classification of DESs as ILs is a topic of continuous discussion within the scholarly community. The structural distinctions between the two are the main source of contention: While DESs are mixes that may contain many anionic and/or cationic agents, as well as non-ionic agents, ILs are usually salts made up of a single distinct anion and cation. Notwithstanding these distinctions, DESs and ILs have a lot in common as appealing solvents, such as being adjustable.

DESs offer a chance to create low-cost, eco-friendly solvents because of the variety of combinations that can be used to prepare them. As a result, it is becoming more plausible to forecast that DESs will contribute significantly to the manufacture of biodiesel and may even displace ILs.

REMOVAL OF GLYCEROL FROM CRUDE BIODIESEL

The purification process, which is both economically expensive and process-intensive, is one of the most important stages in the manufacturing of biodiesel. The fuel must fulfill the requirements set forth by international standards, including ASTM D6751 and EN 14214, before it can be categorized as biodiesel, (European Committee for Standardization, 2003). The quality requirements for biodiesel, which account for 60–80% of the total production costs, are established by these guidelines, (Atadashi, Aroua, & Aziz, 2011). Glycerol and other undesirable byproducts and contaminants are eliminated during the purifying process. In diesel engines, glycerol buildup can cause a number of problems, such as clogged fuel systems, injector fouling, corrosion, fuel tank buildup, valve deposits, and the release of toxic acrolein, (Hayyan et al., 2010).

Glycerol can be extracted from biodiesel using conventional techniques such as membrane extraction and dry and water washing, (Santori et al., 2012). But these methods have a lot of drawbacks, especially because they are expensive and complicated. Because of their non-toxic, affordable, and eco-friendly qualities, deep eutectic solvents (DESs) have drawn interest as an alternative extraction medium. Glycerol and quaternary ammonium salts were combined to show that DESs could efficiently extract glycerol from biodiesel (Abbott et al., 2007). Furthermore, Hayyan et al. (2010) discovered that the most efficient ratios for removing glycerol were 1:1 molar for DES to biodiesel and 1:1 molar for salt to glycerol.

In a similar vein, Shahbaz et al. (2010) used two distinct DESs to successfully recover glycerol from biodiesel derived from palm oil. Choline chloride, a quaternary ammonium salt, was coupled with 2,2,2-trifluoroacetamide and ethylene glycol, two hydrogen-bond donors, to create these DESs. This procedure was further enhanced by Shahbaz et al. (2010), who demonstrated that binary DESs composed of ethylene glycol and methyl triphenylphosphonium bromide could eliminate all free glycerol from biodiesel obtained from palm oil. Also, Abbott et al. (2007), they found that there was no discernible improvement after the ideal molar ratio of 10:10 (DES:biodiesel).

DEEP EUTECTIC SOLVENTS AS CATALYSTS

Different deep eutectic solvents (DESs) are produced by researchers used as catalysts/co-solvents in the manufacture of biodiesel. The application of an ammonium-based DES as a new catalyst for low-grade crude palm oil (LGCPO) transesterification by a large amount of free fatty acids (FFA) was examined by Hayyan and Hashim (2013b). N-diethyl ethanol ammonium chloride (salt) and p-toluene sulfonic acid monohydrate (hydrogen bond donor) made up the DES utilized in this investigation. Under ideal circumstances, the FFA concentration decreased significantly from 9.5% to less than 1%, according to the results. In a different work, Hayyan and Hashim (2013b) investigated a two-step method for esterifying LGCPO that involved an alkali treatment after a phosphonium-based DES. By using this method, the FFA portion was lowered to levels appropriate trans-esterification of alkaline.

With the extra advantage of DES recyclability, the study showed that LGCPO could be converted to biodiesel efficiently while still remaining active after four uses.

To trans-esterify soybean oil, Long, et al. (2010) make use of Lewis acidic catalyst that consisted of $ZnCl_2$ and Cl_2I_2 in CH_3OH at a molar ratio of 2:3. Following 72 hours of ideal circumstances (10% catalyst, CH_3OH -to-oil molar ratio of 17:1.7, and 69.9°C), their results demonstrated a 55% conversion. Choline chloride and metal chloride ILs were employed as catalysts for palm oil transesterification in a related work by Isahak et al. (2011), who obtained ester yields of 69.9% and 68.01%, respectively. By adding 95%vol sulfuric acid, the yields increased to 91.9%.

The use of DESs in conjunction with other catalysts to enhance the generation of biodiesel has also been investigated in a number of studies. The use of a DES prepared from choline chloride and glycerol (2:3 molar ratio) as a solvent in CaO-catalyzed rapeseed oil transesterification was examined by Huang et al. (2013). It was discovered that the DES efficiently activated CaO by clearing the catalyst surface of inactive calcium hydroxide and carbonate layers. In comparison to just 4.0% without the DES, the data demonstrated a high ester yield of 92%.

IONIC LIQUIDS AS BIODIESEL CATALYSTS

One of the most often employed ILs as single catalyst in the transesterification and esterification procedures for the synthesis of biodiesel is Brønsted acidic ionic liquids (ILs), (Vafaeezadeh & Alinezhad, 2016). The covalently bound sulfonic acid group (-SO3H) or a Brønsted acidic counteranion, like hydrogen sulfate (HSO4−) or dihydrogen phosphate (H2PO4−), are responsible for the acidic characteristics of these ILs, (Troter et al., 2016). Brønsted acidic ionic liquids have a number of benefits over traditional acid catalysts, such as a decreased risk to the environment and less equipment damage. Notably, ILs based on pyridine or imidazole have garnered a lot of interest due to their possible application in the manufacturing of biodiesel.

In an esterification reaction of oleic acid with ethanol, it has been discovered that the combination of ILs and bentonite performs better than bentonite treated with sulfuric acid. There have also been studies on other combinations of ionic liquids and silica materials (Troter et al., 016), including the immobilization of ILs onto iron- and magnetite-silica (Zhang et al., 2012), ILs supported on SiO2 modified by sulfhydryl groups (Zhen et al., 2014), or ILs distributed over silica gel with a large surface area through physisorbed confinement (Vafaeezadeh & Hashemi, 2014) (Figure 4). This research provides efficient catalytic solutions for the synthesis of biodiesel, where the catalyst's surface area is increased by supporting the ionic liquid.

Figure 4. Transesterification reaction of triglycerides in methanol

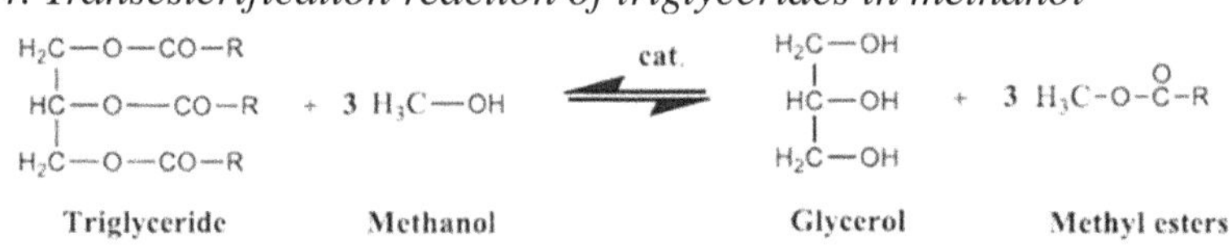

On porous matrices like silica, polymeric ionic liquids—that get prepared by polymerizing ionic liquid monomers—can also be sustained. Core-shell structured Fe_3O_4-SiO_2 has been successfully used to support polymeric acidic Brønsted ionic liquids, exhibiting high catalytic efficiency for the esterification and trans-esterification of fatty acids and vegetable oils, (Mir et al., 2022). This method used radical grafting copolymerization to fix the ionic liquid 1-vinyl-3-(3-sulfopropyl) imidazolium hydrogen sulfate onto the support (Figure 5). The Fe_3O_4-SiO_2 catalyst support's magnetic characteristics make it simple to separate from the reaction by the use of external magnet, allowing for additional reuse. Furthermore, the polymeric ionic liquid is strongly attached thanks to this immobilization approach, which improves its stability and recyclability.

ENZYMATIC BIODIESEL PRODUCTION BY IONIC LIQUIDS

Through the esterification and transesterification of fatty acids and triglycerides, enzymes like lipase can serve as catalysts for lipid-based biodiesel manufacturing process. Low temperatures and pressures can be used using this technique, which lowers the process's energy requirements, (Hama & Kondo, 2013). Enzymatic trans-esterification does have certain drawbacks, though, such as the deactivation of the enzyme, especially when methanol is present, and the blocking effect that results from glycerol deposition on the active sites of the enzyme, (Andreani & Rocha, 2012). Furthermore, the high cost of enzymes has prevented the large-scale implementation of enzymatic biodiesel synthesis to date, (Taher et al., 2017; Ribeiro et al., 2014). These issues can be resolved by using ionic liquids as enzyme immobilizing media, which improves enzyme stability and guarantees long-term activity.

The commercial Novozym 435 lipase (immobilized Candida antarctica lipase B) is utilized in early attempts to produce biodiesel in a variety of ionic liquids, (Arai et al., 2010). These investigations demonstrated that, in comparison to organic solvents or conditions devoid of ionic liquids, the reaction yield was noticeably better when ionic liquids were present. Additionally, the biocatalyst could be recycled. Later, Arai et al. (2010) used the same feedstock and fungal entire cells in ionic liquids that contained the cations 1-ethyl-3-methylimidazolium and 1-butyl-3-methylimidazolium along with the tetrafluoroborate anion. According to their research, the biocatalyst needed to be cross-linked with glutaraldehyde in order to be reusable.

Figure 5. Bronsted acidic IL and i-vinyl-3(3-sulfopropyle) imidazolium hydrogen sulfate undergo radical grafting copolymerization

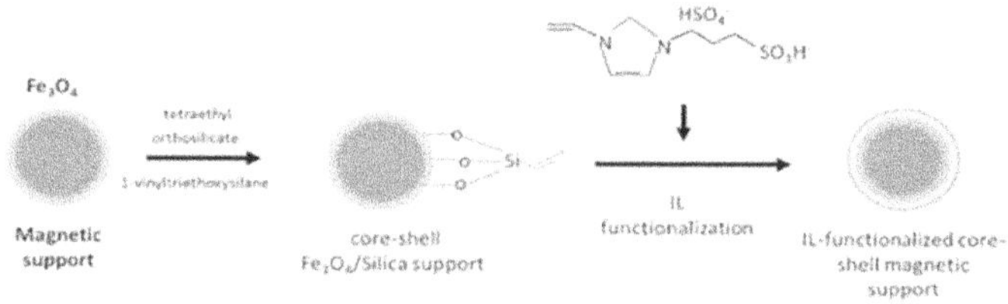

(Mir et al., 2022)

The ultimate biodiesel yields and enzyme stability are greatly treated by the kind of enzyme and the media of reaction. For example, the solvent used can affect how stable an enzyme is in the presence of methanol. The working potential of many lipases under various reaction circumstances was compared by Zhang et al. (2011). Penicillium expansum lipase, Novozym 435 lipase, and Lipozyme TLIM were studied for transesterification of vegetable oil with CH_3OH in two distinct solvents: C_6H_6 and the ionic liquid 1-butyl-3-methylimidazolium hexafluorophosphate. Their findings demonstrated that, depending on the solvent, each lipase responded differently to methanol. Only Penicillium expansum lipase was strict and showed nice activity in both solvents (ionic liquid or hexane), but Lipozyme TLIM remained stable in the ionic liquid and Novozym 435 exhibited methanol resistance in hexane.

Another important factor in assessing the ionic liquids' appropriateness for high-yield procedures is their structure. For instance, because lipids (triglycerides) are poorly soluble in ionic liquids containing short-chain cations, like 1,3-dialkylimidazolium cations, a number of studies have documented low enzymatic activity when utilizing these media, (De Diego et al., 2011). However, de Diego et al. (2011) demonstrated that Novozym® 435 produced good synthetic activity and stability when used in ionic liquids with long alkyl-chain cations. In particular, the enzyme was able to function continuously at high methanol amount because the ionic liquids 1-hexadecyl-3-methylimidazolium triflimide and 1-octadecyl-3-methylimidazolium triflimide provided an appropriate microenvironment for it.

Additionally, De los Ríos et al. (2011) found that while activity improved when the anion's nucleophilicity reduced, enzyme activity increased with higher cation hydrophobicity.

Enzymatic synthesis has also been used to create third-generation biodiesel from microalgae, (Ahmad et al., 2011). Because they can be grown quickly and have a high oil content without competing for land or water, microalgae are becoming more and more popular as a sustainable feedstock for biodiesel, (Lai et al., 2012). The performance of Novozym® 435 and Penicillium expansum lipase for producing biodiesel from different microalgal oils (Botryococcus braunii, Chlorella vulgaris, and Chlorella pyrenoidosa) in 2 systems of solvent—tert-butanol and the ionic liquid 1-butyl-3-methylimidazolium hexafluorophosphate—was compared by Lai et al. (2012).

According to the study, Penicillium expansum lipase performed marginally better than Novozym® 435, and the product was much more in the ionic liquid medium than in the organic solvent. These findings align with research utilizing feedstocks derived from vegetable oils, (Bauer et al., 2017).

Although two-stage enzymatic biodiesel manufacturing, which involves oil extraction from microalgae and transesterification, has been the subject of numerous investigations, combining these two processes into a single step would be more economical. In this sense, ionic liquids are advantageous because they can increase enzyme activity for transesterification and extract algal oil without causing cell disruption. Bauer et al. (2017) recently investigated a one-step transesterification procedure employing Novozym® 435 and a combination of entire Chromochloris zofingiensis microalgae cells with ionic liquids. They obtained a 64% yield by using a zwitterionic ionic liquid.

CONCLUSION

In the manufacturing of biodiesel, ionic liquids (ILs) have become a game-changing option, providing a sustainable substitute for traditional solvents and catalysts. Their use in transesterification, esterification, lipid extraction, and biodiesel purification is made possible by their special physicochemical characteristics, which include low volatility, high thermal stability, and tunability. While deep eutectic solvents (DESs) have further improved process sustainability by providing affordable and ecologically friendly options, acidic and basic ILs have shown remarkable catalytic efficiency. Notwithstanding these benefits, issues including expensive manufacturing costs, possible toxicity, and difficult recycling procedures still stand in the way of broad industrial implementation. The commercialization of low-cost, biodegradable ILs will depend on ongoing research into their development and optimized process integration. Future developments in the synthesis of IL-based biodiesel are probably going to concentrate on increasing production for industrial uses, improving recyclability, and boosting catalytic performance. ILs and DESs have the potential to completely transform the biodiesel sector if these issues are resolved, increasing the productivity, sustainability, and economic viability of biofuel production.

REFERENCES

Abbott, A. P., Cullis, P. M., Gibson, M. J., Harris, R. C., & Raven, E. (2007). Extraction of glycerol from biodiesel into a eutectic based ionic liquid. *Green Chemistry*, *9*(8), 868–872. DOI: 10.1039/b702833d

Ahmad, A. L., Yasin, N. H. M., Derek, C. J. C., & Lim, J. K. (2011). Microalgae as a sustainable energy source for biodiesel production: A review. *Renewable & Sustainable Energy Reviews*, *15*(1), 584–593. DOI: 10.1016/j.rser.2010.09.018

Akoh, C. C., Chang, S.-W., Lee, G.-C., & Shaw, J.-F. (2007). Enzymatic approach to biodiesel production. *Journal of Agricultural and Food Chemistry*, *55*(22), 8995–9005. DOI: 10.1021/jf071724y PMID: 17902621

Andreani, L., & Rocha, J. D. (2012). Use of ionic liquids in biodiesel production: A review. *Brazilian Journal of Chemical Engineering*, *29*(1), 1–13. DOI: 10.1590/S0104-66322012000100001

Arai, S., Nakashima, K., Tanino, T., Ogino, C., Kondo, A., & Fukuda, H. (2010). Production of biodiesel fuel from soybean oil catalyzed by fungus whole-cell biocatalysts in ionic liquids. *Enzyme and Microbial Technology*, *46*(1), 51–55. DOI: 10.1016/j.enzmictec.2009.08.008

Atadashi, I. M., Aroua, M. K., & Aziz, A. A. (2011). Biodiesel separation and purification: A review. *Renewable Energy*, *36*(2), 437–443. DOI: 10.1016/j.renene.2010.07.019

Bauer, G., Lima, S., Chenevard, J., Sugnaux, M., & Fischer, F. (2017). Biodiesel via in situ wet microalgae biotransformation: Zwitter-type ionic liquid supported extraction and transesterification. *ACS Sustainable Chemistry & Engineering*, *5*(2), 1931–1937. DOI: 10.1021/acssuschemeng.6b02665

Bligh, E. G., & Dyer, W. J. (1959). A rapid method of total lipid extraction and purification. *Canadian Journal of Biochemistry and Physiology*, *37*(8), 911–917. DOI: 10.1139/o59-099 PMID: 13671378

Cheong, L.-Z., Tan, C.-P., & Cheah, Y.-K. (2011). Extraction and enrichment of n-3 polyunsaturated fatty acids and ethyl esters through reversible π–π complexation with aromatic rings containing ionic liquids. *Journal of Agricultural and Food Chemistry*, *59*(16), 8961–8967. DOI: 10.1021/jf202043w PMID: 21790198

Choi, S.-A., Oh, Y.-K., Lee, J., Sim, S. J., & Park, J. Y. (2014). Effects of ionic liquid mixtures on lipid extraction from *Chlorella vulgaris*. *Renewable Energy*, *65*, 169–174. DOI: 10.1016/j.renene.2013.08.015

De Diego, T., Manjón, A., Lozano, P., & Iborra, J. L. (2011). A recyclable enzymatic biodiesel production process in ionic liquids. *Bioresource Technology*, *102*(10), 6336–6339. DOI: 10.1016/j.biortech.2011.02.071 PMID: 21392972

De Los Ríos, A. P., Hernández Fernández, F. J., Gómez, D., Rubio, M., & Víllora, G. (2011). Biocatalytic transesterification of sunflower and waste cooking oils in ionic liquid media. *Process Biochemistry (Barking, London, England)*, *46*(7), 1475–1480. DOI: 10.1016/j.procbio.2011.03.021

European Committee for Standardization. (2003). *Automotive fuels—Fatty acid methyl esters (FAME) for diesel engines—Requirements and test methods* (EN 14214).

Fang, D., Yang, J., & Jiao, C. (2011). Dicationic ionic liquids as environmentally benign catalysts for biodiesel synthesis. *ACS Catalysis*, *1*(1), 42–47. DOI: 10.1021/cs100026q

Farra, A., Fratini, E., & Baglioni, P. (2015). Green solvents in carbohydrate chemistry: From raw materials to fine chemicals. *Chemical Reviews*, *115*(14), 6811–6853. DOI: 10.1021/cr500719h PMID: 26121409

Guo, F., Fang, Z., Tian, X.-F., Long, Y.-D., & Jiang, L.-Q. (2011). One-step production of biodiesel from Jatropha oil with high-acid value in ionic liquids. *Bioresource Technology*, *102*(11), 6469–6472. DOI: 10.1016/j.biortech.2011.02.079 PMID: 21420854

Ha, S. H., Lan, M. N., Lee, S. H., Hwang, S. M., & Koo, Y.-M. (2007). Lipase-catalyzed biodiesel production from soybean oil in ionic liquids. *Enzyme and Microbial Technology*, *41*(4), 480–483. DOI: 10.1016/j.enzmictec.2007.03.017

Hama, S., & Kondo, A. (2013). Enzymatic biodiesel production: An overview of potential feedstocks and process development. *Bioresource Technology*, *135*, 386–395. DOI: 10.1016/j.biortech.2012.08.014 PMID: 22985827

Han, M., Yi, W., Wu, Q., Liu, Y., Hong, Y., & Wang, D. (2009). Preparation of biodiesel from waste oils catalyzed by a Brønsted acidic ionic liquid. *Bioresource Technology*, *100*(7), 2308–2310. DOI: 10.1016/j.biortech.2008.10.046 PMID: 19071017

Harris, J., Viner, K., & McCabe, B. K. (2018). Advances in microalgal lipid extraction for biofuel production: A review. *Biofuels, Bioproducts & Biorefining*, *12*(6), 1118–1135. DOI: 10.1002/bbb.1923

Hayyan, A., Hashim, M. A., Mjalli, F. S., & AlNashef, I. M. (2013b). A novel phosphonium-based deep eutectic catalyst for biodiesel production from industrial low-grade crude palm oil. *Chemical Engineering Science*, *92*, 81–88. DOI: 10.1016/j.ces.2012.12.024

Hayyan, M., Mjalli, F. S., Hashim, M. A., & AlNashef, I. M. (2010). A novel technique for separating glycerine from palm oil-based biodiesel using ionic liquids. *Fuel Processing Technology*, *91*(1), 116–120. DOI: 10.1016/j.fuproc.2009.09.002

Huang, W., Tang, S., Zhao, H., & Tian, S. (2013). Activation of commercial CaO for biodiesel production from rapeseed oil using a novel deep eutectic solvent. *Industrial & Engineering Chemistry Research*, *52*(34), 11943–11947. DOI: 10.1021/ie401292w

Isahak, W. N. R. W., Ismail, M., Yarmo, M. A., Jahangirian, H., & Eshraghi, M. J. (2011). Transesterification of palm oil by using ionic liquids as a new potential catalyst. *Trends in Applied Sciences Research*, *6*(9), 1055–1062. DOI: 10.3923/tasr.2011.1055.1062

Lai, J. Q., Hu, Z. L., Wang, P. W., & Yang, Z. (2012). Enzymatic production of microalgal biodiesel in ionic liquid [BMIm] [PF6]. *Fuel*, *95*, 329–333. DOI: 10.1016/j.fuel.2011.11.001

Li, K.-X., Chen, L., Yan, Z.-C., & Wang, H.-L. (2010). Application of pyridinium ionic liquid as a recyclable catalyst for acid-catalyzed transesterification of Jatropha oil. *Catalysis Letters*, *139*(3–4), 151–156. DOI: 10.1007/s10562-010-0409-x

Liang, J., Ren, X., Wang, J., Jiang, M., & Li, Z. (2010). Preparation of biodiesel by transesterification from cottonseed oil using the basic dication ionic liquids as catalysts. *Journal of Fuel Chemistry & Technology*, *38*(3), 275–280. DOI: 10.1016/S1872-5813(10)60033-3

Liu, C.-Z., Wang, F., & Ou-Yang, F. (2015). An overview of the role of ionic liquids in biodiesel reactions. *Journal of Industrial and Engineering Chemistry*, *21*, 1–10. DOI: 10.1016/j.jiec.2014.01.046

Long, T., Deng, Y., Gan, S., & Chen, J. (2010). Application of choline chloride·x-ZnCl2 ionic liquids for preparation of biodiesel. *Chinese Journal of Chemical Engineering*, *18*(2), 322–327. DOI: 10.1016/S1004-9541(08)60359-6

Luo, H., Fan, W., Li, Y., & Nan, G. (2013). Biodiesel production using alkaline ionic liquid and adopted as lubricity additive for low-sulfur diesel fuel. *Bioresource Technology*, *140*, 337–341. DOI: 10.1016/j.biortech.2012.11.112 PMID: 23708846

Manic, M. S., Najdanovic-Visak, V., da Ponte, M. N., & Visak, Z. P. (2011). Extraction of free fatty acids from soybean oil using ionic liquids or poly(ethylene glycol)s. *AIChE Journal. American Institute of Chemical Engineers*, *57*(5), 1344–1355. DOI: 10.1002/aic.12349

Mata, T. M., Martins, A. A., & Caetano, N. S. (2010). Microalgae for biodiesel production and other applications: A review. *Renewable & Sustainable Energy Reviews*, *14*(1), 217–232. DOI: 10.1016/j.rser.2009.07.020

Mir, M. A., & Banik, B. K. (2025). Sustainable healing: Natural compounds facilitating the future cancer treatment. *World Development Sustainability*, *6*, 100215. Advance online publication. DOI: 10.1016/j.wds.2025.100215

Mir, M. A., Chang, S. K., & Hefni, D. (2024). A comprehensive review on challenges and choices of food waste in Saudi Arabia: Exploring environmental and economic impacts. *Environmental Systems Research*, *13*(1), 40. Advance online publication. DOI: 10.1186/s40068-024-00364-5

Mir, M. A., Chang, S. K., & Hefni, D. (2024). Saudi Arabia e-waste management strategies, challenges and opportunities, effect on health and environment: A strategic review. *Emerging Contaminants*, *10*(4), 100357. Advance online publication. DOI: 10.1016/j.emcon.2024.100357

Mir, M. A., Hefni, D., & Chauhan, N. (2025). Heritage to health: The socio-cultural and health impacts of traditional ethnic foods in Saudi Arabian society. *Food Chemistry Advances*, *6*, 100927. Advance online publication. DOI: 10.1016/j.focha.2025.100927

Mir, M. A., Khan, M. A. A., Banik, B. K., Hasnain, S. M., Alzayer, L., Andrews, K., & Abba, S. I. (2025). Microplastics in food products: Prevalence, artificial intelligence-based detection, and potential health impacts on humans. *Emerging Contaminants*, *11*(4), 100477. Advance online publication. DOI: 10.1016/j.emcon.2025.100477

Mir, M. A., Kumar, A., Madwal, S. P., Chauhan, R., & Kumar, T. (2022). Synthesis and spectral analysis of the Cr (II) lapachol chelate crystal complexes using DMC, H2O mixture. *Inorganic and Nano-Metal Chemistry*, *54*(7), 621–626. DOI: 10.1080/24701556.2022.2081202

Moniruzzaman, M., Nakashima, K., Kamiya, N., & Goto, M. (2010). Recent advances of enzymatic reactions in ionic liquids. *Biochemical Engineering Journal*, *48*(3), 295–314. DOI: 10.1016/j.bej.2009.10.002

Mudhaffar, B., & Salimon, J. (2010). Optimization of process variables using D-optimal design for separating linoleic acid in *Jatropha curcas* seed oil by urea complex fractionation. *Biotechnology (Faisalabad)*, *9*(3), 362–367. DOI: 10.3923/biotech.2010.362.367

Niedermeyer, H., Hallett, J. P., Villar-Garcia, I. J., Hunt, P. A., & Welton, T. (2013). A step towards the a priori design of ionic liquids. *Physical Chemistry Chemical Physics*, *15*(27), 11566–11577. DOI: 10.1039/c3cp50521a PMID: 23752557

Passos, H., Freire, M. G., & Coutinho, J. A. P. (2014). Ionic liquid solutions as extractive solvents for value-added compounds from biomass. *Green Chemistry*, *16*(12), 4786–4815. DOI: 10.1039/C4GC00236A PMID: 25516718

Pham, P. J., Li, M., Pittman, C. U., & Li, T. (2009). Selective extraction of polyunsaturated triacylglycerols using a novel ionic liquid precursor immobilized on a mesoporous complexing adsorbent. *Biotechnology Progress*, *25*(5), 1419–1426. DOI: 10.1002/btpr.219 PMID: 19882672

Ribeiro, L. M. O., Silva, A. E., Silva, M. C. S., & Almeida, G., & RMR. (. (2014). A study on ethanolysis and methanolysis of coconut oil for enzymatically catalyzed production of biodiesel. *Journal of Sustainable Bioenergy Systems*, *4*(4), 223–232. DOI: 10.4236/jsbs.2014.44020

Santori, G., Di Nicola, G., Moglie, M., & Polonara, F. (2012). A review analyzing the industrial biodiesel production practice starting from vegetable oil refining. *Applied Energy*, *92*, 109–132. DOI: 10.1016/j.apenergy.2011.10.031

Seddon, K. R. (1997). Ionic liquids for clean technology. *Journal of Chemical Technology and Biotechnology*, *68*(4), 351–356. DOI: 10.1002/(SICI)1097-4660(199704)68:4<351::AID-JCTB613>3.0.CO;2-4

Shahbaz, K., Mjalli, F. S., Hashim, M. A., & AlNashef, I. M. (2010). Using deep eutectic solvents for the removal of glycerol from palm oil-based biodiesel. *Journal of Applied Sciences (Faisalabad)*, *10*(24), 3349–3354. DOI: 10.3923/jas.2010.3349.3354

Sunitha, S., Kanjilal, S., Reddy, P. S., & Prasad, R. B. N. (2007). Ionic liquids as a reaction medium for lipase-catalyzed methanolysis of sunflower oil. *Biotechnology Letters*, *29*(12), 1881–1885. DOI: 10.1007/s10529-007-9471-x PMID: 17634866

Tadesse, H., & Luque, R. (2011). Advances on biomass pretreatment using ionic liquids: An overview. *Energy & Environmental Science*, *4*(10), 3913–3929. DOI: 10.1039/c0ee00667j

Taher, H., Nashef, E., Anvar, N., & Al-Zuhair, S. (2017). Enzymatic production of biodiesel from waste oil in ionic liquid medium. *Biofuels, 10*(4), 463–472. DOI: 10.1080/17597269.2017.1316145

Tang, S., Baker, G. A., & Zhao, H. (2012). Ether- and alcohol-functionalized task-specific ionic liquids: Attractive properties and applications. *Chemical Society Reviews, 41*(10), 4030–4066. DOI: 10.1039/c2cs15362a PMID: 22456483

Troter, D. Z., Todorović, Z. B., Đokić-Stojanović, D. R., Stamenković, O. S., & Veljković, V. B. (2016). Application of ionic liquids and deep eutectic solvents in biodiesel production: A review. *Renewable & Sustainable Energy Reviews, 61*, 473–500. DOI: 10.1016/j.rser.2016.04.011

Vafaeezadeh, M., & Alinezhad, H. (2016). Brønsted acidic ionic liquids: Green catalysts for essential organic reactions. *Journal of Molecular Liquids, 218*, 95–105. DOI: 10.1016/j.molliq.2016.02.017

Vafaeezadeh, M., & Hashemi, M. M. (2014). Efficient fatty acid esterification using silica supported Brønsted acidic ionic liquid catalyst: Experimental study and DFT modeling. *Chemical Engineering Journal, 250*, 35–41. DOI: 10.1016/j.cej.2014.04.001

Vancov, T., McIntosh, S., Curtis, J., & Yu, J. (2012). Use of ionic liquids in converting lignocellulosic material to biofuels. *Renewable Energy, 45*, 1–6. DOI: 10.1016/j.renene.2012.02.033

Walden, P. (1914). Molecular weights and electrical conductivity of several fused salts. *Bulletin de l'Academie Imperiale des Sciences de St. Petersburg*, 405, 42.

Wasserscheid, P., & Welton, T. (Eds.). (2002). *Ionic liquids in synthesis*. Wiley-VCH., DOI: 10.1002/3527600701

Wu, Q., Chen, H., Han, M., Wang, D., & Wang, J. (2007). Transesterification of cottonseed oil to biodiesel catalyzed by highly active ionic liquids. *Industrial & Engineering Chemistry Research, 46*(24), 7955–7960. DOI: 10.1021/ie070678o

Xu, L., Li, W., He, J., Liu, Z., & Zhou, Y. (2011). Assessment of a dry and a wet route for the production of biofuels from microalgae: Energy balance analysis. *Bioresource Technology, 102*(8), 5113–5122. DOI: 10.1016/j.biortech.2011.01.066 PMID: 21334889

Yang, H., Chen, L., Zhou, Y., & Wang, H. (2017). New algal lipid extraction procedure using an amphiphilic amine solvent and ionic liquid. *Biomass and Bioenergy, 100*, 108–115. DOI: 10.1016/j.biombioe.2017.03.018

Yang, Z., Huang, Z.-L., Hu, Z.-L., Zhang, K.-P., & Wang, P.-W. (2010). Both hydrolytic and transesterification activities of *Penicillium expansum* lipase are significantly enhanced in ionic liquid [BMIm] [PF6]. *Journal of Molecular Catalysis. B, Enzymatic*, *63*(1–2), 23–30. DOI: 10.1016/j.molcatb.2009.11.014

Yoo, C. G., Pu, Y., & Ragauskas, A. J. (2017). Ionic liquids: Promising green solvents for lignocellulosic biomass utilization. *Current Opinion in Green and Sustainable Chemistry*, *5*, 5–11. DOI: 10.1016/j.cogsc.2017.03.003

Zhang, K. P., Lai, J. Q., Huang, Z. L., & Yang, Z. (2011). *Penicillium expansum* lipase-catalyzed production of biodiesel in ionic liquids. *Bioresource Technology*, *102*(3), 2767–2772. DOI: 10.1016/j.biortech.2010.11.057 PMID: 21138789

Zhang, L., Cui, Y., Zhang, C., Wang, L., Wan, H., & Guan, G. (2012). Biodiesel production by esterification of oleic acid over Brønsted acidic ionic liquid supported onto Fe-incorporated SBA-15. *Industrial & Engineering Chemistry Research*, *51*(50), 16590–16596. DOI: 10.1021/ie302419y

Zhao, H., Zhang, C., & Crittle, T. D. (2013). Choline-based deep eutectic solvents for enzymatic preparation of biodiesel from soybean oil. *Journal of Molecular Catalysis. B, Enzymatic*, *85–86*, 243–247. DOI: 10.1016/j.molcatb.2012.09.003

Zhen, B., Jiao, Q., Wu, Q., & Li, H. (2014). Catalytic performance of acidic ionic liquid-functionalized silica in biodiesel production. *Journal of Energy Chemistry*, *23*(1), 97–104. DOI: 10.1016/S2095-4956(14)60122-4

Zhou, J. H., Wu, L. M., Wang, J., & Zhang, L. (2011). Preparation of biodiesel from Tung oil catalyzed by sulfonic-functional Brønsted acidic ionic liquids. *Advanced Materials Research*, *314–316*, 1459–1462. . DOI: 10.4028/www.scientific.net/AMR.314-316.1459

Zhou, S., Liu, L., Wang, B., Xu, F., & Sun, R. C. (2012). Biodiesel preparation from transesterification of glycerol trioleate catalyzed by basic ionic liquids. *Chinese Chemical Letters*, *23*(4), 379–382. DOI: 10.1016/j.cclet.2012.01.034

Chapter 6
Ionic Liquids as Sustainable Solvents for Eco-Friendly Industrial Applications

Muhammad Waqar Ashraf
https://orcid.org/0000-0002-7087-9714
Prince Mohammad Bin Fahd University, Saudi Arabia

Mohammad Amin Mir
Prince Mohammad Bin Fahd University, Saudi Arabia

Syed Hasnain
https://orcid.org/0000-0002-5102-3101
Prince Mohammad Bin Fahd University, Saudi Arabia

Kim Andrews
https://orcid.org/0000-0002-0304-9630
Prince Mohammad Bin Fahd University, Saudi Arabia

ABSTRACT

Ionic liquids (ILs) are a family of adaptable, adjustable solvents that show a lot of promise for industrial and sustainable chemistry uses. They are excellent candidates to replace traditional volatile organic solvents due to their special physicochemical characteristics, which include low volatility, good thermal stability, and tunable solubility. The structure, production, and several uses of ILs are examined in this work, with a focus on their use in metal ion extraction, organic synthesis, catalysis, and nanoparticle synthesis. ILs' promise as green solvents is highlighted in this critical evaluation of their environmental impact, which also acknowledges issues

DOI: 10.4018/979-8-3373-1245-3.ch006

2026, IGI Global Scientific Publishing. Copying or distributing in print or electronic forms without written permission of IGI Global Scientific Publishing is prohibited. Use of this chapter to ative artificial intelligence (AI) technologies is expressly prohibited. The publisher reserves all rights to license its use for generative AI training and machine learning model development.

with toxicity, difficulty of synthesis, and sustainability of the life cycle. This study emphasizes the importance that ILs play in promoting more sustainable industrial processes by incorporating them into chemical education and green chemistry practices.

INTRODUCTION

Ionic liquids (ILs) are a unique class of liquid materials that may be adjusted and have a remarkable set of features. An IL is defined historically and conventionally as a salt that stays molten at temperatures lower than 100 °C. Of these, room-temperature ionic liquids (RTILs) are especially pertinent for real-world uses because their melting points are lower than 25 °C. They differ from traditional molecular solvents in that they have a distinct set of properties due to their unique ion composition. Notably, ILs may be used as electrolytes in a variety of electrochemical systems due to their intrinsic conductivity and wide electrochemical window. Supercapacitors, batteries, and dye-sensitized solar cells (DSSCs) are a few examples of these uses. DSSCs are essential for energy conversion and storage, particularly as the use of renewable energy sources increases, (Rauber et al., 2018). ILs' low vapor pressure, which inhibits evaporation into the environment and lowers operator exposure through gaseous emissions, is one of its main advantages. Furthermore, ILs can operate efficiently at high temperatures due to their generally great thermal stability. By reducing the possibility of igniting or explosion in the event of failure, this feature improves the safety of devices like batteries.

The capacity to modify the characteristics of ILs by adding functional groups or choosing particular cations and anions is one of their most amazing features. They are sometimes referred to as "task-specific fluids" or "designer solvents" (Plechkova & Seddon, n.d.) because of their versatility. Their phase behavior, which is frequently used in IL recycling procedures, can vary from extremely hydrophilic to extremely lipophilic. ILs are sometimes seen as a link between high-temperature molten salts and traditional molecular liquids due to their adaptability. ILs were commonly considered "green solvents" in the past (and sometimes even now), especially in contrast to conventional molecular solvents that were categorized as volatile organic compounds (VOCs). Their safety, non-volatility, and recyclability in chemical reactions served as the foundation for this impression. Even though these characteristics fit the guidelines for sustainable solvents, new research shows that, when considering their entire ecological impact, ILs are not always a more environmentally friendly option than VOCs, (Mir et al., 2025). Later on, this subject will be covered in more detail. However, the planned and cautious use of low-temperature molten salts can help make processes, devices, or reactions that use ILs more sustainable.

Even though certain early claims regarding their sustainability have not been fully fulfilled, IL research is still a vibrant and promising field within green chemistry, despite current controversies, (Ranke et al., 2007). ILs are rarely used in chemistry education, despite the fact that they have received a lot of attention in many scientific domains, including chemistry, since the introduction of air- and moisture-stable ILs in the 1990s, (Stark et al., 2010). This absence might result from the comparatively high cost of ILs as well as the requirement for sophisticated knowledge of chemistry and associated fields. However, by fusing state-of-the-art research with recent social and technical developments, their integration in curricula could provide special learning opportunities. In general, student motivation is increased when they are exposed to pertinent, real-world subjects.

Furthermore, by showcasing real-world applications rather than just theoretical ideas, connecting ILs with green chemistry allows for a more hands-on approach to teaching sustainability. Students can become more conscious of sustainable development with focused instruction. Fundamental chemical concepts like structure-property connections and the slow transition from molecular to ionic compounds can also be taught using ILs. Given the environmental problems brought on by human activity, developing greener and more sustainable chemistry is essential for future generations. These difficulties include global problems like climate change, ozone depletion, and marine pollution, as well as local contamination by persistent, harmful compounds (from mining and industrial chemical manufacture, for example). Innovative solutions are needed to address these issues, and ILs have a big role to play in creating a more sustainable future.

STRUCTURE OF IONIC LIQUIDS

Ionic liquids (ILs) are made up of different cationic and anionic components, just like regular salts. However, because of their large and uneven cationic structures, ILs are less likely to crystallize than regular salts, (Flieger & Flieger, 2020). Precise customisation of IL characteristics is made feasible by the large number of potential cation-anion combinations. Cations affect melting temperatures and solubility in organic solvents, but anions have a major effect on air and water stability, (Kaur, Kumar, & Singla, 2022).

ILs, sometimes known as "designer solvents," have the benefit of being able to customize certain properties to fit specific uses. By combining big positively charged cations like alkylimidazolium, alkylpyridinium, alkylpyrrolidinium, alkylphosphonium, or alkylmorpholinium with small negatively charged anions like [Tf2N]−, PF6−, or PF4−, researchers can create ILs, (Mir, Chang, & Hefni, 2024). These specially made ILs are frequently used to dissolve specific substances or remove contaminants from solutions in a targeted manner. The production of ILs with characteristics especially suited to a range of commercial and scientific applications is made possible by the capacity to alter their molecular architectures. Even though IL cations, anions, and their interactions have been well studied, research is still being done to find new liquid salts that are stable at room temperature, (Choi et al., 2023). The main characteristics of ILs that set them apart from traditional solvents are their amphiphilic composition, which results in complex molecular correlations across longer distances, and their coulombic interactions, which create local ordering. The arrangement of cation-cation and anion-anion domains, among other structural characteristics of ILs, create patterns that interact with the distribution of cation-anion pairs to shape the ILs' overall characteristics. Superacidity, basicity, hydrophilicity, water miscibility, water immiscibility, and hydrophobicity are some of these characteristics.

SYNTHESIS OF IONIC LIQUIDS

Ionic liquids (ILs) are created using a variety of methods that combine particular cations and anions to produce customized characteristics. Several popular techniques are as follows:

Neutralization Reaction

In this method, an IL is created by reacting a basic molecule with an acidic equivalent. For instance, imidazole (a base) and hydrochloric acid combine to form imidazolium chloride. The resultant cation and anion are determined by the choice of acid and base, which has a direct impact on the properties of the produced IL.

Metathesis Reaction

This process creates a new IL by ion exchange between two salts. Researchers can create ILs with certain characteristics by carefully choosing precursor salts that contain the desired cations and anions. This method offers versatility in adjusting the composition of IL.

Ion Exchange

This procedure creates an IL with desired characteristics by swapping out the ions in a salt for new ones. To create ILs with particular cation-anion combinations, for example, halide salts can go through ion exchange. Depending on the choice of precursor salts, this method enables customisation.

Quaternization

This process creates a quaternary ammonium salt by reacting a tertiary amine with an alkyl halide. An IL is then produced by combining this molecule with the proper anion. The final IL's distinct properties are influenced by the presence of alkylated nitrogen cations.

Functionalization

Using chemical reactions to add functional groups, this technique alters already-existing ILs. This approach improves efficiency and adaptability by modifying the features of existing ILs to satisfy certain needs rather than creating new ones from start. These synthesis methods give scientists a wide range of resources to create ILs with specific cation-anion pairings. Each approach offers unique benefits that influence aspects like cost-effectiveness, synthesis complexity, and exact control over the final IL structure, enabling optimization in a range of applications.

PHYSICAL PROPERTIES OF IONIC LIQUIDS

Distinctive Physical Properties of Ionic Liquids (ILs)

Ionic liquids (ILs) differ from traditional solvents due to their distinct physico-chemical characteristics. Among these attributes are:

Low Volatility

ILs have a very low vapor pressure because of their ionic composition, which results in strong electrostatic interactions between cations and anions. Because of this feature, they are perfect for processes that need to have little solvent loss, like closed systems or ecologically delicate operations, (Greer, Jacquemin, & Hardacre, 2020).

High Thermal Stability

ILs can withstand high temperatures without decomposing significantly because they don't contain any volatile molecular components. In heat-intensive applications, such as thermal energy storage and catalysis, this stability is very advantageous, (Mir, Hefni, & Chauhan, 2025).

Low Melting Temperatures

Because of their low melting temperatures, a large number of ILs are still liquid at normal temperature, (Rothe et al., 2021). This characteristic increases their adaptability in a variety of applications by permitting their use over a wide temperature range.

Tunable Solvating Ability

Researchers may alter the polarity and interaction strength of ILs, making them efficient solvents for a variety of molecules, by choosing particular cations and anions, (Boamah et al., 2023).

Density and Viscosity Control

Applications involving fluid dynamics and mass transport are made more flexible by the ability to modify the density and viscosity of ILs by selecting the right ions, (Boualem et al., 2022).

Broad Liquid Range

ILs are appropriate for procedures with a range of thermal conditions because they preserve a stable liquid phase across a wide temperature range, (Mir & Chang, 2024).

Non-Flammable Nature

Due to their ionic structure and low vapor pressure, many ILs are non-flammable. By lowering the risk of fire and explosion, this characteristic improves safety in industrial settings, (Greer, Jacquemin, & Hardacre, 2020).

Water Miscibility and Hydrophobicity

ILs can be either extremely water-soluble or hydrophobic, depending on the cation-anion combination that is selected. Because of this characteristic, they can be used in procedures that call for either total water miscibility or water exclusion, (Freire et al., 2007).

In conclusion, ILs' unique qualities make them extremely adaptable for use in electrochemistry, green chemistry, energy storage, and catalysis. The potential of ILs is expanded across numerous scientific and industrial fields by ongoing study.

Green Characteristics of Ionic Liquids

ILs are regarded as environmentally benign because of a number of important characteristics:

Customizable Composition

By choosing appropriate cation-anion combinations, ILs can be specially made to satisfy the needs of specific applications, (Seddon, 1997).

Multicomponent Systems

ILs' usefulness can be further increased by combining them with inorganic salts to create multicomponent liquid systems.

Modifiable Physical Properties

For best results, properties including density, hydrophobicity, miscibility, and viscosity can be adjusted.

Near-Zero Vapor Pressure

During handling and transportation, dangerous volatile organic compounds (VOCs) are not released due to the strong electrostatic interactions between ions, which provide a negligible vapor pressure, (Suresh & Sandhu, 2011).

Non-Flammable and Non-Explosive

ILs are intrinsically safer than many conventional solvents because they do not oxidize readily.

Increased Reaction Rates

ILs are useful in catalytic applications because of their ionic composition, which speeds up reaction kinetics.

Long-Term Stability

ILs don't break down when kept in storage for long periods of time.

Control of Stereoselectivity

They are helpful in reactions that need to produce stereoselective results, (Orha, Akien, & Horvath, 2012).

No Coordination with Metal Complexes

ILs do not create coordinate bonds with metal complexes, in contrast to certain traditional solvents.

Broad Thermal Stability

ILs don't change in temperature across a broad range.

Common Cations and Anions. Nitrate, phosphate, tetrafluoroborate, hydrogen sulfate, acetate, and halides are among the anions that are frequently found in ILs, along with organic cations such imidazolium, pyrrolidinium, ammonium, phosphonium, and pyridinium, (Handy & Zhang, 2001).

Versatility in Dissolution

ILs can efficiently dissolve both organic and inorganic compounds due to their high organic cation and modest inorganic anion composition.

Because of these characteristics, ILs are seen as possible substitutes for traditional solvents in a variety of fields, including environmental protection, green chemistry, and sustainable industrial operations.

APPLICATIONS OF IONIC LIQUIDS (ILS)

Organic Synthesis and ILs

Green chemistry encourages the application of less harmful and sustainable techniques in an effort to reduce the influence on the environment. The use of ionic liquids (ILs) in place of traditional volatile organic solvents is a notable development in this regard. They are perfect substitutes for environmentally friendly organic synthesis due to their adjustable physicochemical characteristics and low vapor pressure. In order to optimize reaction conditions, ILs can be created as task-specific solvents by altering their cationic and anionic components. In a variety of organic transformations, such as the Michael reaction, the Biginelli & Hantzsch reaction, the Aldol condensation, the Knoevenagel reaction, and the Doebner condensation, ILs operate as both catalysts and reaction media. ILs affect these processes' stereoselectivity in addition to increasing reaction speeds, (Orha, Akien, & Horvath, 2012).

One well-known instance is the Stille reaction, as shown in figure 1which occurs when an electrophilic reagent and an organotin molecule interact with a palladium catalyst to create carbon-carbon bonds. Iodocyclohexanone and vinyl tributyl tin reacted in this reaction using [bmim]BF4 (1-butyl-3-methylimidazolium tetrafluoroborate) as the IL medium. Ethylene was used to extract the product, but the catalyst and IL were stable in moisture and air, allowing them to be used again in other processes, (Suresh & Sandhu, 2011). Their relevance in efficient and sustainable organic synthesis is further supported by the stability and recyclability of ILs in catalytic methods. As seen in Figure 2 (Suresh & Sandhu, 2011), numerous more reactions, including the modified Mannich process, used green solvent [bmim] [NTf2].

Figure 1. Stille reaction in ionic liquids

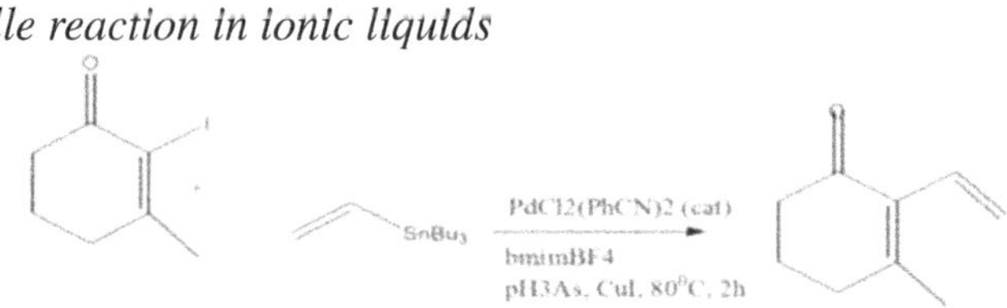

Figure 2. [bmim][NTf2] modified mannich reaction

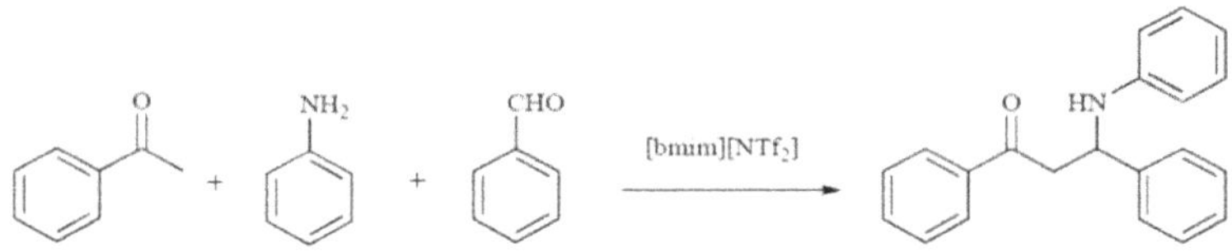

ILs as Solvents and Catalysts in Chemical Reactions

Ionic liquids (ILs) have highly adjustable characteristics like density, viscosity, and solvating power that allow them to function as both solvents and catalysts in a variety of chemical reactions. ILs can be customized for certain processes by choosing the right cation-anion combinations, increasing sustainability and efficiency.

Heterocyclic Compound Fluorination

When fluorinating heterocyclic compounds, ILs like [bmim]BF4 (1-butyl-3-methylimidazolium tetrafluoroborate) work well as solvents, improving reaction efficiency and selectivity, (Takeuchi, Tarui, & Shibata, 2000).

Opening Ring Reactions

In ring-opening events, ILs also function as reaction media, facilitating the production of compounds with biological activity. Because of their excellent yields and ability to continue at low temperatures, ILs are a useful substitute for conventional solvents, (Wasserscheid & Welton, 2008).

Aldehyde to Alcohol Reduction

Even at room temperature, ILs help reduce aldehydes to alcohols. ILs that provide gentle reaction conditions and improved yields, such as [bmim]BF4, [emim]BF4, and [emim]PF6, are frequently utilized for this transformation, (Reynolds, Erdner, & Jones, 2002).

Reactions of Suzuki-Miyaura Coupling

ILs play an important role in the Suzuki-Miyaura coupling process, which creates biaryls from aryl halides and aryl boronic acids in the presence of a base and a Pd catalyst. The primary benefit is that, unlike in traditional reaction medium, the Pd catalyst is easily recovered and reused because it is held in the IL phase, (Mathews, Smith, & Welton, 2000).

Reactions of Hydrogenation

ILs enhance catalyst retention in hydrogenation processes, which involve the addition of hydrogen across a carbon-carbon double bond. Transition metals or their derivatives are typically used as catalysts in these processes, although recovering them and separating the products can be difficult. When carried out in ILs, the product separates into the alcoholic phase and the catalyst stays in the IL phase, making recovery simple through decantation and catalyst reuse, (Dupont, Consorti, & Spencer, 2000).

Reformatsky Responses

Traditionally, volatile organic solvents such as dichloromethane and ethereal substances are used in the Reformatsky reaction as shown in figure 3, which creates β-hydroxyesters via condensation of an aldehyde or ketone with an α-halo ester in the presence of Zn dust. Although these solvents function as Lewis bases, their emissions of volatile organic compounds (VOCs) are a problem for the environment. These solvents have been effectively replaced with ILs, which have a very low vapor pressure, making the process more environmentally friendly and sustainable, (Suresh & Sandhu, 2011). In conclusion, by increasing reaction efficiency, facilitating catalyst recovery, mitigating environmental effect, and offering improved control over reaction conditions, ILs are essential to organic synthesis. They are a vital instrument for green chemistry and sustainable chemical processes because of their adaptability.

Figure 3. The reformatsky reaction modified in Illinois

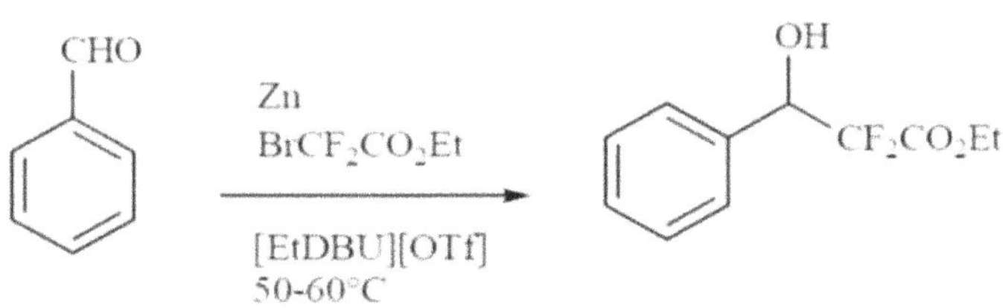

(Suresh & Sandhu, 2011)

ROLE OF IONIC LIQUIDS IN NANOPARTICLE SYNTHESIS

The low vapor pressure of ionic liquids (ILs) prevents the emission of volatile organic compounds (VOCs) into the atmosphere, making them environmentally benign solvents, (Kim et al., 2006). During the creation of nanoparticles, they can act as stabilizing and reducing agents due to their adjustable characteristics, (Kim et al., 2006). By creating a stabilizing layer surrounding the metal ions, the strong ionic interactions in ILs reduce aggregation of nanoparticles and provide better solvation and protection of metal ions than traditional solvents, (Suresh & Sandhu, 2011).

Because of their superior reducing capabilities, alcohol-based ionic liquids (AILs) are especially useful for the synthesis of gold nanoparticles (AuNPs), (Kim et al., 2006). Research has demonstrated that the length of the cation's alkyl chain affects the size of the nanoparticles in the ionic liquid (HEMMor)(BF4), with smaller nanoparticles resulting from longer alkyl chains, (Dupont & Scholten, 2010).

ILs also act as environmentally friendly catalysts for the production of nanoparticles. For example, it has been discovered that imidazolium-based ILs, like 1-n-butyl-3-methylimidazolium, function as effective catalysts in multiphase processes, (Dupont & Scholten, 2010).

Extraction of Lignocellulosic Materials

Optimizing reaction conditions to improve yield and selectivity is made possible by the design of ILs. By dissolving enormous volumes of cellulose, ILs have shown great promise for cellulose extraction. In particular, during pulsed microwave heating, ILs with 1-butyl-3-methylimidazolium cations ($[C4mim]^+$) and chloride anions can dissolve up to 25 weight percent cellulose, outperforming conventional binary solvent systems as DMAC/LiCl, (Swatloski et al., 2002). These ILs allow cellulose to dissolve by breaking down and penetrating its robust hydrogen bonding network. The activity of chloride ions, which interact with cellulose's hydroxyl groups, is closely related to how effective these solvents are, (Pinkert et al., 2009).

ILs Have a Number of Benefits Over Binary Solvent Systems, Including:

By altering the composition of ILs, the mobility and reactivity of chloride ions can be precisely controlled; the concentration of chloride ions in ILs is not constrained by salt solubility limits, (Pinkert et al., 2009). The development of novel IL systems for the dissolving of cellulose is the subject of ongoing study, especially those that use methyl imidazolium (MIM) and methyl pyridinium (MPy) as cations, (Pinkert et al., 2009). The best anions for dissolving cellulose have been found to be chlorides and other hydrogen bond acceptors. But there are still issues that prevent ILs from being fully green solvents, including as their sensitivity to water contaminants and worries about their toxicity and environmental effects.

Because of the functional groups on their cations or anions, ILs are also known as Task-Specific Ionic Liquids (TSILs), which enable them to operate as solvents, reagents, or catalysts in chemical reactions, (Davis, 2004). ILs with amine-functionalized groups are used for CO_2 capture, and ILs with large aromatic head groups are used for aromatic extraction in aqueous biphasic systems, (Davis, 2004).

Using hydroxyl-functionalized ILs for phase-transfer catalysis in the production of ethoxy benzene, (Davis, 2004). Sulfonic acid-functionalized ILs are used in esterification and acid-catalyzed processes (Welton, 2004), and ILs that can form metal complexes are used to remove metal ions from aqueous solutions, (Welton, 2004).

Furthermore, in a variety of organic processes, ILs can operate as an acid, base, or organo catalyst.

Ionic Liquids as Acid Catalysts

In many synthetic reactions, hazardous mineral acids are gradually being replaced with Task-Specific Ionic Liquids (TSILs), especially reusable ones, which are becoming increasingly effective catalytic systems. In the esterification of carboxylic acids and alcohols, for example, the Bronsted acid ionic liquid N-methyl-2-pyrrolidinium methyl sulphonate has been effectively employed as an alternative to halogenated mineral acids, (Xing et al., 2005). Under microwave circumstances, 1-benzyl-3-methylimidazolium hydrogensulphate ([bnmim][HSO4]) has demonstrated potential as an efficient catalyst in the condensation of indoles with benzaldehyde, (Zhang et al., 2009). Particularly in tumor chemotherapeutic applications, the products of this reaction have substantial medical value, (Yannai et al., 1996).

While Bronsted acidic 1-methyl-3-propanesulphonic imidazolium hydrogen sulfate ([mimps][HSO_4]) has been used to synthesize xanthene, which has numerous uses in laser technology as well as in the biological and pharmaceutical domains, Lewis acidic ionic liquids, such as [bmim]Cl·$AlCl_3$, have been employed in Friedel-Crafts acylation reactions, (Gong et al., 2009). Additionally, ILs that include sulfonyl are efficient recyclable catalysts, especially when it comes to the synthesis of 3-vinyl indoles from indoles and ketones, (Das et al., 2010). In this instance, the simultaneous presence of sulfonyl and sulfonic acid groups in the same ionic liquid increases the catalytic activity.

Furthermore, the selective alkylation of phenol and anisole has been accomplished using Bronsted acidic triflate IL [MIMBS][OTf] in a biphasic medium, which has an advantage over traditional techniques in that it does not require neutralizing excess acid by-products, (Titze-Frech et al., 2013). The reduced oxophilicity of ILs in comparison to conventional mineral acids is responsible for the higher selectivity seen in this procedure.

Ionic Liquids as Base Catalysts

For some processes, ionic liquids are also very effective base catalysts. They have advantages over mixes of inorganic bases and ionic liquids, such as easy recyclability and increased catalytic efficiency. For instance, the Michael addition of N-heterocycles to α,β-unsaturated molecules at ambient temperature has been carried out using the basic IL ([bmim]OH) as a catalyst and reaction media, (Xu et al., 2007). An ethanol-functionalized TSIL performs several roles in the Pd-catalyzed Heck process, including that of a base, ligand, and reaction media. It also offers the advantage of catalyst recyclability, (Wang, Li, & li, 2009).

Secondary and tertiary nitrogens found in BrTBDPEG150TBDBr can be used to activate methanol in transesterification processes. Together, cycloaddition and transesterification are made easier by this activation. Additionally, soybean oil can be converted into biodiesel when a basic IL catalyst, like choline hydroxide, is present. It has also been demonstrated that basic ILs, like ([bmim]OH), are effective catalysts in the synthesis of substituted ureas, (Jiang et al., 2008). These techniques have the following main benefits:

- Reaction conditions without a solvent.
- The removal of water byproducts without the need for a dehydrating agent the catalyst's recyclable nature and ease of use.

Ionic Liquids as Organocatalysts

Because of their capacity for hydrogen bonding, ionic liquids (ILs) have demonstrated notable success as organocatalysts, especially in Diels-Alder cycloaddition reactions. For instance, low-coordinating bistriflimide anions and ILs with cations from hydrogen-bond-rich D-glucopyranoside derivatives have been successfully employed as organocatalysts in the Diels-Alder reaction of different dienes and dienophiles, (Erfirt et al., 2014).

Ionic Liquids in Metal Ion Removal

Extractants that increase the concentration of metal ions in a hydrophobic phase are usually used in traditional methods for removing metal ions from solutions. These extractants facilitate the transport of metal ions into the extracting phase by creating a hydrophobic environment, (Mir & Banik, 2025). Task-Specific Ionic

Liquids (TSILs), on the other hand, have attached groups that serve as essential parts of the extracting solvent. TSILs can therefore operate as the extractant and extracting solvent at the same time. For example, crown ethers in combination with ILs containing PF_6 or Tf_2N- anions have been successfully used to remove strontium nitrate, a byproduct of radioactive materials, (Dai, Ju, & Barnes, 1999). TSILs have also been used to remove mercury and cadmium from tainted water.

PF_6 anions or mixes with [bmim]BF_4 were used with imidazolium cations, which have been modified with functional groups that operate as ligands for binding to these metals, making them hydrophobic solvents and dual-purpose extractants, (Mir, 2024). TSILs offer a practical way to handle nuclear materials since they can be very successful in removing radioactive lanthanides and actinides. [Cnmim][PF_6] was used to successfully extract the Na+ and Cs+ ions (n=4, 6, 8), (Visser et al., 2000). Several functional groups can be added to the cation's side chains as complexing agents to further adjust the extraction capacity of ILs. The following are some crucial elements for metal ion extraction process optimization:

- as employed in ILs, crown ethers improve the selectivity of metal ion extraction due to their thermodynamically advantageous solvation as compared to organic solvents, (Chun, Dzyuba, & Bartsch, 2001).
- By changing the structure of ILs, the hydrophobicity can be modified, improving the metal ion partition coefficients, (Mir, 2024).
- ILs and extractants can be used to optimize the extraction of metal ions by adjusting the pH of the system.

Ionic Liquids in Biological Systems

Enzymes in ILs have been effectively used for the extraction of amino acids and the production of medicinal compounds because of their stability (Zhao, Xia, & Ma, 2005) as mentioned in Table 1. Despite being readily available and renewable energy sources for the chemical industry, carbohydrates frequently have limited potential to be converted into useful compounds because of their poor solubility in organic solvents. The high solubility of carbohydrates in ILs has been used to overcome this issue, (Liu et al., 2005). For instance, an IL was used to acetylate glucose in the presence of lipase enzymes, which enhanced the reaction rate and regioselectivity, (Park & Kazlauskas, 2001). In the same way, sucrose and other carbohydrates that are insoluble in a variety of organic solvents can be esterified with dodecanoic acid in the IL [bmim][dca], where they show excellent solubility, (Liu et al., 2005). In biphasic whole-cell biocatalytic processes, water-immiscible ILs such as [bmim]PF_6 and [bmim][Tf_2N] can function as substrate reservoirs and in situ product extractors since they do not damage the cells of E. coli and S. cerevisiae. Because toxic chemical solvents are incompatible with biological systems, their usage for these procedures has decreased, (Pfruender, Jones, & Weuster-Botz, 2006).

Table 1. Examples of IL extractions of metal ions [49]

	Substances	IL	Extractant/ligand/metal chelator
Alkali metals	Li^+, Na^+, K^+, Rb^+, Cs^+	[C*n*MIM][PF_6] (*n* = 4–9)	$DC_{18}C_6$
	Na^+, Cs^+	[C*n*MIM][PF_6] (*n* = 4, 6, 8)	$18C_6$, $DC_{18}C_6$, $Dtb_{18}C_6$
	Cs^+	[C*n*MIM][Tf_2N] (*n* = 2, 3, 4, 6, 8)	BOBCalixC_6
	Na^+, K^+, Cs^+	[C*n*MIM][Tf2N] (*n* = 2, 4, 6, 8)	$DC_{18}C_6$, *N*-alkyl aza-18-crown-6 ethers
Alkaline earth metals	Mg^{2+}, Ca^{2+}, Sr^{2+}, Ba^{2+}	[C*n*MIM][PF_6] (*n* = 4–9)	$DC_{18}C_6$
	Sr^{2+}	[C*n*MIM][PF_6] (*n* = 4, 6, 8)	$18C_6$, $DC_{18}C_6$, $Dtb_{18}C_6$
	Sr^{2+}	[R_1R_2MeIM][PF_6], [R_1R_2MeIM][Tf_2N]	$DC_{18}C_6$

Note: C*n*MIM = 1-alkyl-3-methylimidazolium; DC18C6 = dicyclohexano-18-crown-6; 18C6 = 18-crown-6; Dtb18C6 = 4,4_-(5_)-di-(*tert* butyl cyclo hexano)- 18-crown-6; BOBCalixC6 = calix (Ranke et al., 2007) arene-bis(*tert*-octylbenzo-crown-6); Tf2N = bis[(trifluoromethyl)sulfonyl] amide; R1R2MeIM = 1-R1-2-R2-3-methylimidazolium (R^1 =Bu, Et, or Pr; R^2 = H, or Me).

Ionic Liquids and Green Chemistry Principles

The United Nations' 2015 report "Transforming our World: The 2030 Agenda for Sustainable Development" (United Nations General Assembly, 2015) emphasizes the rising need for green and sustainable chemistry. In this sense, sustainability includes the interconnected ecological, economic, and social facets. This translates to lowering expenses in the field of chemistry by using more effective procedures and using less energy and chemicals. It also highlights the growing usage of chemicals and renewable feedstocks that present few hazards to the environment and operators. Waste production and chemical emissions into the environment can be reduced by implementing efficient recycling techniques.

A thorough understanding of ionic liquids' (ILs') basic characteristics is necessary to comprehend how they affect new, environmentally friendly technologies, especially when compared to conventional solvents. Understanding this is essential for determining how ILs might support greener practices. Figure 4 shows some of the main molecular structures of ILs, such as their cations and anions, as well as the characteristics that emerge from these structures.

Figure 4. Molecular structures of some important cations and anions used for the formation of ionic liquids and notable characteristics in the context of green and sustainable chemistry

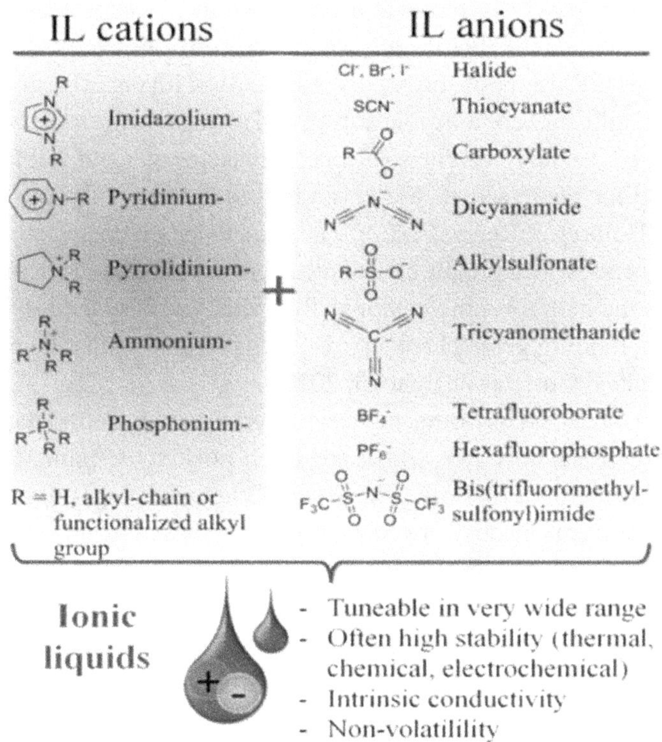

Solvents and Their Reduction in the Chemical Industry

Both the chemical industry and scientific research are becoming increasingly interested in the usage of solvents and their decrease. In order to regulate heat, produce uniform reaction conditions, and direct organic reactions toward the intended products, solvents are essential, (Reichardt & Welton, 2010). The environmental impact of solvents, however, usually outweighs that of the chemical process itself since the amounts of solvents employed are frequently greater than those of the actual chemicals. Beyond chemical reactions, solvent consumption includes procedures like extraction and purification, which can call for significant solvent usage. According to recent estimates, around 20 million metric tons of organic solvents are produced annually worldwide, (Clark et al., 2015). Volatile organic compounds

(VOCs), which have a major influence on pollution and energy consumption, make up a large number of these solvents.

The production of solvents from fossil fuels and the energy required to remove them are two ways that VOCs impact climate change and contribute to environmental deterioration by contaminating the air, water, and soil. Unsustainable chemistry is demonstrated by the history of solvent use, especially with regard to non-renewable, poisonous, volatile, and combustible solvents. This emphasizes the need for more environmentally friendly solvent solutions, which the general public frequently ignores even if the chemical sector plays a part in creating solutions to deal with these issues, (Holbrey & Rogers, 2002). The need to lessen the negative impacts of traditional solvents, which greatly contribute to environmental and health problems, is what drives the hunt for more ecologically friendly solvents. As a result, one of the most active fields of green chemistry research is the investigation of sustainable solvent substitutes, (Anastas & Eghbali, 2010).

Ionic liquids (ILs), among other neoteric solvents, are the most adaptable. Compared to conventional solvents, which are often poisonous, flammable, volatile, and generated from fossil fuels, they have significant advantages. By offering safe, biodegradable solvents made from renewable materials and requiring less energy to produce, neoteric solvents aim to eradicate these adverse consequences. Commonly used tactics include cutting back on solvent usage or switching out hazardous solvents for safer ones (such as ethanol for chlorinated solvents). For many reactions, however, such substitutions can be economically impractical or chemically limited.

In contrast to the green imperative that regulates the use of conventional solvents, which calls for a more interdisciplinary approach to producing green and sustainable chemistry, which is still maturing, the recycling of neoteric solvents is usually motivated by economic considerations, (Clarke et al., 2018). ILs are frequently referred to as "green solvents" since they satisfy a wide range of sustainability requirements. Among their many benefits is their non-volatility, which shields operators from exposure and environmental evaporation. Furthermore, because of their low vapor pressure and excellent thermal stability, they are less likely to cause mishaps like explosions or thermal runaways, which are frequent with volatile solvents. The variety of ILs and their adaptable qualities make them perfect for recycling procedures, which can lower the total amount of solvent used.

ILs have the potential to number in the trillions, providing countless modification options to meet particular requirements, whereas organic solvents are few, (Plechkova & Seddon, 2008). Because of their dual organic and ionic properties, ILs can effectively dissolve both organic and inorganic substances that would otherwise be intractable in conventional solvents. This is especially helpful for processing and isolating biological macromolecules that come from renewable sources, such as cellulose and lignin.

ILs are perfect for multipurpose applications since they also have "beyond solvent" qualities like catalytic activity or magnetic qualities. In some reactions, they can also act as catalysts and solvents, significantly lowering the need for resources. Anastas and Warner's twelve principles of green chemistry are met by a large number of reactions involving ILs, (Mir, 2023). In Table 2 there are highlighted beneficial influence of ILs in achieving these green chemistry principles, (Zhao, Liao, & Zhang, 2007; Marr & Marr, 2016). A framework for attaining more sustainable chemistry is provided by these ideas, which include:

- Waste Prevention: Reduce energy and chemical waste since disposing of garbage requires more resources.
- Optimizing Atom Economy: Plan reactions to use as few starting elements as possible in the finished product.
- Lowering Synthesis Risks: Make use of materials with minimal toxicity and possible risks.
- Creating Safer Chemical Products: Create compounds that are less harmful while maintaining their effectiveness.
- Safer Auxiliaries and Solvents: When it is not possible to avoid using non-essential chemicals, utilize the safest reagents and solvents.
- Energy-Efficient Design: Prefer reactions that take place in ambient settings and optimize chemical processes to use the least amount of energy.
- Use of Renewable Feedstocks: Avoid depleting non-renewable resources by obtaining materials from renewable sources.
- Derivatives reduction: Steer clear of extra chemical processes, which typically call for more energy and chemicals.
- Selective catalysts are preferred over stoichiometric reagents in catalytic reactions.
- Product Degradability: To prevent environmental persistence, chemicals should be designed to decompose into innocuous compounds.
- Real-Time Analysis for Pollution Prevention: Use real-time process monitoring to stop dangerous compounds from forming.
- Inherently Safer Chemistry: Select materials and reaction configurations that reduce the possibility of mishaps, detonations, and other dangers.

Table 2. Highlights the beneficial influence of ILs in achieving these green chemistry principles

Green Chemistry Principle	**Contribution of Ionic Liquids**
Waste Prevention	- Ionic liquids (ILs) do not evaporate, allowing for high recycling potential by adjusting their properties. - Their high stability often eliminates the need for additional cleaning steps, facilitating complete recycling. - ILs can enhance reaction rates, selectivity, and yields by stabilizing or destabilizing intermediates, thus promoting desired reaction pathways. - Designed for long-term use, ILs are integrated into applications like batteries and supercapacitors, where recycling is incorporated.
Atom Economy	- ILs typically do not impact atom economy negatively, and are generally not used as starting materials in reactions.
Reduce Hazards	- ILs eliminate risks associated with volatile compounds, such as inhalation and flammability. - They can be engineered to be non-toxic.
Safer Products	- Their non-volatile nature makes ILs safer for transportation and use, as compared to volatile, flammable solvents. - ILs are stable and thermally resistant, producing lower pressure during incidents, which is crucial for applications like energy storage. - ILs can be combined with other materials, creating novel, safer compounds such as ionic liquid crystals or polymerized ionic liquids.
Reduced Auxiliaries / Safer Solvents	- ILs can dissolve a wide range of substances, minimizing the need for additional solvents or reaction/extraction steps. - Multifunctional ILs can reduce the overall chemical demand in processes. - In general, ILs are safer than conventional molecular solvents.
Energy Efficiency	- New, energy-efficient processes become feasible with ILs, such as multiphasic catalysis or reactions at elevated temperatures without pressurizing the reaction vessel.
Usage of Renewable Feedstocks	- ILs offer efficient pathways for utilizing renewable resources like lignin and cellulose due to their strong solvation properties. - ILs can also be derived from renewable feedstocks.
Reduction of Derivatives	- ILs do not interfere with derivatization, making them neutral in this regard.
Catalytic Reactions	- Certain ILs can catalyze specific reactions (especially those involving Lewis or Brønsted acid/base behavior). - ILs can activate catalysts and stabilize catalytically active species.
Degradability of Products	- ILs can be designed to degrade biologically. - Some ILs are naturally occurring and biodegradable.
Real-Time Analysis	- ILs do not interfere with real-time analytical methods, which can be applied to reactions involving ILs.
Safer Chemistry	- ILs make chemical processes safer by eliminating the risks associated with volatile, flammable solvents. - They have high thermal stability, preventing accidents in case of operational errors or technical failures.

(Zhao, Liao, & Zhang, 2007; Marr & Marr, 2016)

It is often known that ionic liquids (ILs) are safer and more recyclable than conventional organic solvents, especially when it comes to green chemistry. They have a number of advantageous qualities, including excellent thermal stability, non-volatility, and application-specific tuning. Notwithstanding its encouraging characteristics, a number of important factors make it difficult to present ILs as wholly "green."

Key Considerations in the Sustainability of ILs

Synthesis Complexity

When compared to traditional solvents, ILs are more complicated compounds. Multiple steps are usually involved in their synthesis, which increases energy usage and chemical use. Concerns regarding their total environmental impact are raised by the possibility that their complexity will result in waste being generated during the production phase.

Change in the Effect on the Environment

According to Life Cycle Assessments (LCAs), the use of ILs may only transfer the environmental burden to the production phase, even if they are typically thought to lessen environmental harm during the consumption stage (e.g., by replacing volatile, hazardous solvents). The potential advantages in use may be outweighed by the complexity of IL synthesis, particularly in light of the energy and chemical waste associated with their manufacturing, (Zhang, Bakshi, & Demessie, 2008).

Non-Renewable Feedstocks and Toxic Chemicals

A large number of the chemicals employed in the synthesis of IL, including alkylation agents, are toxic and frequently come from non-renewable feedstocks. Because these precursor compounds might not be in line with the objectives of sustainable chemistry, this raises possible environmental and sustainability concerns.

Renewable Feedstocks

By creating ILs from renewable feedstocks, like chemicals generated from biomass, attempts are being made to increase their sustainability. This method tackles some of the issues related to the use of non-renewable resources in IL synthesis and lessens dependency on fossil fuels, (Handy, 2003). Furthermore, using solvent-free techniques for IL synthesis has been investigated as a way to reduce the production's negative environmental effects, (Varma & Nambocdiri, 2001).

Considerations and Limitations of Ionic Liquids (ILs) in Green Chemistry

Although ionic liquids (ILs) offer a number of benefits in terms of fulfilling the requirements of green chemistry, such as being safer and recyclable solvents in contrast to conventional organic solvents, there are significant factors that cast doubt on the idea that ILs are entirely "green." A thorough evaluation of their sustainability must consider a number of criteria beyond their immediate environmental advantages, despite the fact that they have many advantages, including non-volatility, high thermal stability, and adjustable features. The energy and chemical consumption required for the production of ILs is one of the main obstacles. Although ILs can be used in place of traditional solvents, their manufacture usually entails more intricate synthetic processes that use more energy, chemicals, and produce waste.

Although the use of some ILs can lessen the environmental impact during the operational stage (by replacing hazardous solvents), life cycle assessments (LCAs) of these chemicals have demonstrated that the overall environmental burden may simply shift to the production phase, where more resources are used, (Zhang, Bakshi, & Demessie, 2008). This problem is made worse by the fact that a large number of the chemicals required to synthesize IL, including alkylation agents, are dangerous and come from non-renewable resources. However, the sustainability of ILs may be enhanced by using renewable feedstocks. Some ILs, for instance, are made from sugars, which are renewable raw materials. This provides a means to increase the "greenness" of ILs by lowering their need on fossil fuels.

Furthermore, it has been shown that using solvent-free synthesis techniques for ILs can help lessen some of the negative environmental effects related to their manufacture, (Handy, 2003; Varma & Nambocdiri, 2001).

Reliance on Fossil Fuels

The sustainability of ILs is limited because they are mostly made from materials sourced from fossil fuels. Their "greenness" may be improved by creating ILs from renewable resources, although this change hasn't been widely adopted yet.

New Chemicals with Unknown Environmental Effects

Because ILs are still in their infancy, more research is needed to determine their effects on the environment, human health, and safety before they can be extensively used. Because of their intricate chemical makeup, little is known about how they will affect ecosystems and human health in the long run.

Possible Health and Environmental Risks

Because of their structural resemblance to growth regulators and herbicides, some ILs may pose health and environmental risks. To ascertain if ILs are safe, particularly in situations involving prolonged exposure, their toxicity and ecotoxicity must be carefully assessed.

Residue in Polymers

Polymers produced with ILs present may still include sizable quantities of IL residues. Because the leftover ILs may be harmful to health, this can be an issue when these polymers are utilized in products like food packaging or personal care items.

Solubility in Water and Environmental Durability

Certain ILs, especially those that dissolve in water, can find their way into aquatic habitats through unintentional spills or effluents. It has been demonstrated that common ILs like [bmim][PF6] and [bmim][BF4] break down in the presence of water, producing harmful byproducts including phosphoric acid and hydrofluoric acid. Concerns regarding their toxicity and environmental permanence may arise from this breakdown. Long-term ecological concerns may arise from the high persistence of some ILs in the environment, which means they may not break down readily.

Ecotoxicity and Toxicology

For ILs to be completely approved as "green solvents," a thorough evaluation of their toxicity, eco-toxicity, and degradability is required. This entails being aware of how ILs decompose in various settings, their potential for bioaccumulation, and their effects on ecosystems.

Despite the fact that ionic liquids have a number of benefits over conventional organic solvents, including non-volatility, recyclability, and tunability, there are still important issues that need to be resolved to guarantee their sustainability. To avoid simply moving the environmental burden from one step of the product life cycle to another, it is important to carefully assess the synthesis's complexity, possible environmental effects, and safety considerations. ILs made from renewable resources require greater study and development, as well as more environmentally friendly synthesis techniques, to reach their full potential as "green solvents."

CONCLUSION

Ionic liquids (ILs) have shown a great deal of promise as eco-friendly solvents that help bridge the gap between industrial applications and sustainability. They are desirable substitutes for conventional volatile organic solvents due to their special qualities, which include low volatility, excellent thermal stability, and adjustable solubility. ILs are widely used in a variety of processes, including metal ion extraction, biomass processing, organic synthesis, catalysis, and nanoparticle creation. Their significance in green chemistry is demonstrated by their capacity to improve catalyst recovery, increase reaction efficiency, and lessen environmental effect. But even with these benefits, the sustainability of ILs is still a complicated matter. Their synthesis presents several difficulties that need to be carefully considered, such as significant energy consumption, the use of hazardous precursors, and possible environmental persistence. According to life cycle studies, the overall environmental impact of ILs is dependent on their production processes and long-term stability, even though they may lower emissions and dangers while in use. To create ILs from renewable feedstocks, enhance biodegradability, and improve solvent-free or energy-efficient synthesis methods, more research is needed. Going forward, innovation in sustainable chemistry can be stimulated by the incorporation of ILs into chemical education and industrial processes. ILs can be extremely important in the shift to more sustainable, greener chemical processes if their advantages are balanced with environmental concerns. Unlocking their full potential and guaranteeing their ethical use in a variety of scientific and industrial applications will need ongoing multidisciplinary research and technical breakthroughs.

REFERENCES

Anastas, P. T., & Eghbali, N. (2010). Green chemistry: Principles and practice. *Chemical Society Reviews*, *39*(1), 301–312. DOI: 10.1039/B918763B PMID: 20023854

Boamah, P. O., Wang, L., Shen, W., Tang, S., & Lee, H. K. (2023). Applications of ionic liquids in the microextraction of pesticides: A mini-review. *Journal of Chromatography Open*, *4*, 100090. Advance online publication. DOI: 10.1016/j.jcoa.2023.100090

Boualem, A. D., Argoub, K., Benkouider, A. M., Yahiaoui, A., & Toubal, K. (2022). Viscosity prediction of ionic liquids using NLR and SVM approaches. *Journal of Molecular Liquids, 368*(Part A). *Journal of Molecular Liquids, 120610*, 120610. Advance online publication. DOI: 10.1016/j.molliq.2022.120610

Choi, O. K., Lim, H. C., Cho, Y., Cho, A. E., Al Abdulgader, H., Lithoxoos, G. P., Aljama, H. A., Choi, S.-H., Choi, Y. C., & Lee, J. W. (2023). Room-temperature ionic liquids as candidate materials for produced water desalination: Experiments and molecular dynamic analysis. *Desalination*, *557*, 116608. Advance online publication. DOI: 10.1016/j.desal.2023.116608

Chun, S., Dzyuba, S. V., & Bartsch, R. A. (2001). Influence of structural variation in room-temperature ionic liquids on the selectivity and efficiency of competitive alkali metal salt extraction by a crown ether. *Analytical Chemistry*, *73*(15), 3737–3741. DOI: 10.1021/ac010061v PMID: 11510842

Clark, J. H., Farmer, T. J., Hunt, A. J., & Sherwood, J. (2015). Opportunities for bio-based solvents created as petrochemical and fuel products transition towards renewable resources. *International Journal of Molecular Sciences*, *16*(8), 17101–17159. DOI: 10.3390/ijms160817101 PMID: 26225963

Clarke, C. J., Tu, W.-C., Levers, O., Bröhl, A., & Hallett, J. P. (2018). Green and sustainable solvents in chemical processes. *Chemical Reviews*, *118*(2), 747–800. DOI: 10.1021/acs.chemrev.7b00571 PMID: 29300087

Dai, S., Ju, Y. H., & Barnes, C. E. (1999). Solvent extraction of strontium nitrate by a crown ether using room-temperature ionic liquids. *Journal of the Chemical Society, Dalton Transactions: Inorganic Chemistry*, (8), 1201–1202. DOI: 10.1039/a809672d

Das, S., Rahman, M., Kundu, D., Majee, A., & Hajra, A. (2010). Task-specific ionic-liquid-catalyzed efficient synthesis of indole derivatives under solvent-free conditions. *Canadian Journal of Chemistry*, *88*(2), 150–154. DOI: 10.1139/V09-154

Davis, J. H.Jr. (2004). Task-specific ionic liquids. *Chemistry Letters, 33*(9), 1072–1077. DOI: 10.1246/cl.2004.1072

Dupont, J., Consorti, C. S., & Spencer, J. (2000). Room temperature molten salts: Neoteric "green" solvents for chemical reactions and processes. *Journal of the Brazilian Chemical Society, 11*(4), 337–344. DOI: 10.1590/S0103-50532000000400002

Dupont, J., & Scholten, J. D. (2010). On the structural and surface properties of transition-metal nanoparticles in ionic liquids. *Chemical Society Reviews, 39*(5), 1780–1804. DOI: 10.1039/b822551f PMID: 20419219

Erfirt, K., Wandzik, I., Walczak, K., Matuszek, K., & Chrobok, A. (2014). Hydrogen-bond-rich ionic liquids as effective organocatalysts for Diels–Alder reactions. *Green Chemistry, 16*(7), 3508–3514. DOI: 10.1039/C4GC00380B

Flieger, J., & Flieger, M. (2020). Ionic liquids toxicity—Benefits and threats. *International Journal of Molecular Sciences, 21*(17), 6267. Advance online publication. DOI: 10.3390/ijms21176267 PMID: 32872533

Freire, M. G., Neves, C. M. S. S., Carvalho, P. J., Gardas, R. L., Fernandes, A. M., Marrucho, I. M., Santos, L. M. N. B. F., & Coutinho, J. A. P. (2007). Mutual solubilities of water and hydrophobic ionic liquids. *The Journal of Physical Chemistry B, 111*(45), 13082–13089. DOI: 10.1021/jp076271e PMID: 17958353

Gong, K., Fang, D., Wang, H. L., Zhou, X. L., & Liu, Z. L. (2009). The one-pot synthesis of 14-alkyl- or aryl-14H-dibenzo[a,j]xanthenes catalyzed by task-specific ionic liquid. *Dyes and Pigments, 80*(1), 30–33. DOI: 10.1016/j.dyepig.2008.02.011

Greer, A. J., Jacquemin, J., & Hardacre, C. (2020). Industrial applications of ionic liquids. *Molecules (Basel, Switzerland), 25*(21), 5207. Advance online publication. DOI: 10.3390/molecules25215207 PMID: 33182328

Handy, S. T. (2003). Greener solvents: Room temperature ionic liquids from biorenewable sources. *Chemistry (Weinheim an der Bergstrasse, Germany), 9*(13), 2938–2944. DOI: 10.1002/chem.200304799

Handy, S. T., & Zhang, X. (2001). Organic synthesis in ionic liquids: The Stille coupling. *Organic Letters, 3*(2), 233–236. DOI: 10.1021/ol0068849 PMID: 11430042

Holbrey, J. D., & Rogers, R. D. (2002). Green chemistry and ionic liquids: Synergies and ironies. In Rogers, R. D., & Seddon, K. R. (Eds.), *Ionic liquids: Industrial applications for green chemistry* (pp. 2–14). American Chemical Society., DOI: 10.1021/bk-2002-0818.ch001

Jiang, T., Ma, X., Zhou, Y., Liang, S., Zhang, J., & Han, B. (2008). Solvent-free synthesis of substituted ureas from CO2 and amines with a functional ionic liquid as the catalyst. *Green Chemistry*, *10*(4), 465–469. DOI: 10.1039/b717868a

Kaur, G., Kumar, H., & Singla, M. (2022). Diverse applications of ionic liquids: A comprehensive review. *Journal of Molecular Liquids*, *351*, 118556. Advance online publication. DOI: 10.1016/j.molliq.2022.118556

Kim, K. S., Choi, J. H., Yeon, S. H., & Lee, H. (2006). Facile one-pot synthesis of gold nanoparticles using alcohol ionic liquids. *Journal of Materials Chemistry*, *16*(14), 1315–1317. DOI: 10.1039/b601478j

Liu, Q., Janssen, M. H. A., van Rantwijk, F., & Sheldon, R. A. (2005). Room-temperature ionic liquids that dissolve carbohydrates in high concentrations. *Green Chemistry*, *7*(1), 39–42. DOI: 10.1039/b412848f

Marr, P. C., & Marr, A. C. (2016). Ionic liquid gel materials: Applications in green and sustainable chemistry. *Green Chemistry*, *18*(1), 105–128. DOI: 10.1039/C5GC02277K

Mathews, C. J., Smith, P. J., & Welton, T. (2000). Palladium catalyzed Suzuki cross-coupling reactions in ambient temperature ionic liquids. *Chemical Communications (Cambridge)*, (14), 1249–1250. DOI: 10.1039/b002755n

Mir, M. A. (2023). DFT, Hirshfeld surface, molecular docking and drug likeness studies of medicinally important coumarin molecule. *Arabian Journal for Science and Engineering*, *48*(6), 7445–7462. DOI: 10.1007/s13369-022-07476-z

Mir, M. A. (2024). Biogenic amines: Catalysis, quality, and safety aspects of food items consumed in Saudi Arabia. *Current Organocatalysis*, *11*(4), 310–320. DOI: 10.2174/0122133372285097240220062811

Mir, M. A. (2024). Utilization of medicinally important plants as antidiabetic medicines in a sustainable manner. In Banik, B. K. (Ed.), *Advances in green and sustainable chemistry: Green approaches in medicinal chemistry for sustainable drug design* (2nd ed., Vol. 1, pp. 397–410). Elsevier., DOI: 10.1016/B978-0-443-16166-7.00017-7

Mir, M. A., & Banik, B. K. (2025). Synthesis of Schiff base ligands under environment-friendly conditions: A systematic review. *Inorganic Chemistry Communications*, *174*(Part 1), 113987. Advance online publication. DOI: 10.1016/j.inoche.2025.113987

Mir, M. A., & Chang, S. K. (2024). Saudi Arabia e-waste management strategies, challenges and opportunities, effect on health and environment: A strategic review. *Emerging Contaminants*, *10*(4), 100357. Advance online publication. DOI: 10.1016/j.emcon.2024.100357

Mir, M. A., Chang, S. K., & Hefni, D. (2024). A comprehensive review on challenges and choices of food waste in Saudi Arabia: Exploring environmental and economic impacts. *Environmental Systems Research*, *13*(1), 40. Advance online publication. DOI: 10.1186/s40068-024-00364-5

Mir, M. A., Hefni, D., & Chauhan, N. (2025). Heritage to health: The socio-cultural and health impacts of traditional ethnic foods in Saudi Arabian society. *Food Chemistry Advances*, *6*, 100927. Advance online publication. DOI: 10.1016/j.focha.2025.100927

Mir, M. A., Khan, M. A. A., Banik, B. K., Hasnain, S. M., Alzayer, L., Andrews, K., & Abba, S. I. (2025). Microplastics in food products: Prevalence, artificial intelligence-based detection, and potential health impacts on humans. *Emerging Contaminants*, *11*(2), 100477. Advance online publication. DOI: 10.1016/j.emcon.2025.100477

Orha, L., Akien, G. R., & Horvath, I. T. (2012). Ionic liquids as green solvents for nucleophilic substitution reactions. In P. T. Anastas (Ed.), *Handbook of green chemistry: Vol. 7. Green synthesis* (pp. 93–120). Wiley-VCH.

Park, S., & Kazlauskas, R. J. (2001). Improved preparation and use of room-temperature ionic liquids in lipase-catalyzed enantio- and regioselective acylations. *The Journal of Organic Chemistry*, *66*(25), 8395–8401. DOI: 10.1021/jo015761e PMID: 11735517

Pfruender, H., Jones, R., & Weuster-Botz, D. (2006). Water immiscible ionic liquids as solvents for whole cell biocatalysis. *Journal of Biotechnology*, *124*(1), 182–190. DOI: 10.1016/j.jbiotec.2005.12.004 PMID: 16413078

Pinkert, A., Marsh, K. N., Pang, S., & Staiger, M. P. (2009). Ionic liquids and their interaction with cellulose. *Chemical Reviews*, *109*(12), 6712–6728. DOI: 10.1021/cr9001947 PMID: 19757807

Plechkova, N. V., & Seddon, K. R. (2008). Applications of ionic liquids in the chemical industry. *Chemical Society Reviews*, *37*(1), 123–150. DOI: 10.1039/B006677J PMID: 18197338

Ranke, J., Stolte, S., Störmann, R., Arning, J., & Jastorff, B. (2007). Design of sustainable chemical products—The example of ionic liquids. *Chemical Reviews*, *107*(6), 2183–2206. DOI: 10.1021/cr050942s PMID: 17564479

Reichardt, C., & Welton, T. (2010). *Solvents and solvent effects in organic chemistry* (4th ed.). Wiley-VCH. DOI: 10.1002/9783527632220

Reynolds, J. L., Erdner, K. R., & Jones, P. B. (2002). Photoreduction of benzophenones by amines in room-temperature ionic liquids. *Organic Letters, 4*(6), 917–919. DOI: 10.1021/ol017290o PMID: 11893185

Rothe, M., Müller, E., Denk, P., & Kunz, W. (2021). Ionic liquids based on the concept of melting point lowering due to ethoxylation. *Molecules (Basel, Switzerland), 26*(13), 4034. Advance online publication. DOI: 10.3390/molecules26134034 PMID: 34279374

Seddon, K. R. (1997). Ionic liquids for clean technology. *Journal of Chemical Technology and Biotechnology, 68*(4), 351–356. DOI: 10.1002/(SICI)1097-4660(199704)68:4<351::AID-JCTB613>3.0.CO;2-4

Stark, A., Ott, D., Kralisch, D., Kreisel, G., & Ondruschka, B. (2010). Ionic liquids and green chemistry: A lab experiment. *Journal of Chemical Education, 87*(2), 196–201. DOI: 10.1021/ed8000396

Suresh, & Sandhu, J. S. (2011). Recent advances in ionic liquids: Green unconventional solvents of this century: Part 1. *Green Chemistry Letters and Reviews, 4*(4), 289–310. DOI: 10.1080/17518253.2011.572295

Suresh, & Sandhu, J. S. (2011). Recent advances in ionic liquids: Green unconventional solvents of this century: Part II. *Green Chemistry Letters and Reviews, 4*(4), 311–320. DOI: 10.1080/17518253.2011.572296

Swatloski, R. P., Spear, S. K., Holbrey, J. D., & Rogers, R. D. (2002). Dissolution of cellulose with ionic liquids. *Journal of the American Chemical Society, 124*(18), 4974–4975. DOI: 10.1021/ja025790m PMID: 11982358

Takeuchi, Y., Tarui, T., & Shibata, N. (2000). A novel and efficient synthesis of 3-fluorooxindoles from indoles mediated by Selectfluor. *Organic Letters, 2*(5), 639–642. DOI: 10.1021/ol991400y PMID: 10814398

Titze-Frech, K., Ignatiev, N., Uerdingen, M., Schulz, P. S., & Wasserscheid, P. (2013). Highly selective aromatic alkylation of phenol and anisole by using recyclable Brønsted acidic ionic liquid systems. *European Journal of Organic Chemistry, 2013*(30), 6961–6966. DOI: 10.1002/ejoc.201300579

United Nations General Assembly. (2015). *Transforming our world: The 2030 Agenda for Sustainable Development*. United Nations. https://sdgs.un.org/2030agenda

Varma, R. S., & Namboodiri, V. V. (2001). An expeditious solvent-free route to ionic liquids using microwaves. *Chemical Communications (Cambridge)*, (7), 643–644. DOI: 10.1039/b101375k

Visser, A. E., Swatloski, R. P., Reichert, W. M., Griffin, S. T., & Rogers, R. D. (2000). Traditional extractants in nontraditional solvents: Groups 1 and 2 extraction by crown ethers in room-temperature ionic liquids. *Industrial & Engineering Chemistry Research, 39*(10), 3596–3604. DOI: 10.1021/ie000426m

Wang, L., Li, H., & Li, P. (2009). Task-specific ionic liquid as base, ligand and reaction medium for the palladium-catalyzed Heck reaction. *Tetrahedron, 65*(1), 364–368. DOI: 10.1016/j.tet.2008.10.042

Wasserscheid, P., & Welton, T. (Eds.). (2008). *Ionic liquids in synthesis* (2nd ed.). Wiley-VCH.

Welton, T. (2004). Ionic liquids in catalysis. *Coordination Chemistry Reviews, 248*(21–24), 2459–2477. DOI: 10.1016/j.ccr.2004.04.015

Xing, H., Wang, T., Zhou, Z., & Dai, Y. (2005). Novel Brønsted-acidic ionic liquids for esterifications. *Industrial & Engineering Chemistry Research, 44*(11), 4147–4150. DOI: 10.1021/ie0488703

Xu, J. M., Wu, Q., Zhang, Q. Y., Zhang, F., & Lin, X. F. (2007). Basic ionic liquid as catalyst and reaction medium: A rapid and facile protocol for aza-Michael addition reactions. *European Journal of Organic Chemistry, 2007*(11), 1798–1802. DOI: 10.1002/ejoc.200600999

Yannai, X., Ge, S., Rennert, G., Gruener, N., & Fares, F. A. (1996). 3,3′-Diindolylmethane induces apoptosis in human cancer cells. *Biochemical and Biophysical Research Communications, 228*(1), 153–158. DOI: 10.1006/bbrc.1996.1631 PMID: 8912651

Zhang, Y., Bakshi, B. R., & Demessie, E. S. (2008). Life cycle assessment of an ionic liquid versus molecular solvents and their applications. *Environmental Science & Technology, 42*(5), 1724–1730. DOI: 10.1021/es0713983 PMID: 18441827

Zhao, D., Liao, Y., & Zhang, Z. (2007). Toxicity of ionic liquids. *Clean (Weinheim), 35*(1), 42–48. DOI: 10.1002/clen.200600015

Zhao, H., Xia, S., & Ma, P. (2005). Use of ionic liquids as 'green' solvents for extractions. *Journal of Chemical Technology and Biotechnology, 80*(10), 1089–1096. DOI: 10.1002/jctb.1333

Chapter 7
IoNanofluids: Bridging Efficiency and Sustainability in Chemistry

Harshit Gouri
https://orcid.org/0009-0006-4507-8989
Department of Chemistry, University of Delhi, Delhi, India

ABSTRACT

This chapter explores the role of ionanofluids—hybrid materials combining ionic liquids with nanoparticles—in advancing sustainable chemical applications. IoNanofluids offer enhanced thermal stability, tunable viscosity, and superior catalytic properties, making them valuable in green chemistry, catalysis, and industrial processes. The chapter is structured into three key sections: (1) their use in sustainable chemistry, focusing on reducing environmental impact and improving reaction efficiency, (2) their role in catalytic applications, where they enhance reaction rates, selectivity, and catalyst longevity, and (3) their integration into process chemistry, including heat transfer, separation, and electrochemical applications. By addressing both scientific advancements and industrial scalability, this chapter highlights the transformative potential of ionanofluids in modern chemical engineering and concludes with future research directions.

INTRODUCTION

Ionanofluids, an advanced class of hybrid materials that integrate the intrinsic properties of ionic liquids with the enhanced functionalities of nanoparticles. Ionic liquids, defined as salts in the liquid state at room temperature or slightly elevated temperatures, are composed of organic cations and inorganic or organic anions(de Castro et al., 2013; Murshed et al., 2012). These liquids are characterized by their

DOI: 10.4018/979-8-3373-1245-3.ch007

© 2026, IGI Global Scientific Publishing. Copying or distributing in print or electronic forms without written permission of IGI Global Scientific Publishing is prohibited. Use of this chapter to train generative artificial intelligence (AI) technologies is expressly prohibited. The publisher reserves all rights to license its use for generative AI training and machine learning model development.

remarkable thermal stability, low vapor pressure, and high ionic conductivity. The inclusion of nanoparticles into these ionic liquids forms ionanofluids, yielding materials with synergistic physicochemical properties that surpass those of their individual components(Almutairi, 2023; Murshed et al., 2012).

A distinctive feature of ionanofluids is their high thermal stability, which enables their application in processes requiring exposure to extreme temperatures without significant degradation(Moosavi et al., 2024). This property is particularly advantageous in thermal management systems and high-temperature chemical reactions. In addition, ionanofluids exhibit enhanced thermal conductivity, attributed to the efficient heat transfer facilitated by the dispersed nanoparticles(Pereira et al., 2024). The tunable viscosity of ionanofluids can be modulated by altering the nanoparticle type, size, or concentration, thereby enabling customization for specific industrial processes.

Figure 1. The role of ionanofluids in chemistry

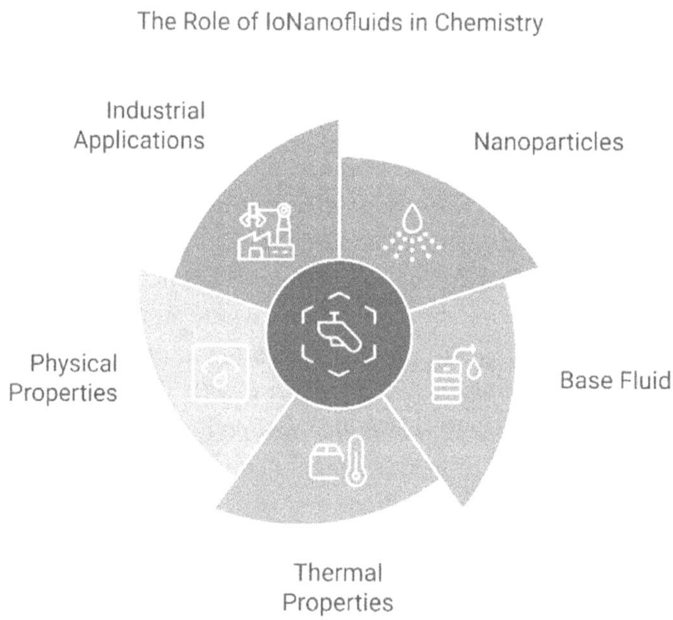

UTILIZATION OF IONANOFLUIDS FOR SUSTAINABLE CHEMISTRY

Ionanofluids exhibit strong alignment with the principles of sustainable chemistry, serving as a cornerstone for eco-friendly scientific and industrial advancements. The foundation of sustainable chemistry is rooted in minimizing environmental and health hazards, a principle well-addressed by ionanofluids(He & Alexandridis, 2017). Unlike traditional organic solvents, ionic liquids—core components of ionanofluids—demonstrate significantly lower toxicity(Ogodo & Abosede, 2025).

Biodegradability a critical aspect of sustainable chemistry that is increasingly incorporated into the design of ionanofluids. Although conventional ionic liquids are not inherently biodegradable, substantial progress has been made in developing bio-compatible and eco-friendly variants(Ogodo & Abosede, 2025). When combined with nanoparticles, the resulting ionanofluids can be engineered to maintain high performance while minimizing ecological disruption. This ensures a balance between industrial functionality and environmental stewardship(Bharathi et al., 2025).

The reusability and recyclability of ionanofluids address another vital tenet of sustainability: waste reduction. These fluids can often be recovered and recycled with minimal degradation in their physicochemical properties. This recyclability not only extends the lifecycle of the material but also minimizes the environmental burden associated with waste disposal(Beil et al., 2021; Jastorff et al., 2005).

Efficiency, a hallmark of sustainable chemistry, is inherent to the performance of ionanofluids. Enhanced thermal conductivity, coupled with tunable viscosity, enables highly efficient energy transfer and utilization in thermal and catalytic systems. In industrial applications, this translates to reduced energy consumption and improved process kinetics. For example, the superior heat transfer properties of ionanofluids optimize thermal management in high-temperature processes, while their electrical conductivity supports advanced electrochemical applications such as energy storage systems(Singh et al., 2022).

Moreover, ionanofluids facilitate reaction optimization in catalytic processes by providing a highly stable and efficient reaction medium(Sakhahari, 2025). The ability to tailor both the ionic liquid matrix and the dispersed nanoparticles allows for precise control over chemical reactivity and selectivity. This adaptability ensures that industrial processes utilizing ionanofluids achieve maximum efficiency with minimal resource consumption.

Environmental Benefits of IoNanofluids

A major advantage of ionanofluids in sustainable chemistry is their ability to significantly reduce waste production and environmental impact. Traditional chemical processes rely heavily on organic solvents, which contribute to air pollution, hazardous waste accumulation, and high energy consumption due to their volatility and the need for strict containment measures. Many of these solvents also pose health risks and require expensive disposal methods, further straining both industry and the environment. In contrast, ionanofluids—characterized by their low vapor pressure, excellent thermal stability, and recyclability—offer a cleaner, more efficient, and more sustainable alternative. Their unique physicochemical properties support safer reaction conditions, improved catalyst stability, and enhanced selectivity while also minimizing the release of volatile organic compounds (VOCs) into the atmosphere(-Clarke et al., 2018; Ogodo & Abosede, 2025; Welton, 2015).

One of the most significant environmental benefits of ionanofluids is their ability to replace hazardous organic solvents in catalytic and synthetic processes, thereby reducing waste generation(Jhanji, 2023; Welton, 2015). Their tunable solubility and ionic nature enable effective dissolution of reactants and catalysts, leading to higher reaction efficiencies and fewer toxic byproducts(Rastogi et al., 2025). By shifting toward solvent-free or reduced-solvent systems, these fluids help decrease the volume of chemical waste that requires treatment or disposal, making industrial operations more sustainable(Granger & Parvulescu, 2020). Additionally, their high recyclability and reusability enhance both environmental and economic benefits. Unlike conventional solvents that degrade over time and require frequent replacement, ionanofluids retain their stability and performance across multiple reaction cycles. Their ability to be efficiently separated from reaction products and reused in subsequent processes reduces resource consumption, cuts hazardous waste disposal costs, and improves the overall sustainability of chemical manufacturing.

Practical applications have demonstrated the potential of ionanofluids to lower the environmental footprint of industrial chemical reactions. For example, in hydroformylation and hydrogenation processes—widely used in fine chemical and pharmaceutical production—ionanofluids enable phase separation-based catalyst recycling. This advancement not only improves catalyst recovery rates but also minimizes metal waste and solvent emissions, thereby reducing the environmental impact of these reactions(Durga & Mishra, 2016; Holbrey, 2005). Similarly, in polymerization processes, ionanofluids enhance efficiency by stabilizing catalysts, allowing for better-controlled polymer architectures while eliminating the need for toxic organic solvents. This has significant implications for sustainable materials production, facilitating the development of high-performance polymers with a lower environmental impact.

Applications in Green Synthesis

Ionanofluids, a hybrid of ionic liquids (ILs) and nanoparticles, are advanced materials that have garnered significant attention in green chemistry. Their unique physicochemical properties, such as low volatility, high thermal stability, and tunable polarity, make them highly suitable for promoting sustainable and environmentally friendly chemical processes(Jhanji, 2023; Murshed et al., 2012).

FACILITATION OF SOLVENT-FREE REACTIONS

In solvent-free reactions, ionanofluids play a pivotal role as multifunctional materials. Their ionic nature creates a highly polar environment that stabilizes reaction intermediates and facilitates interactions between reactants. This property enhances reaction rates and selectivity, eliminating the need for traditional volatile organic solvents(Tomar & Jain, 2022). Additionally, the nanoparticles dispersed in ionanofluids often exhibit catalytic activity, further accelerating reaction kinetics. The thermal and ionic conductivity of ionanofluids ensures that heat and charge are efficiently transferred within the system, enabling reactions to proceed at lower energy inputs. This characteristic not only improves energy efficiency but also aligns with the principles of green chemistry by reducing the carbon footprint of chemical processes(Valli Sowbhagyam, 2024).

PROMOTION OF ATOM-ECONOMICAL PROCESSES

A critical aspect of green chemistry is atom economy, which aims to maximize the incorporation of reactant atoms into the final product while minimizing waste. Ionanofluids contribute significantly to this goal by enabling selective catalytic processes that favor desired reaction pathways. Their highly tunable composition allows precise control over reaction conditions, enhancing product yields and reducing byproducts. Moreover, ionanofluids are often recyclable, with both the ionic liquid and the embedded nanoparticles recoverable for reuse. This recyclability reduces the need for fresh materials in subsequent reactions, thereby conserving resources and minimizing waste generation. The combination of high reactant utilization and selective catalysis ensures that ionanofluids support atom-economical processes effectively(R. A. Sheldon, 2000; Trost, 1991).

Examples of Green Syntheses Enabled by Ionanofluids

The utility of ionanofluids in green chemistry is exemplified by their applications in diverse synthetic processes. In the production of biopolymers, ionanofluids offer a sustainable alternative to conventional solvents. For instance, cellulose, a widely available biopolymer, can be dissolved and chemically modified in ionanofluids without the need for harsh solvents(Polesca et al., 2022). This approach is instrumental in the development of bioplastics and biofuels, contributing to the advancement of renewable materials. Similarly, chitosan, another biopolymer, can be processed and functionalized using ionanofluids for applications in pharmaceuticals and biodegradable packaging(Aziz et al., 2024).

In the realm of organic synthesis, ionanofluids have demonstrated efficacy in several key reactions. Aldol condensations, which are fundamental in forming carbon-carbon bonds, can be catalyzed by ionanofluids under solvent-free conditions. This eliminates the dependence on volatile organic compounds (VOCs) and reduces environmental pollution(Alqahtani, 2024). Furthermore, ionanofluids have been employed in palladium-catalyzed cross-coupling reactions, such as Suzuki-Miyaura coupling, to achieve high yields and excellent selectivity. These solvent-free protocols not only improve process sustainability but also enhance reaction efficiency.

Nanomaterial synthesis is another area where ionanofluids have proven their versatility. The ionic liquid component of ionanofluids can stabilize nanoparticles during in situ synthesis, ensuring uniform size distribution and functionality. These nanoparticles serve as catalytic supports, enabling efficient hydrogenation and oxidation reactions. By combining the stabilizing properties of ionic liquids with the catalytic potential of nanoparticles, ionanofluids create a synergistic platform for advanced materials development(Dutta & Sit, 2024; Pinkert et al., 2009).

Environmental and Economic Advantages of Ionanofluids

The environmental and economic advantages of ionanofluids are substantial, making them highly attractive for sustainable industrial applications. Their unique physicochemical properties, including low volatility and non-flammability, significantly reduce the risks associated with traditional organic solvents. Many conventional solvents are volatile organic compounds (VOCs), which contribute to air pollution and pose health and safety hazards due to their flammability and toxicity. In contrast, ionanofluids exhibit negligible vapor pressure, eliminating the risk of solvent evaporation and atmospheric contamination. This not only enhances workplace safety but also reduces the need for costly ventilation and containment systems in industrial settings(Ribeiro et al., 2011).

In addition to their safety benefits, ionanofluids offer exceptional thermal stability and high thermal conductivity, which contribute to more energy-efficient operations. Unlike many organic solvents that degrade or evaporate at high temperatures, ionanofluids maintain their structural integrity even under extreme thermal conditions, making them ideal for high-temperature catalytic reactions, heat transfer applications, and energy storage systems. Their superior thermal conductivity enables more efficient heat dissipation, reducing overheating risks and enhancing the performance of chemical reactions that require precise temperature control. By optimizing thermal management, ionanofluids help lower overall energy consumption, leading to significant cost savings and a reduced carbon footprint in industrial processes(Murshed et al., 2012).

From an economic perspective, one of the most compelling advantages of ionanofluids is their recyclability and reusability, which significantly reduces operational costs. Traditional solvents are often consumed in large quantities and require frequent replenishment due to losses from evaporation, degradation, or contamination(Karadaghi et al., 2023). In contrast, ionanofluids can be recovered and reused multiple times without substantial loss of performance. Their chemical stability allows for efficient recycling processes, reducing waste generation and minimizing the need for fresh solvent input. This translates into lower raw material costs, reduced waste disposal expenses, and improved overall process sustainability.

The ability of ionanofluids to enhance catalyst stability and longevity further contributes to their economic benefits. In catalytic applications, ionanofluids help prevent catalyst deactivation, which is a major challenge in industrial chemistry(Granger & Parvulescu, 2020). By stabilizing active catalytic species and reducing unwanted side reactions, ionanofluids extend catalyst lifespan, minimizing the need for frequent catalyst replacement and lowering operational costs. This advantage is particularly valuable in high-value catalytic processes, such as pharmaceutical synthesis, fine chemical production, and petrochemical refining, where catalyst efficiency directly impacts production costs and product quality(Valli Sowbhagyam, 2024).

Additionally, the use of ionanofluids can lead to lower maintenance and equipment costs due to their non-corrosive nature. Many traditional solvents and process chemicals cause significant wear and degradation of industrial equipment over time, leading to frequent maintenance, downtime, and replacement expenses. Ionanofluids, on the other hand, exhibit excellent chemical stability and compatibility with various materials, reducing the risk of equipment corrosion and extending the operational lifespan of reactors, pipelines, and storage systems. This reduction in maintenance costs adds another layer of economic viability for industries adopting ionanofluid-based processes(Ferdous et al., 2024; R. Sheldon, 2001).

UTILIZATION OF IONANOFLUIDS FOR CATALYTIC PROCESSES

Introduction to Catalysis and IoNanofluids

Catalysis plays a fundamental role in chemical processes by enhancing reaction rates without being consumed in the process. Catalysts function by lowering activation energy, enabling chemical transformations under milder conditions and improving efficiency (Abhijeet Kr Shrivastava, 2024). Catalysis is broadly classified into homogeneous and heterogeneous catalysis:

Figure 2. Homogenous v. heterogeneous catalysts

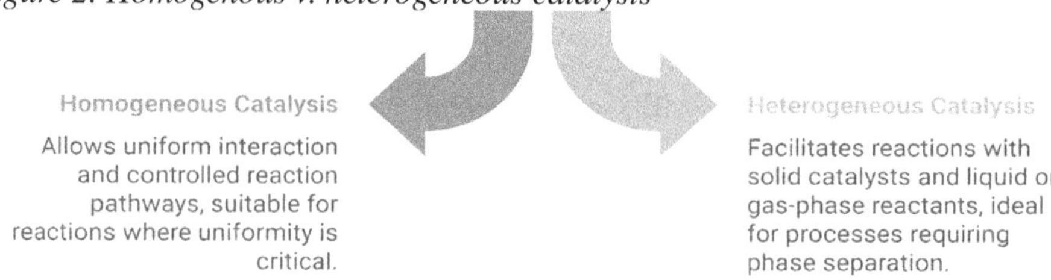

Role of IoNanofluids in Catalytic Reactions & Enhancing Catalysis

Ionanofluids, which are ionic liquids infused with nanoparticles, serve as promising media for catalytic processes due to their unique properties like, enhanced solubility, thermal and chemical stability, tunable viscosity and polarity, reusability.

Improved Dispersion of Nanoparticles

One of the major advantages of ionanofluids in catalysis is their ability to disperse nanoparticles uniformly(Guan et al., 2025). In catalytic applications, well-dispersed nanoparticles enhance active surface area exposure, leading to higher catalytic efficiency and selectivity.

Stabilization of Active Catalytic Sites

Catalyst deactivation due to aggregation, leaching, or sintering is a common challenge in catalytic processes(Zvereva et al., 2017). Ionanofluids help stabilize catalytic sites by:

- **Preventing Aggregation:** Their ionic nature provides electrostatic stabilization, reducing nanoparticle clustering.
- **Minimizing Leaching:** Strong interactions between nanoparticles and the ionic liquid matrix enhance catalyst longevity.
- **Enhancing Metal Catalyst Performance:** Noble metal catalysts such as palladium and platinum show increased stability and efficiency when incorporated into ionanofluids.

Influence on Reaction Kinetics and Selectivity (Hawker et al., 2018)

Ionanofluids influence reaction pathways, improving reaction kinetics and selectivity by:

- **Modulating Reactant Solubility:** Improved solubility of reactants leads to enhanced reaction rates.
- **Tuning Microenvironment:** The ionic environment of ionanofluids impacts reaction pathways, favoring selective transformations.
- **Controlling Diffusion Rates:** Optimized viscosity levels regulate mass transfer, avoiding side reactions and increasing product purity.

Applications in Homogeneous and Heterogeneous Catalysis

Ionanofluids, a class of ionic liquid-based nanofluids, have emerged as highly effective media for catalytic processes due to their unique combination of ionic conductivity, thermal stability, and tunable physicochemical properties(Basu & Banik, 2024). These fluids not only improve catalyst solubility and dispersion but also enhance reaction selectivity, efficiency, and recyclability. Their ability to act as structured reaction media makes them invaluable in both homogeneous and heterogeneous catalysis.

Homogeneous Catalysis

Figure 3. Ionanofluids in catalysis

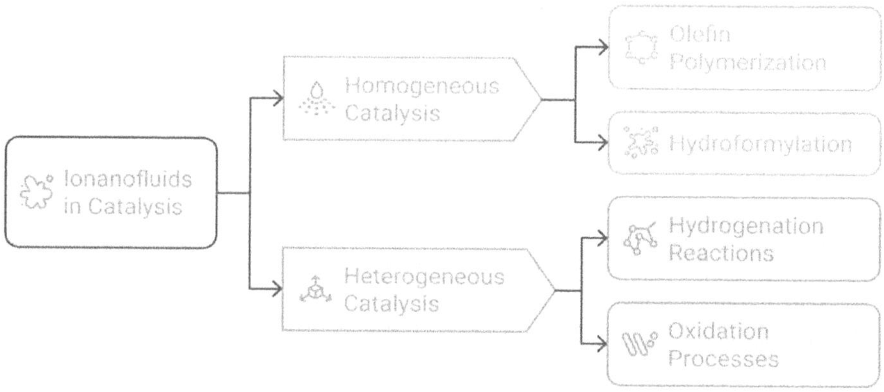

Homogeneous catalysis involves catalysts that are dissolved in the same phase as the reactants, often leading to higher reaction rates and selectivities. However, traditional homogeneous catalytic systems suffer from difficulties in catalyst separation and recycling. Ionanofluids address these challenges by providing a stable ionic environment that enhances catalyst solubility, reduces catalyst deactivation, and facilitates recovery(Marullo et al., 2021). Key applications include:

Olefin Polymerization: Ionanofluids serve as an ideal solvent system for olefin polymerization, offering precise control over polymer architecture, molecular weight distribution, and overall polymer properties. Their ionic nature enhances catalyst solubility, particularly for metallocenes and Ziegler–Natta catalysts, leading to improved polymerization efficiency. Additionally, ionanofluids influence chain propagation and termination reactions, enabling fine-tuned control over polymer microstructures(Shamiri et al., 2014). By stabilizing growing polymer chains, they help minimize chain transfer reactions, reducing unwanted side reactions that could compromise polymer yield and quality. Moreover, their negligible volatility and high thermal stability make them a sustainable alternative to conventional solvents, contributing to greener polymerization processes with lower solvent emissions and reduced catalyst losses(Kubisa, 2005).

Hydroformylation: Hydroformylation, or the oxo process, is a crucial industrial carbonylation reaction that converts olefins into aldehydes using syngas ($CO + H_2$) and transition metal catalysts. Ionanofluids enhance this process by improving catalyst stability, particularly for rhodium-based complexes, which are prone to deactivation due to leaching or aggregation. Their ionic environment also influences regioselectivity, allowing for greater control over product distribution and selectivity for either linear or branched aldehydes (Punzalan et al., 2023). Additionally, ionanofluids facilitate efficient catalyst recycling by enabling easy phase separation, reducing operational costs and improving process sustainability. In some cases, they serve as a solvent-free alternative to volatile organic solvents (VOCs), significantly lowering the environmental footprint of hydroformylation reactions.

Heterogeneous Catalysis: Heterogeneous catalysis involves catalysts that exist in a different phase from the reactants, typically solid catalysts facilitating reactions in gaseous or liquid phases (W. Zhang & Zhang, 2024). Ionanofluids significantly enhance these catalytic processes by preventing metal nanoparticle aggregation, thereby maintaining a high catalytic surface area and improving catalyst dispersion. Their tunable viscosity and polarity enhance mass transport, allowing for better diffusion of reactants and products, which optimizes reaction efficiency(Bartlewicz et al., 2020). Additionally, ionanofluids provide exceptional thermal and chemical stability, ensuring sustained catalytic performance even under harsh reaction conditions, making them a valuable alternative to conventional solvents in heterogeneous catalysis. Applications include:

Hydrogenation Reactions: Hydrogenation is a fundamental reaction in petrochemical refining, pharmaceutical synthesis, and food processing, and ionanofluids play a crucial role in optimizing its efficiency. By stabilizing metal nanoparticles such as palladium, platinum, and nickel, ionanofluids prevent catalyst aggregation, ensuring uniform catalytic activity and improved dispersion(Chacón & Dupont, 2019). Additionally, some ionanofluids are specifically designed to enhance hydrogen solubility, increasing reaction rates compared to conventional solvents. Their ionic nature also facilitates better interaction between reactants and catalyst surfaces, reducing mass transfer limitations and accelerating hydrogenation kinetics(Canton et al., 2017). Furthermore, ionanofluids protect active metal sites from poisoning and deactivation, extending catalyst lifetimes and reducing replacement costs, making them a highly effective medium for sustainable hydrogenation processes.

Oxidation Processes: Oxidation reactions are essential for the production of fine chemicals, pharmaceuticals, and specialty materials, and ionanofluids play a significant role in enhancing their efficiency. By stabilizing metal oxide catalysts such as RuO_2, MnO_2, and TiO_2, ionanofluids help maintain high catalytic activity over extended reaction times, preventing deactivation(Wang et al., 2024). Their ability to control the electronic environment of catalysts improves selectivity, ensuring the targeted oxidation of desired compounds while minimizing unwanted byproducts. Additionally, ionanofluids serve as environmentally friendly alternatives to hazardous organic solvents, reducing waste generation and improving process sustainability(Bakthavatchalam et al., 2020a). Their high thermal conductivity also facilitates efficient heat transfer, dissipating excess heat from exothermic oxidation reactions and preventing catalyst degradation, making them a valuable tool in oxidation processes.

Table 1. Comparative analysis: IoNanofluids vs. traditional solvents

Property	**Traditional Solvents**	**IoNanofluids**
Volatility	High	Low
Thermal Stability	Moderate	High
Catalyst Recovery	Difficult	Facilitated
Environmental Impact	High	Low
Selectivity	Moderate	Enhanced

Industrial-Scale Applications in Catalytic Process

The application of ionanofluids in catalysis has revolutionized multiple industrial sectors by improving reaction efficiency, selectivity, and sustainability(Dupont et al., 2024). Their unique physicochemical properties—such as high thermal stability, tunable viscosity, and enhanced ionic conductivity—enable superior catalytic performance across various large-scale chemical processes.

Fuel Processing

Ionanofluids play a critical role in fuel refining and processing by improving catalytic efficiency in key processes such as hydrocracking and catalytic reforming. In hydrocracking, ionanofluids stabilize metal catalysts, such as molybdenum and nickel-based systems, preventing deactivation due to coking and extending catalyst lifespan(X. Zhang et al., 2025). Their ability to enhance hydrogen solubility improves reaction rates, leading to higher yields of desirable fuel fractions, such as diesel and gasoline, from heavy crude oil and biomass-derived feedstocks. In catalytic reforming, ionanofluids assist in upgrading low-octane hydrocarbons into high-octane gasoline components by optimizing reaction conditions, improving catalyst dispersion, and minimizing side reactions(Luska et al., 2015). Additionally, ionanofluids contribute to more sustainable fuel processing by reducing energy consumption and solvent waste, making them an attractive option for modern refineries.

Fine Chemical Production

The fine chemicals industry, including pharmaceuticals, agrochemicals, and specialty chemicals, benefits significantly from the use of ionanofluids in catalytic transformations. In pharmaceutical synthesis, ionanofluids improve reaction efficiency in key processes such as asymmetric hydrogenation, oxidation, and carbonylation, leading to high-value intermediates with superior selectivity and yield(Lantos et al., 2024). Their ability to stabilize chiral catalysts ensures enantioselective transformations, which are essential for producing active pharmaceutical ingredients (APIs) with precise stereochemistry. In specialty chemical production, ionanofluids facilitate the synthesis of complex molecules by enhancing catalyst stability, increasing reaction rates, and enabling solvent-free or low-waste processes(Compagno et al., 2023). Their role in green chemistry initiatives is particularly notable, as they help reduce reliance on hazardous organic solvents, minimize byproduct formation, and enable catalyst recycling, leading to more cost-effective and environmentally friendly production processes.

Integration with Renewable Energy Sources

The drive toward sustainable industrial processes has led to the exploration of innovative solutions for reducing environmental impact and improving energy efficiency(Guzmán-Lucero et al., 2021). By combining the unique properties of ionanofluids with the clean energy potential of solar, bio-based, and other renewable sources, catalytic processes can be made more energy-efficient and environmentally friendly.

Photocatalytic Applications

Ionanofluids are increasingly being utilized in photocatalytic applications, which harness solar energy to drive chemical reactions. In solar-driven catalytic processes, the absorption of light is crucial for initiating reactions, particularly in environmental remediation, hydrogen production, and carbon dioxide reduction. Ionanofluids improve the efficiency of these processes by enhancing light absorption and promoting better charge separation in photocatalysts, such as titanium dioxide (TiO_2) and zinc oxide (ZnO)(Han et al., 2023). Their ionic nature can optimize the electronic properties of these catalysts, increasing the number of active sites and improving the overall rate of photocatalytic reactions. By providing a stable and efficient medium, ionanofluids help boost the performance of solar-driven reactions, making them a vital component in developing renewable energy technologies aimed at reducing the carbon footprint of industrial processes (Khan et al., 2019). Moreover, ionanofluids help to avoid issues of catalyst aggregation and deactivation, which are common challenges in conventional photocatalytic systems, thus improving long-term stability and reusability.

Bio-Based Ionic Fluids

In line with the push for greener and more sustainable processes, bio-based ionic fluids are gaining significant attention. These ionic liquids are derived from renewable biomass sources such as plant oils, lignocellulosic materials, or natural sugars, making them an eco-friendly alternative to traditional petrochemical-based solvents and catalysts (Hulsbosch et al., 2016). The development of bio-based ionic fluids for catalytic applications can reduce reliance on fossil fuels, lower the environmental impact of chemical processes, and promote the circular economy. Ionanofluids, when designed with bio-based ionic liquids, enhance catalysis by providing a non-volatile, thermally stable, and environmentally benign solvent system. They have shown promise in a range of catalytic processes, including biomass conversion, biodiesel production, and CO_2 capture, by improving reaction rates, selectivity, and catalyst stability.

IONANOFLUIDS IN PROCESS CHEMISTRY

Process chemistry is a specialized field that focuses on the development, optimization, and scaling of chemical reactions for industrial applications. It plays a crucial role in various industries, including pharmaceuticals, petrochemicals, food production, and material synthesis. Ionanofluids, a subclass of nanofluids, have emerged as promising materials in process chemistry due to their unique physicochemical properties(Bakthavatchalam et al., 2020b). These fluids offer enhanced thermal stability, tunable viscosity, and superior solvent capabilities which aligns with the objective of process chemistry to enhance reaction efficiency, maximize product yield, and ensure sustainability while minimizing waste generation and energy consumption. Ionanofluids integration into process chemistry applications has opened new avenues for improving efficiency, energy conservation, and sustainability.

Figure 4. Unique properties of ionanofluids in process chemistry

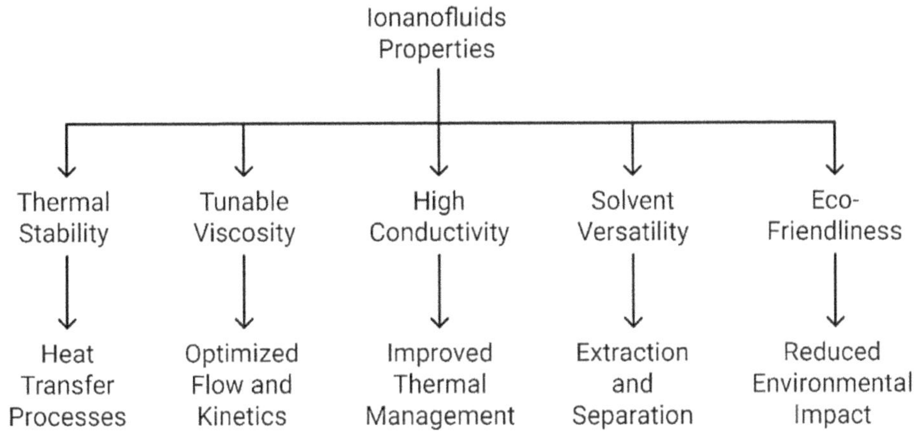

Applications of IoNanofluids in Process Chemistry

Heat Transfer Applications

One of the primary applications of ionanofluids in process chemistry is in heat transfer systems. Their enhanced thermal conductivity, coupled with their ability to remain stable under high temperatures, makes them highly effective in thermal management systems used in industrial processes. Heat transfer efficiency is crucial for optimizing energy consumption in chemical processes. Ionanofluids contribute significantly to improving heat exchange in reaction vessels, distillation columns, and cooling systems(Mohite et al., 2024). Their use in heat exchangers and reactors helps in maintaining uniform temperature distribution, thus preventing hotspots and improving reaction yield. In chemical synthesis, maintaining precise temperature control is essential. Ionanofluids have been employed in industrial reactors to improve heat transfer efficiency, reducing energy consumption and improving product yield(Khan et al., 2019). Industries dealing with large-scale heat exchange operations, such as the petrochemical sector, have successfully utilized ionanofluids to enhance the efficiency of their cooling and heating systems.

Separation and Extraction Processes

The solvent properties of ionanofluids enable their use in separation and extraction processes. Unlike traditional organic solvents, ionanofluids can be customized to exhibit specific solubility and selectivity characteristics, making them ideal for various separation techniques. Industrial wastewater often contains toxic heavy metals, which pose environmental and health risks. Ionanofluids have been developed to selectively bind and extract heavy metals such as lead, mercury, and cadmium from aqueous solutions(Yudaev & Chistyakov, 2022). The ability of ionanofluids to dissolve organic contaminants makes them effective in removing pollutants from industrial effluents, leading to cleaner production processes. Many industrial processes generate byproducts that contain valuable materials. Ionanofluids facilitate the extraction and recovery of rare earth elements, pharmaceutical intermediates, and other high-value chemicals. (Regel-Rosocka & Wisniewski, 2011)By offering selective solubility and high efficiency in separations, ionanofluids contribute to greener and more cost-effective process chemistry applications.

Electrochemical and Photochemical Applications

Ionanofluids have gained attention for their role in electrochemical processes due to their high ionic conductivity and electrochemical stability. These properties make them well-suited for applications in batteries, capacitors, and fuel cells. The stability and conductivity of ionanofluids improve battery performance by enhancing charge transfer efficiency and reducing degradation over time (Pereira et al., 2023). Ionanofluids can be used to modify electrodes, improving electrochemical reactions and increasing the lifespan of energy storage devices.

Photochemical processes rely on light-driven reactions, and ionanofluids have shown potential in enhancing the efficiency of such reactions. Ionanofluids are used to support photocatalytic degradation of pollutants, enabling more efficient environmental remediation processes(Regel-Rosocka & Wisniewski, 2011). In solar-driven chemical processes, ionanofluids contribute to improved light absorption and energy conversion, facilitating sustainable energy storage solutions.

Process Optimization and Energy Efficiency

Process optimization is a critical aspect of industrial chemistry, and ionanofluids play a key role in improving reaction kinetics. Their ability to stabilize reactive intermediates, control solvation dynamics, and enhance catalyst performance contributes to faster and more efficient chemical reactions. The use of ionanofluids in heat transfer and solvent applications significantly reduces energy consumption(Khan et al., 2019). By improving heat exchange and reaction efficiency, these fluids help industries cut down on operational costs while enhancing sustainability.

Figure 5. Applications of ionanofluids in process chemistry

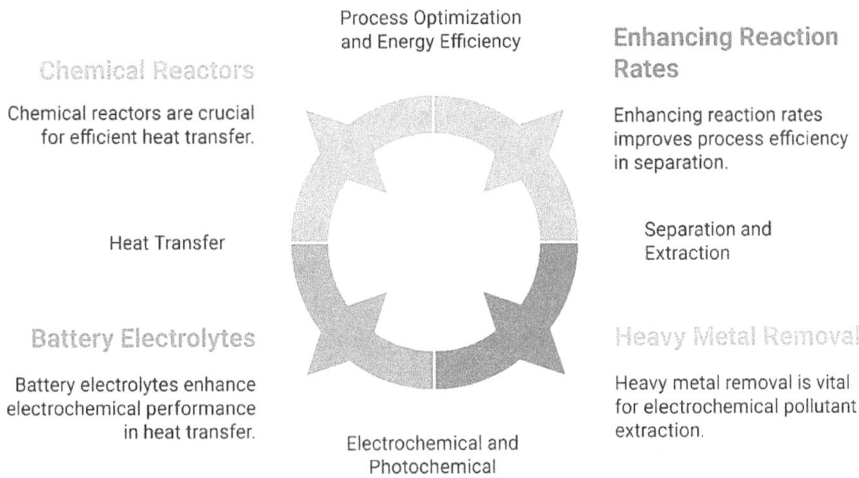

Real-World Applications of IoNanofluids

Industries have successfully integrated ionanofluids into a variety of process chemical applications, capitalizing on their unique properties to improve efficiency, sustainability, and product quality across multiple sectors.

Petrochemical Industry

In the petrochemical industry, ionanofluids have been employed in processes such as catalytic cracking and refining(Berton et al., 2022). Their ability to enhance thermal stability and improve catalyst performance has led to more efficient fuel production and a reduction in energy consumption(Khan et al., 2019). By stabilizing metal catalysts and reducing the formation of coke deposits, ionanofluids contribute to higher reaction yields and lower emissions, making the refining process more environmentally friendly(Akhtar et al., 2024). This efficiency gain not only reduces operational costs but also supports the industry's shift toward more sustainable production practices, reducing the environmental footprint of traditional fuel refining methods.

Pharmaceuticals

In the pharmaceutical industry, ionanofluids have revolutionized drug synthesis and purification processes. Their selective solubility and ability to dissolve both organic and inorganic compounds make them ideal for efficiently separating active pharmaceutical ingredients (APIs) from reaction mixtures. This results in improved yields, lower solvent waste, and enhanced drug purity(Gazal et al., 2021). Ionanofluids also enable more precise control over reaction conditions, ensuring that pharmaceutical processes are both cost-effective and environmentally sustainable. By minimizing the need for toxic solvents and reducing byproducts, the use of ionanofluids not only improves the economics of drug production but also contributes to safer, more eco-friendly pharmaceutical manufacturing(Zhuo et al., 2024).

Materials Science

In materials science, ionanofluids have shown great promise in polymerization and nanomaterial production. These fluids enhance process control, ensuring that polymerization reactions occur under optimal conditions. Ionanofluids also improve the mechanical and thermal characteristics of final products, such as advanced polymers, composites, and nanomaterials(Almutairi, 2023). For instance, in the production of high-performance carbon nanotubes or graphene composites, ionanofluids can improve the dispersion of nanoparticles, leading to stronger and more durable materials. This capability has made them invaluable in industries requiring materials with specialized properties, such as aerospace, automotive, and electronics.

Food and Beverage Industry

In the food and beverage industry, ionanofluids have been explored for applications in food extraction and preservation. Their unique ability to enhance the efficiency of natural compound separation makes them ideal for extracting bioactive compounds, such as antioxidants, vitamins, and flavors, from plants or food waste(Cañadas et al., 2020). By facilitating more efficient extractions, ionanofluids help preserve the integrity of these compounds, ensuring that food products retain their nutritional and sensory qualities. Moreover, their use in food preservation processes, such as pasteurization or sterilization, can help reduce the reliance on harsh chemicals, improving both food safety and sustainability(Chiozzi et al., 2022).

Textile Industry

In the textile industry, ionanofluids have been incorporated into dyeing and fabric treatment processes to improve the overall sustainability of textile production. Their ability to enhance dye solubility and fabric penetration enables more uniform and efficient dyeing processes, resulting in better color consistency and less waste(Hynes et al., 2020). Additionally, ionanofluids can help reduce wastewater generation by enabling the reuse of dyes and reducing the need for excessive water during treatment processes. By improving the sustainability of dyeing and fabric treatment, ionanofluids contribute to reducing the environmental impact of textile manufacturing, making the industry more eco-friendly, (Van der Bruggen et al., 2017).

Economic Viability and Sustainability

Ionanofluids represent a significant advancement in process chemistry, offering a unique blend of high efficiency, sustainability, and versatility that is driving their adoption across various industrial sectors. Their integration into industrial applications delivers not only economic benefits but also supports sustainable practices, aligning with both cost-effectiveness and environmental responsibility (Berton et al., 2022).

From an economic perspective, ionanofluids contribute to substantial cost savings by optimizing processes and reducing the need for traditional, expensive, and hazardous organic solvents. Their ability to reduce energy consumption through enhanced thermal conductivity and catalyst stabilization directly impacts operational costs, especially in energy-intensive industries like petrochemicals and pharmaceuticals(Mukherjee et al., 2024). Additionally, their effectiveness in improving process efficiency leads to higher yields, reducing waste and material costs(Khan et al., 2019). Ionanofluids also extend the lifespan of catalytic systems and reduce equipment maintenance due to their ability to prevent catalyst deactivation and corrosion. This reduction in maintenance costs is a key factor in making ionanofluids a financially attractive option for long-term industrial applications. Moreover, their recyclability and reusability further enhance their economic viability, making them a sustainable option for large-scale operations, where the ability to recover and reuse fluids leads to lower overall expenses.(Karthikeyan et al., 2022)

On the environmental side, the benefits of ionanofluids are even more pronounced. By reducing the reliance on volatile organic compounds (VOCs), ionanofluids help minimize harmful emissions that contribute to air pollution and health hazards, making them a greener alternative to traditional solvents(Valli Sowbhagyam, 2024). Their efficient use of energy and reduction in waste generation results in lower carbon emissions, supporting industries in meeting sustainability targets and reducing their environmental footprint. Furthermore, ionanofluids contribute to the principles of green chemistry by enabling solvent-free reactions and more atom-efficient processes, which align with global efforts to adopt more eco-friendly industrial practices(Gazal et al., 2021).

DISCUSSION

The development and application of ionanofluids represent a transformative advancement in chemical science, particularly in the fields of sustainable chemistry, catalysis, and industrial process applications. Their unique combination of ionic liquids and nanoparticles provides a multifunctional platform that enhances reaction selectivity, efficiency, and recyclability, all while reducing environmental impact. Unlike conventional solvents, ionanofluids exhibit low volatility, high thermal stability, and excellent ionic conductivity, making them superior alternatives for green chemistry applications. Their ability to replace volatile organic solvents (VOCs), improve energy efficiency, and enhance catalyst stability has made them increasingly attractive for large-scale industrial use.

A major advantage of ionanofluids is their enhanced thermal conductivity and stability, which enables efficient heat transfer and improved process efficiency in catalytic and electrochemical reactions. Their recyclability and reusability further contribute to their economic and environmental viability, ensuring cost-effective and sustainable industrial operations. Industries such as petrochemicals, pharmaceuticals, materials science, food and beverage processing, and textiles have successfully adopted ionanofluids, leveraging their superior solvent properties, tunability, and catalytic efficiency.

However, despite these advantages, challenges remain in the large-scale implementation of ionanofluids. The high cost of synthesis and the complexity of recycling ionic liquids and nanoparticles pose economic barriers. While ionanofluids offer promising sustainability benefits, their widespread adoption depends on further research into cost-effective production techniques and improved recovery and separation processes. Additionally, regulatory and safety considerations need to be addressed, as the long-term environmental impact of some ionic liquids remains under investigation. Understanding their biodegradability, toxicity, and potential environmental accumulation is critical for ensuring safe and sustainable use in industrial applications.

FUTURE TREND

The future of ionanofluids is poised for significant advancements, driven by ongoing research in materials science, catalysis, and green chemistry, with several key trends shaping their evolution and industrial adoption. A major focus is the development of bio-based and biodegradable ionic fluids, where researchers are exploring ionic liquids derived from renewable sources such as lignocellulosic biomass, plant oils, and natural polysaccharides. These eco-friendly alternatives will help mitigate environmental concerns associated with synthetic ionic liquids, making ionanofluids a more sustainable option for industrial applications. Additionally, the emergence of smart and responsive ionanofluids is expected to revolutionize chemical processes, as these fluids can adapt to changes in temperature, pH, or external stimuli, enabling self-regulating catalytic systems, enhanced reaction selectivity, and improved process efficiency. Such innovations will be particularly beneficial in biocatalysis, pharmaceutical synthesis, and precision material fabrication, where fine control over reaction conditions is essential. Another transformative trend is the integration of ionanofluids with renewable energy technologies, where their unique properties are being harnessed in solar-driven photocatalysis, bio-based chemical production, fuel cells, CO_2 capture, hydrogen production, and battery systems. Their ability to enhance energy storage and conversion technologies will play a crucial role in supporting global sustainability initiatives. However, for ionanofluids to achieve widespread industrial adoption, research efforts must address cost and scalability challenges by developing more efficient synthesis, recovery, and recycling methods. Advancements in continuous-flow processing, scalable manufacturing, and improved separation techniques will be key to making ionanofluid-based technologies more economically viable for large-scale applications. Finally, regulatory and environmental assessments will be critical to ensuring the safe and responsible use of ionanofluids. Future research must evaluate their biodegradability, toxicity, and long-term ecological impact, while governments and regulatory agencies may establish green certification standards to promote eco-friendly industrial adoption. As these trends continue to evolve, ionanofluids are set to become a cornerstone of sustainable chemistry, enabling cleaner, more efficient, and environmentally responsible industrial processes.

CONCLUSION

Ionanofluids have emerged as versatile and sustainable materials with significant potential to revolutionize chemical processes across multiple industries. Their high thermal stability, recyclability, and ability to enhance catalytic efficiency make them an attractive alternative to traditional solvents and process fluids. By replacing hazardous organic solvents and improving energy efficiency, ionanofluids contribute to reducing the environmental footprint of chemical manufacturing, aligning with global sustainability initiatives. Their successful application in catalysis, polymerization, heat transfer, separation processes, and electrochemical systems highlights their transformative potential. However, challenges related to cost, large-scale production, and regulatory considerations must be addressed to facilitate their widespread industrial adoption. The development of bio-based ionic fluids, smart and adaptive ionanofluids, and integration with renewable energy technologies will further enhance their sustainability and economic viability. As research continues to refine their formulation, efficiency, and scalability, ionanofluids are expected to play an increasingly crucial role in the future of sustainable chemistry. Their adoption in industrial processes will not only improve reaction efficiency and economic feasibility but also support global efforts toward greener, safer, and more sustainable chemical manufacturing.

REFERENCES

Abhijeet Kr Shrivastava. Dr. (2024). Nanocatalysis: new dimensions in catalytic industry. In *Futuristic Trends in Chemical Material Sciences & Nano Technology Volume 3 Book 14* (pp. 137–148). Iterative International Publishers, Selfypage Developers Pvt Ltd. DOI: 10.58532/V3BECS14P4CH1

Akhtar, M. S., Ali, S., & Zaman, W. (2024). Recent Advancements in Catalysts for Petroleum Refining. *Catalysts*, *14*(12), 841. DOI: 10.3390/catal14120841

Almutairi, R. (2023). *Ionanofluids: A review on its properties and thermal applications*. 16–37. DOI: 10.21741/9781644902790-3

Alqahtani, A. S. (2024). Indisputable roles of different ionic liquids, deep eutectic solvents and nanomaterials in green chemistry for sustainable organic synthesis. *Journal of Molecular Liquids*, *399*, 124469. DOI: 10.1016/j.molliq.2024.124469

Aziz, T., Li, W., Zhu, J., & Chen, B. (2024). Developing multifunctional cellulose derivatives for environmental and biomedical applications: Insights into modification processes and advanced material properties. *International Journal of Biological Macromolecules*, *278*, 134695. DOI: 10.1016/j.ijbiomac.2024.134695 PMID: 39151861

Bakthavatchalam, B., Habib, K., Saidur, R., Saha, B. B., & Irshad, K. (2020a). Comprehensive study on nanofluid and ionanofluid for heat transfer enhancement: A review on current and future perspective. *Journal of Molecular Liquids*, *305*, 112787. DOI: 10.1016/j.molliq.2020.112787

Bakthavatchalam, B., Habib, K., Saidur, R., Saha, B. B., & Irshad, K. (2020b). Comprehensive study on nanofluid and ionanofluid for heat transfer enhancement: A review on current and future perspective. *Journal of Molecular Liquids*, *305*, 112787. DOI: 10.1016/j.molliq.2020.112787

Bartlewicz, O., Dąbek, I., Szymańska, A., & Maciejewski, H. (2020). Heterogeneous Catalysis with the Participation of Ionic Liquids. *Catalysts*, *10*(11), 1227. DOI: 10.3390/catal10111227

Basu, S., & Banik, B. K. (2024). Nanoparticles as Catalysts: Exploring Potential Applications. *Current Organocatalysis*, *11*(4), 265–272. DOI: 10.2174/01221333 72285610231227094959

Beil, S., Markiewicz, M., Pereira, C. S., Stepnowski, P., Thöming, J., & Stolte, S. (2021). Toward the Proactive Design of Sustainable Chemicals: Ionic Liquids as a Prime Example. *Chemical Reviews*, *121*(21), 13132–13173. DOI: 10.1021/acs.chemrev.0c01265 PMID: 34523909

Berton, P., Abidi, N., & Shamshina, J. L. (2022). Ionic liquids: Implementing objectives of sustainability for the next generation chemical processes and industrial applications. *Current Opinion in Green and Sustainable Chemistry*, *35*, 100625. DOI: 10.1016/j.cogsc.2022.100625

Bharathi, C., Rajeswari, R., Janaki, P., Sivamurugan, A. P., Senthil, G. K., Radhamani, S., Sathya Priya, R., Sangeetha, M., Thenmozhi, S., Chiranjeevirajan, N., Sharmila, R., Ramya, B., & Balamurugan, R. (2025). Bio-mediated synthesis of nanoparticles: A new paradigm for environmental sustainability. *Plant Science Today*, *12*(sp1). Advance online publication. DOI: 10.14719/pst.5192

Cañadas, R., González-Miquel, M., González, E. J., Díaz, I., & Rodríguez, M. (2020). Overview of neoteric solvents as extractants in food industry: A focus on phenolic compounds separation from liquid streams. *Food Research International*, *136*, 109558. DOI: 10.1016/j.foodres.2020.109558 PMID: 32846603

Canton, P., Perosa, A., & Selva, M. (2017). Metal Nanoparticles Stabilized in Ionic Liquids for Catalytic Multiphase Reactions. *Current Organic Chemistry*, *21*(24). Advance online publication. DOI: 10.2174/1385272820666161009222123

Chacón, G., & Dupont, J. (2019). Arene Hydrogenation by Metal Nanoparticles in Ionic Liquids. *ChemCatChem*, *11*(1), 333–341. DOI: 10.1002/cctc.201801363

Chiozzi, V., Agriopoulou, S., & Varzakas, T. (2022). Advances, Applications, and Comparison of Thermal (Pasteurization, Sterilization, and Aseptic Packaging) against Non-Thermal (Ultrasounds, UV Radiation, Ozonation, High Hydrostatic Pressure) Technologies in Food Processing. *Applied Sciences (Basel, Switzerland)*, *12*(4), 2202. DOI: 10.3390/app12042202

Clarke, C. J., Tu, W.-C., Levers, O., Bröhl, A., & Hallett, J. P. (2018). Green and Sustainable Solvents in Chemical Processes. *Chemical Reviews*, *118*(2), 747–800. DOI: 10.1021/acs.chemrev.7b00571 PMID: 29300087

Compagno, N., Profeta, R., & Scarso, A. (2023). Recent advances in the synthesis of active pharmaceutical and agrochemical ingredients in micellar media. *Current Opinion in Green and Sustainable Chemistry*, *39*, 100729. DOI: 10.1016/j.cogsc.2022.100729

de Castro, C. N., & Ribeiro, C. A. P., I.C., S., P. Franca, J. M., J.V., M., V., F., M.S, S., Goodrich, P., & Hardacre, C. (2013). Synthesis, Properties and Physical Applications of IoNanofluids. In *Ionic Liquids - New Aspects for the Future*. InTech. DOI: 10.5772/52596

Dupont, J., Leal, B. C., Lozano, P., Monteiro, A. L., Migowski, P., & Scholten, J. D. (2024). Ionic Liquids in Metal, Photo-, Electro-, and (Bio) Catalysis. *Chemical Reviews*, *124*(9), 5227–5420. DOI: 10.1021/acs.chemrev.3c00379 PMID: 38661578

Durga, G., & Mishra, A. (2016). Ionic Liquids: Industrial Applications. In *Encyclopedia of Inorganic and Bioinorganic Chemistry* (pp. 1–13). Wiley. DOI: 10.1002/9781119951438.eibc2434

Dutta, D., & Sit, N. (2024). A comprehensive review on types and properties of biopolymers as sustainable bio-based alternatives for packaging. *Food Biomacromolecules*, *1*(2), 58–87. DOI: 10.1002/fob2.12019

Ferdous, A. R., Shah, S. N. A., Shah, S. S., & Aziz, M. (2024). Advancements in nanotechnology applications: Transforming catalysts, sensors, and coatings in petrochemical industries. *Fuel*, *371*, 132020. DOI: 10.1016/j.fuel.2024.132020

Gazal, U., Khan, I., Bhat, A. H., & Pathak, V. (2021). Ionic liquids as green solvents in the pharmaceutical industry. In *Green Sustainable Process for Chemical and Environmental Engineering and Science* (pp. 41–54). Elsevier., DOI: 10.1016/B978-0-12-821885-3.00002-5

Granger, P., & Parvulescu, V. I. (2020). Application of Ionic Liquids for Sustainable Catalysis. In *Heterogeneous Catalysis for Energy Applications* (pp. 304–360). The Royal Society of Chemistry., DOI: 10.1039/9781788019576-00304

Guan, Y., Li, L., & Piao, H. (2025). Factors Influencing Nanoscale Catalysts Prepared Impregnation Method. *Applied and Computational Engineering*, *122*(1), 35–39. DOI: 10.54254/2755-2721/2025.19593

Guzmán-Lucero, D., Guzmán-Pantoja, J., Velázquez, H. D., Likhanova, N. V., Bazaldua-Domínguez, M., Vega-Paz, A., & Martínez-Palou, R. (2021). Isobutane/butene alkylation reaction using ionic liquids as catalysts. Toward a sustainable industry. *Molecular Catalysis*, *515*, 111892. DOI: 10.1016/j.mcat.2021.111892

Han, H., Jang, J. U., Oh, D., Na, K.-H., Choi, W.-Y., Jayakrishnan, N., & Nayak, A. K. (2023). Advances and Perspectives of Titanium-Based Nanocomposites for Energy Generation and Environmental Remediation Applications: A Review. *Energy & Fuels*, *37*(23), 17708–17735. DOI: 10.1021/acs.energyfuels.3c02382

Hawker, R. R., Haines, R. S., & Harper, J. B. (2018). Rational selection of the cation of an ionic liquid to control the reaction outcome of a substitution reaction. *Chemical Communications*, *54*(18), 2296–2299. DOI: 10.1039/C8CC00241J PMID: 29443346

He, Z., & Alexandridis, P. (2017). Ionic liquid and nanoparticle hybrid systems: Emerging applications. *Advances in Colloid and Interface Science*, *244*, 54–70. DOI: 10.1016/j.cis.2016.08.004 PMID: 27567031

Holbrey, J. D. (2005). Industrial Applications of Ionic Liquids. *ChemInform*, *36*(27), chin.200527275. Advance online publication. DOI: 10.1002/chin.200527275

Hulsbosch, J., De Vos, D. E., Binnemans, K., & Ameloot, R. (2016). Biobased Ionic Liquids: Solvents for a Green Processing Industry? *ACS Sustainable Chemistry & Engineering*, *4*(6), 2917–2931. DOI: 10.1021/acssuschemeng.6b00553

Hynes, N. R. J., Kumar, J. S., Kamyab, H., Sujana, J. A. J., Al-Khashman, O. A., Kuslu, Y., Ene, A., & Suresh Kumar, B. (2020). Modern enabling techniques and adsorbents based dye removal with sustainability concerns in textile industrial sector -A comprehensive review. *Journal of Cleaner Production*, *272*, 122636. DOI: 10.1016/j.jclepro.2020.122636

Jastorff, B., Mölter, K., Behrend, P., Bottin-Weber, U., Filser, J., Heimers, A., Ondruschka, B., Ranke, J., Schaefer, M., Schröder, H., Stark, A., Stepnowski, P., Stock, F., Störmann, R., Stolte, S., Welz-Biermann, U., Ziegert, S., & Thöming, J. (2005). Progress in evaluation of risk potential of ionic liquids—Basis for an eco-design of sustainable products. *Green Chemistry*, *7*(5), 362. DOI: 10.1039/b418518h

Jhanji, I. (2023). The Role of Ionic Liquids in Green Solvent Chemistry: Properties and Applications. *International Research Journal of Engineering & Applied Sciences*, *11*(4), 57–67. DOI: 10.55083/irjeas.2023.v11i04009

Karadaghi, L. R., Pan, B., Baddour, F. G., Malmstadt, N., & Brutchey, R. L. (2023). A techno-economic approach to guide the selection of flow recyclable ionic liquids for nanoparticle synthesis. *RSC Sustainability*, *1*(7), 1861–1873. DOI: 10.1039/D3SU00182B

Karthikeyan, A., Raphael, W., & Tavares, J. R. (2022). Nanofluids as heat transfer fluids: Hurdles to industrial application and economic considerations. *Canadian Journal of Chemical Engineering*, *100*(11), 3311–3320. DOI: 10.1002/cjce.24338

Khan, I., Bhat, A. H., Sharma, D. K., Usmani, M. A., & Khan, F. (2019). *Overview of Nanofluids to Ionanofluids: Applications and Challenges* (pp. 199–227). DOI: 10.1007/978-981-13-9833-9_10

Kubisa, P. (2005). Ionic liquids in the synthesis and modification of polymers. *Journal of Polymer Science. Part A, Polymer Chemistry*, *43*(20), 4675–4683. DOI: 10.1002/pola.20971

Lantos, J., Kumar, N., & Saha, B. (2024). A Comprehensive Review of Fine Chemical Production Using Metal-Modified and Acidic Microporous and Mesoporous Catalytic Materials. *Catalysts*, *14*(5), 317. DOI: 10.3390/catal14050317

Luska, K. L., Migowski, P., & Leitner, W. (2015). Ionic liquid-stabilized nanoparticles as catalysts for the conversion of biomass. *Green Chemistry*, *17*(6), 3195–3206. DOI: 10.1039/C5GC00231A

Marullo, S., D'Anna, F., Rizzo, C., & Billeci, F. (2021). Ionic liquids: "normal" solvents or nanostructured fluids? *Organic & Biomolecular Chemistry*, *19*(10), 2076–2095. DOI: 10.1039/D0OB02214D PMID: 33606870

Mohite, D. D., Goyal, A., Singh, A. S., Ansari, M. I., Patil, K. A., Yadav, P. D., Patil, M. J., & Londhe, P. V. (2024). Improvement of thermal performance through nanofluids in industrial applications: A review on technical aspects. *Materials Today: Proceedings*. Advance online publication. DOI: 10.1016/j.matpr.2024.04.083

Moosavi, M., Torkzadeh, M., & Akbarinezhad, Z. (2024). Molecular dynamics investigation of ionanofluids (INFs): Towards a deeper understanding of their thermophysical, structural and dynamical properties. *Journal of Molecular Liquids*, *399*, 124355. DOI: 10.1016/j.molliq.2024.124355

Mukherjee, S., Wciślik, S., Chandra Mishra, P., & Chaudhuri, P. (2024). Nanofluids: Critical issues, economics and sustainability perspectives. *Particuology*, *87*, 147–172. DOI: 10.1016/j.partic.2023.06.021

Murshed, S. M. S., de Castro, C. A. N., Lourenço, M. J. V., Lopes, M. L. M., & Santos, F. J. V. (2012). Current research and future applications of nano- and ionanofluids. *Journal of Physics: Conference Series*, *395*, 012117. DOI: 10.1088/1742-6596/395/1/012117

Ogodo, U. P., & Abosede, O. O. (2025). The Role of Chemistry in Achieving Sustainable Development Goals: Green Chemistry Perspective. *International Research Journal of Pure and Applied Chemistry*, *26*(1), 1–8. DOI: 10.9734/irjpac/2025/v26i1893

Pereira, J., Souza, R., & Moita, A. (2024). A Review of Ionic Liquids and Their Composites with Nanoparticles for Electrochemical Applications. *Inorganics*, *12*(7), 186. DOI: 10.3390/inorganics12070186

Pereira, J., Souza, R., Moita, A., & Moreira, A. (2023). Nanofluids and Ionic Fluids as Liquid Electrodes: An Overview on Their Properties and Potential Applications. *Processes (Basel, Switzerland)*, *11*(11), 3189. DOI: 10.3390/pr11113189

Pinkert, A., Marsh, K. N., Pang, S., & Staiger, M. P. (2009). Ionic Liquids and Their Interaction with Cellulose. *Chemical Reviews*, *109*(12), 6712–6728. DOI: 10.1021/cr9001947 PMID: 19757807

Polesca, C., Passos, H., Coutinho, J. A. P., & Freire, M. G. (2022). Sustainable development of biomaterials using ionic liquids. *Current Opinion in Green and Sustainable Chemistry*, *37*, 100675. DOI: 10.1016/j.cogsc.2022.100675

Punzalan, E., Froese, R. D., & Zimmerman, P. M. (2023). Revealing the Interactions between the Flexible Polymer and Counteranion during Propagation and Termination for Olefin Polymerization with the Ti(IV) Constrained Geometry Catalyst. *Organometallics*, *42*(22), 3236–3248. DOI: 10.1021/acs.organomet.3c00365

Rastogi, S., Sethi, P., Panwar, N., Yadav, P., Verma, R., & Madan, A. (2025). *Harnessing Nanotechnology to Transform Waste Reclamation for Enhanced Efficiency and Environmental Impact.*, DOI: 10.4018/979-8-3693-7282-1.ch016

Regel-Rosocka, M., & Wisniewski, M. (2011). Ionic Liquids in Separation of Metal Ions from Aqueous Solutions. In *Applications of Ionic Liquids in Science and Technology*. InTech., DOI: 10.5772/23909

Ribeiro, A. P. C., & Vieira, C. S. I., M., J., S., C., Langa, E., V. Loureno, M. J., S. Murshed, S. M., & de Castro, C. A. N. (2011). Thermal Properties of Ionic Liquids and Ionanofluids. In *Ionic Liquids: Theory, Properties, New Approaches*. InTech. DOI: 10.5772/13920

Sakhahari, K. V. (2025). Green Chemistry and Catalysis: Innovations toward a cleaner and Sustainable World. *International Journal of Pharmaceutical Research and Applications*, *10*(1), 106–116. DOI: 10.35629/4494-1001106116

Shamiri, A., Chakrabarti, M., Jahan, S., Hussain, M., Kaminsky, W., Aravind, P., & Yehye, W. (2014). The Influence of Ziegler-Natta and Metallocene Catalysts on Polyolefin Structure, Properties, and Processing Ability. *Materials (Basel)*, *7*(7), 5069–5108. DOI: 10.3390/ma7075069 PMID: 28788120

Sheldon, R. (2001). Catalytic reactions in ionic liquids. *Chemical Communications*, *23*(23), 2399–2407. DOI: 10.1039/b107270f PMID: 12239988

Sheldon, R. A. (2000). Atom efficiency and catalysis in organic synthesis. *Pure and Applied Chemistry*, *72*(7), 1233–1246. DOI: 10.1351/pac200072071233

Singh, V., Amirchand, K. D., & Gardas, R. L. (2022). Ionic liquid-nanoparticle based hybrid systems for energy conversion and energy storage applications. *Journal of the Taiwan Institute of Chemical Engineers*, *133*, 104237. DOI: 10.1016/j.jtice.2022.104237

Tomar, P., & Jain, D. (2022). A review of green solvent ionic liquids: as a future solvent. *International Journal of Advances in Scientific Research, 13*(06), 1–16. DOI: 10.55218/JASR.202213601

Trost, B. (1991). The Atom Economy—A Search for Synthetic Efficiency. *Science, 254*(5037), 1471–1477. DOI: 10.1126/science.1962206 PMID: 1962206

Valli Sowbhagyam, D. (2024). Ionic Liquids as Green Solvents: A Comprehensive Review. *International Research Journal on Advanced Engineering Hub (IRJAEH), 2*(02), 220–224. DOI: 10.47392/IRJAEH.2024.0035

Van der Bruggen, B., Canbolat, Ç. B., Lin, J., & Luis, P. (2017). *The Potential of Membrane Technology for Treatment of Textile Wastewater.*, DOI: 10.1007/978-981-10-5623-9_13

Wang, J., Zhang, Y., Guo, X., Tang, J., Chen, Z., Ha, M. N., Cui, P., & Ke, Q. (2024). Manganese oxide-based catalysts for the sustainable synthesis of value-added chemicals through oxidation processes: A critical review and perspectives for the future. *Green Chemistry, 26*(5), 2365–2383. DOI: 10.1039/D3GC04117D

Welton, T. (2015). Solvents and sustainable chemistry. *Proceedings - Royal Society. Mathematical, Physical and Engineering Sciences, 471*(2183), 20150502. DOI: 10.1098/rspa.2015.0502 PMID: 26730217

Yudaev, P. A., & Chistyakov, E. M. (2022). Ionic Liquids as Components of Systems for Metal Extraction. *ChemEngineering, 6*(1), 6. DOI: 10.3390/chemengineering6010006

Zhang, W., & Zhang, S. (2024). Catalysts: Combinatorial heterogeneous catalysis. In *Encyclopedia of Condensed Matter Physics* (pp. 729–737). Elsevier. DOI: 10.1016/B978-0-323-90800-9.00201-8

Zhang, X., Zhang, J., Yin, J., Liu, X., Qiu, W., He, J., Jiang, W., Zhu, L., Li, H., & Li, H. (2025). Mo-MOF-based ionanofluids for highly efficient extraction coupled catalytic oxidative desulfurization. *Separation and Purification Technology, 353*, 128289. DOI: 10.1016/j.seppur.2024.128289

Zhuo, Y., Cheng, H.-L., Zhao, Y.-G., & Cui, H.-R. (2024). Ionic Liquids in Pharmaceutical and Biomedical Applications: A Review. *Pharmaceutics, 16*(1), 151. DOI: 10.3390/pharmaceutics16010151 PMID: 38276519

Zvereva, E. E., Katsyuba, S. A., Dyson, P. J., & Aleksandrov, A. V. (2017). Leaching from Palladium Nanoparticles in an Ionic Liquid Leads to the Formation of Ionic Monometallic Species. *The Journal of Physical Chemistry Letters, 8*(15), 3452–3456. DOI: 10.1021/acs.jpclett.7b01212 PMID: 28692296

Chapter 8
An Insight Into Pharmaceutical and Biomedical Application of Ionic Liquids

M. Amin Mir
https://orcid.org/0000-0003-4689-0507
Prince Mohammad Bin Fahd University, Saudi Arabia

Maythem Mahmoud
Prince Mohammad Bin Fahd University, Saudi Arabia

Syed M Hasnain
Prince Mohammad Bin Fahd University, Saudi Arabia

Kim Andrews
https://orcid.org/0000-0002-0304-9630
Prince Mohammad Bin Fahd University, Saudi Arabia

Lingala Syam
https://orcid.org/0000-0002-0804-2024
Prince Mohammad Bin Fahd University, Saudi Arabia

ABSTRACT

In pharmaceutical applications, ionic liquids (ILs) have become a revolutionary class of materials, especially in drug delivery systems. They are excellent options for boosting transdermal drug transport and boosting bioavailability due to their adjustable physicochemical characteristics, that includes high solubility, thermal stability, and biocompatibility. Each of the four generations of ILs has improved

DOI: 10.4018/979-8-3373-1245-3.ch008

2026, IGI Global Scientific Publishing. Copying or distributing in print or electronic forms without written permission of IGI Global Scientific Publishing is prohibited. Use of this chapter to ative artificial intelligence (AI) technologies is expressly prohibited. The publisher reserves all rights to license its use for generative AI training and machine learning model development.

drug solubility, decreased toxicity, and increased biodegradability. By acting as chemical permeation enhancers, they help transport medications that aren't highly soluble in water, increasing therapeutic effectiveness and reducing systemic side effects. Furthermore, ILs have been extensively investigated as environmentally friendly solvents, catalysts, and stabilizers that can replace volatile organic chemicals in drug synthesis, extraction, and purification procedures. Their versatility in the biological sciences is demonstrated by their promise for antibacterial applications, protein stability, and cancer therapy.

INTRODUCTION

Drug delivery has shown great promise using ionic liquids (ILs), especially when it comes to improving transdermal transport. An important factor in raising skin permeability is their ionic makeup. Compared to traditional solvents, ILs provide a number of benefits, such as stability, non-flammability, non-volatility, and environmental friendliness. ILs are useful as solvents and materials in drug delivery mechanism and effective pharmaceutical ingredient (API) formulations because of their distinct and adaptable physical, chemical, and biological characteristics. ILs can increase drug stability and solubility by modifying their composition, which will increase bioavailability. Their adaptable qualities make it easier to create novel delivery systems, which could increase the effectiveness and safety of treatment. Furthermore, some ILs are suitable for medicinal applications because to their non-toxic and biodegradable nature, which guarantees reduced environmental effect and improved patient easyness. Consequently, ILs have great promise for improving drug delivery systems, (Lu et al., 2022). They can aid in the production of micelles and nanoparticles, which will allow medicinal substances to be delivered to particular body parts with precision. This focused strategy lowers the possibility of adverse effects while increasing medication efficacy, (Kapre, Palakurthi, & Jain, 2024). Furthermore, ILs can be used to create solid dispersions, which enormously increases the solubility and bioavailability of medications that are not very soluble in water. Better drug distribution and absorption follow from this improvement, which eventually improves therapeutic results, (Moshikur & Goto, 2021). As a result, ILs are essential to the creation of innovative pharmaceutical products, which help create safer and more effective treatments for patients.

ILs are categorized into three classes according to their changing characteristics and time of discovery. From the first to the third generation, they have developed further, each with distinct chemical structures and uses. Metal halide anions were mixed with dialkyl imidazolium and alkyl pyridinium cations to create the first generation of ILs, which were mostly used in electroplating. These ILs were potential substitutes for some organic solvents due to their exceptional physical characteristics, which included large liquid portions, small melting points, and efficient thermal stability. Small biodegradability, strong toxicity in aquatic, and expensive formation costs, however, restricted their use. Cations like dialkyl imidazolium, alkyl pyridinium, NH_3, and phosphonium, as well as anions like tetrafluoroborate and hexafluoro-phosphate, were used to create the second generation of ILs, that are strong in both aqueous and air, (Brahma & Gardas, 2024). By altering their anions, cations, and substituents, these ILs provide tunable physical and chemical characteristics viz, solubility, melting point, thermal stability, viscosity, hydrophilicity, toxicity, and biodegradability, (Magina & Barros-Timmons, 2021). However, issues with toxicity, biocompatibility, regulatory obstacles, and environmental effect make it difficult to use them in drug delivery systems, (Shukla, Tiwari, & Verma, 2023). Further limiting their extensive use are formulation and compatibility issues, (Amaral & Pereiro, 2021). The third generation of ILs increased their potential for use in medicinal applications by introducing natural sources for cations (like choline) and anions (like fatty acids, amino acids).

These ILs provided improved biodegradability and decreased toxicity while maintaining promising physical and chemical characteristics. Because of their enhanced safety profiles and environmental compatibility, third-generation ILs have sparked an increasing amount of research into their biological uses, especially in transdermal medication delivery systems. Since ILs have been demonstrated to efficiently promote both transcellular and paracellular transport, their function as chemical permeation enhancers is well known. They accomplish this by avoiding the stratum corneum (SC) by means of processes such lipid fluidization, cell integrity disruption, diffusional pathway creation, and lipid component extraction from the SC. These characteristics of ILs have the potential to increase transdermal medication delivery, which would ultimately result in better patient outcomes and more effective treatments, (Egorova & Gordeev, 2017). In transdermal drug administration, room-temperature ionic liquids are especially adaptable and frequently used as chemical penetration enhancers. Through the disruption of cellular integrity, fluidization of lipids, formation of diffusional routes, and extraction of lipids from the outermost layer of the skin, they enable the movement of substances across cells and cellular barriers. Better clinical results are a result of this remarkable adaptability, which greatly increases the effectiveness of transdermal drug delivery systems, (Sidat, Marimuthu, & Kumar, 2019). A thorough analysis of IL design and development techniques was carried out by Moshikur and Carrier (2023) with an emphasis on biocompatible IL-based formulations and delivery methods.

In formation and delivery of drug methods, they emphasized the uses and benefits of ILs over conventional organic solvents, (Moshikur & Carrier, 2023). The effectiveness of biodegradable ILs to improve transdermal drug delivery systems

(TDDS) was thoroughly reviewed in their review, which also included their functions as active pharmaceutical components, drug modifying agents, and permeation enhancers. In order to maximize drug delivery, the study also recommended using ILs in combination with other chemical imbibition enhancers. The chapter also included important issues with ILs in TDDS, such as their physical characteristics, interactions with membranes and biomolecules, and the overall effect of imbibition improvement. Shukla et al. (2023) investigated the growing function of ILs in drug delivery systems, concentrating on their capacity to improve the solubility of medications that are not particularly soluble in water. They highlighted current developments and the adaptable qualities of ILs, highlighting their forwarded compulsion in biological and pharmacological applications. Ali et al. (2022) investigated the application of ILs in drug administration in a different study, emphasizing their adjustable characteristics that get around the drawbacks of conventional permeation enhancers. The authors highlighted new developments and exciting uses in drug delivery by concentrating on biocompatible IL-based systems that use cutting-edge cations and anions to increase permeability while reducing cytotoxicity and skin irritation. In contrast to the previously stated literature, this chapter focuses on the difficulties that arise when ILs are used in transdermal drug delivery systems (TDDS) and presents creative solutions to these difficulties. In order to enhance the applicability and therapeutic usage via the transdermal method, this chapter also focuses on merging ILs with innovative drug delivery vehicles. The strategy is to fill up current knowledge gaps, offer a thorough grasp of recent developments in this field, and suggest future lines of inquiry for the creation of IL-based transdermal delivery systems of drug. The possible uses of ILs in transdermal delivery are depicted in Figure 1.

Figure 1. Application of ILs in transdermal drug delivery

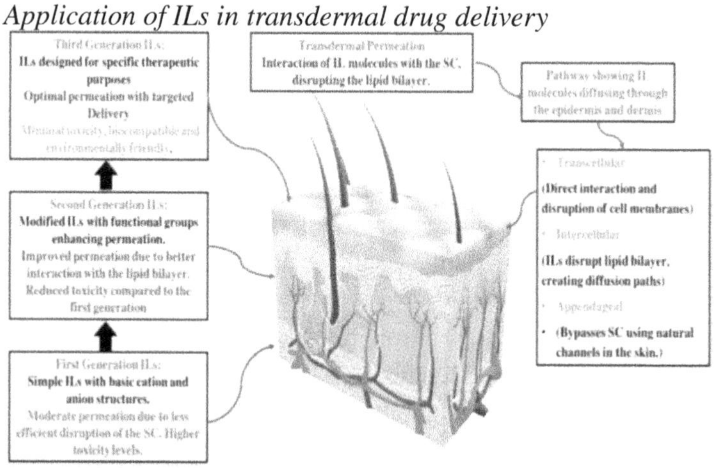

IONIC LIQUID EVOLUTION

There are now four generations of ILs available on the market, each with a distinct function, (MacFarlane et al., 2018).

Generation 1

The low melting point, improved thermal capability, decreased vapor pressure, and enhanced fluidity are some of the noteworthy physical parameters of the first generation of ILs, (Welton, 2018). These ILs have the potential to take the place of environmentally hazardous and combustible organic solvents and agents. Nevertheless, maximum no of first-generation ILs, including $[C_4MIM][BF_4]$ and $[C_4MIM][PF_6]$, are poorly biodegradable and aquatic ecotoxic. Furthermore, trifluoroacetate (TFA), triflate, methyl sulfate, and other weakly coordinating anions are frequently linked to first-generation ILs, (Welton, 2018).

Generation 2

Second-generation ILs are comparatively more helpful in complexation of metal ion, lubricants, and power materials due to their versatile physical and chemical characteristics. The maximum number of biocompatible cations and anions that occur naturally, and with reasonable price, and ecologically benign sources, like amino acids and carbohydrates, are present in these ILs. Cholinium ILs, for instance, can be created by reacting cheap choline hydroxide with naturally occurring carboxylic acids, yielding only carboxylate salts and water as a byproduct.

Generation 3

In addition to their strong chemical and physical characteristics, third-generation ILs have an additional biological trait that makes them ideal for usage in biological and pharmacological properties, (Ferraz et al., 2011). The active biological ions from well-known biocompatible resources are the source of their efficacy in biopharmaceutical products, that includes anesthetics, antifungal, -bacterial, and -cholinergic medications. The pharmaceutically active ingredients or their precursors are used as cations, anions, or a combination of both to synthesis various kinds of ILs, (Dias et al., 2019). Additionally, these ILs are very biodegradable, which makes them perfect for biological applications, (Elgharbawy, Moniruzzaman, & Goto, 2020). 3rd generation ILs have a number of significant benefits over first- and second-generation ILs, including as reduced production costs, easy anion and cation tunability, maintenance of API structural and functional integrity, and controlled polymorphism,

(Wu et al., 2021). Because of their advantages in oral, transdermal, and topical drug delivery as well as their enhanced stability and solubility, the pharmaceutical industry is using ILs more and more. Furthermore, possible issues with solid-state medicines are successfully addressed by transforming solid-state APIs into the IL form, (Shamshina, Barber, & Rogers, 2013). Research papers on the insertion of ILs into therapeutic formulations and drug administration have significantly increased over the last 20 years, (Agatemor et al., 2018). Third-generation ILs are noteworthy because they are extensively to create ILs with desired biological characteristics that can be used as active medicinal components (APIs), (Egorova & Ananikov, 2018).

Generation 4

Ionic liquids (ILs) of the fourth generation were released in 2018. In addition to displaying biocompatibility, these ILs had unusual and distinctive characteristics both in solution and when mixed with other molecular liquids, (Mir & Banik, 2025).

PHARMACEUTICAL APPLICATIONS OF IONIC LIQUIDS

The industrial use of various kinds of ionic liquids (ILs) in the pharmaceutical industry has drawn a lot of attention from around the world in recent years. This advancement attempts to address issues with crystalline medications, such as low bioavailability, instability, poor permeability, limited solubility, and polymorphism. Figure 2 shows examples of the chemical structures of IL cations and anions.

Figure 2. Chemical structures IL cations and anions

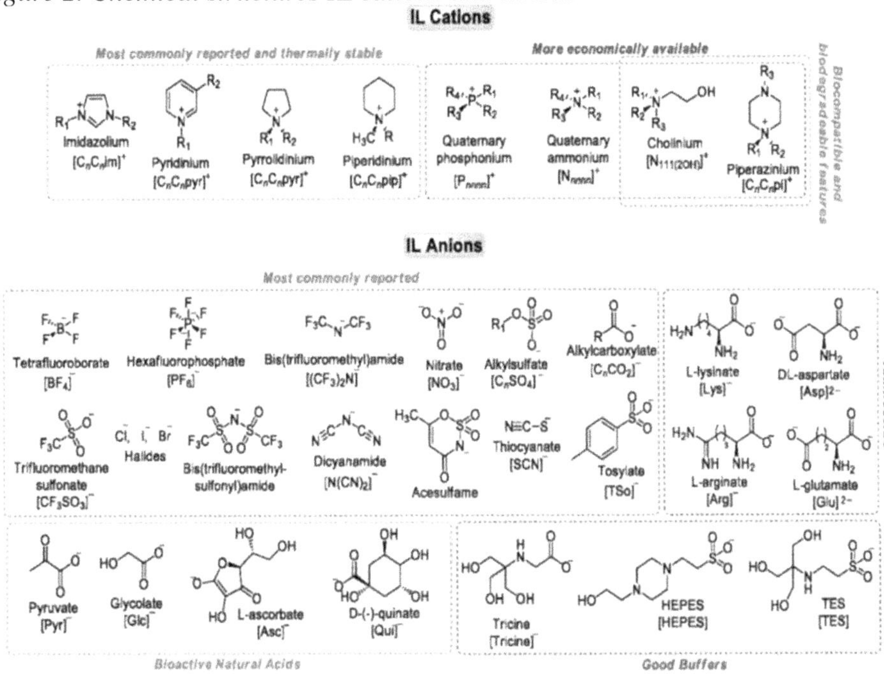

Ionic Liquids Application in Drug Synthesis

Medium and Catalysts in Drug Synthesis

In keeping with the growth of green chemistry, researchers from both domestic and foreign institutions have been very interested in green solvents since the early 1990s. Water, ionic liquids (ILs), low eutectic solvents, and supercritical CO2 are the main types of green solvents. The widespread use of hazardous and volatile organic solvents in drug manufacture processes is a major environmental problem. In mild settings, ILs have recently attracted more interest due to their potential to replace traditional volatile organic solvents in drug production, (Phukan & Sarma, 2021). The manufacture of medicinal substances and drugs is one of the many chemical processes in which ILs are involved. They have numerous benefits, including great solubility, polarity, hydrophilicity, and exceptional thermal and chemical durability. Active groups like hydroxyl and ether groups can be added to ILs to functionalize them to satisfy certain needs. This results in functionalized ILs that are low toxicity and readily degradable, (Mir et al., 2025). Due to their low vapor pressure and low volatility when compared to typical organic solvents, ILs are appropriate for use in separation, purification, and general chemical processes. Ionic liquid-organic systems are created when ILs are employed as solvents in drug manufacturing. Utilizing the small vapor pressure of ILs, vacuum distillation can be used to eliminate volatile organic solvents during the purifying process. Generally, ILs are divided into two groups: hydrophilic and hydrophobic. The reaction products of hydrophilic ILs are extracted using organic solvents, which makes it easier to separate other IL components and recycle the ILs.

Conversely, hydrophobic ILs are insoluble in ether and hexane, two examples of organic solvents. Two phases develop in IL solutions upon the addition of non-miscible organic solvents. IL separation and recovery are made possible by the extraction of hydrophobic products—which are challenging to volatilize—into the organic solvent. Because of their excellent stability, high solubility, recyclability, and resistance to combustion or explosion, ILs are also ideally suited for reactions that need strong vacuum or temperature. Compared to traditional solvents, ILs help reach chemical equilibrium faster, boost reaction speeds, decrease the number of reaction steps, and enhance product yields, efficiency, and selectivity. Their strong polarity and poor bonding properties make them perfect media for reactions involving transition metal catalysts. In the production of several pharmaceutically active ingredients, ILs are recently used as green solvents, catalysts, reagents, and enantioselectivity enhancers, (Algieri et al., 2023). Marcin and colleagues, for example, investigated the effectiveness of ILs containing different cations (ammonium-, piperidinium-, pyridinium-, and imidazolium-based), each with unique alkyl side-chain structures, in combination with the [N(Tf)2] anion as new co-solvents for the moderately induced reaction of acetic anhydride with curcumin. This analysis showed the possibility of using ILs based on bis(trifluoromethylsulfonyl)-imide as recyclable media for curcumin esterification, which would yield curcumin diacetate with no need for extra solvents. The findings showed that [N(Tf)2]-based ILs are efficient solvents for curcumin esterification, with optimal circumstances resulting in a substantial reduction in esterification time to just 15 minutes.

Furthermore, [C4C1im][N(Tf)2] produced the maximum yield of curcumin diacetate (98%) and [C6C1im][N(Tf)2] was recyclable up to three times without significantly losing its catalytic activity. Depressurized filtration can be used to extract the ILs from the reactions, and it was then cleaned with ethyl acetate and dried for use in later cycles, (Gano, Klebeko, & Peech, 2021). Sangwan et al. used ILs such 1-butyl-3-methyl imidazolium tetrafluoroborate (BMIF), 1-butyl-3-methyl imidazolium bromide (BMIB), and 1-butyl-3-methylimidazolium chloride (BMIC) to create 1, 8-dioxooctahydroxanthene derivatives (3a–3k) without the need for a solvent. An outstanding yield of up to 90% was attained using this effective and environmentally friendly procedure. One of the derivatives demonstrated a significant inhibition effect on the A549 cancer cell line against human lung cancer cell lines (A549), (Sangwan et al., 2020). Catalysts are commonly used in industrial pharmaceutical synthesis; their catalytic actions dictate the end product's yield but can also cause organic contamination. For some processes, ILs have catalytic and selective qualities in addition to being reaction solvents. Their exceptional catalytic qualities, environmental advantages, low cost, quick reaction times, recyclability, ease of use, and safe, mild reaction conditions make them popular in a variety of drug synthesis, such as heterocyclic synthesis, alkylation, oxidation, and dehydration, (Das et al., 2020). The intended product with high substrate conversion (93.79%) and product selectivity was produced using this approach. The catalyst was reused up to five times without experiencing a substantial loss of activity, and the remaining methanol was recovered by vacuum distillation, (Mir, Chang, & Hefni, 2024).

Wu et al. used H-bonding and acid co-catalytic pathway to synthesize diarylmethanes, which are common structures in physiologically and biologically active compounds. This reaction was effectively catalyzed by ILs such as 1-propylsulfonic acid-3-methylimidazolium trifluoro methane sulfonate ([SO3H-PMIm][OTf]), which outperformed more conventional acidic catalysts such triflic acids and H_2SO_4. Good to exceptional yields of diarylmethanes were produced by the process. While the anions served as hydrogen bond acceptor for activation of C-H bonds on aromatic hydrocarbons, allowing the benzylation of aromatic compounds, the IL cations used acid and hydrogen bonding to activate benzyl alcohol and generate benzyl cations. Due to the system's distinct phase behavior, the reaction notably took place at the joining face between the aromatic and IL-based phases, and the IL separated spontaneously following the reaction, (Wu et al., 2022).

Ionic Liquid Application in Drug Extraction and Analysis

The particular application in drug separation owing to adjustable structure, ILs have garnered a lot of attention in recent years, (Trujillo- Rodríguez, Pino, & Míró, 2020). A family of secondary metabolites of natural species, including bacteria, plants, animals, and marine life, are known as natural products. They are crucial to human health and have a significant impact on the advancement of human medicine and health. The market for medicinal medications made from natural resources has expanded in recent years. Volatile organic solvents such as Benzine, $C_4H_8O_2$, and C_3H_6O are used in the decoction, reflux, maceration, and Soxhlet extraction procedures in the conventional extraction of natural materials. Low extraction efficiency, contamination of the environment non-renewability, and the creation of readily produced hazardous byproducts are the shortcomings of these techniques.

A family of safe and effective media known as specific ionic liquids that are renewable and their solutions in aqueous form has been found as a promising solvent to address some of the issues related to the usage of organic compounds that are volatile in the isolation and separation mechanisms. Many of the results offer fresh insights into the usage of suitable aqueous IL solutions rather than often employed organic solutions for the extraction of biologically active components from natural goods without the requirement for supplementary product recovery/separation procedures, (Trujillo- Rodríguez, Pino, & Míró, 2020). Two stereochemically distinct C-7 isomers, 7R,8R,8′R)-(-)-allo-hydroxy matairesinol (HMR1) and 7S,8R,8′R)-(−)-7-hydroxymatairesinol (HMR2), combine to form natural 7-hydroxymatairesinol (HMR). Because of its antioxidative and anticarcinogenic qualities, HMR and its derivatives have been thoroughly studied for use in pharmaceutical applications. In place of volatile organic solvents (such as $(CH_3)_2CO$, H_2O– $(CH_3)_2CO$, and H_2O-C_2H_5OH combinations), Ferreira et al. investigated solutions of aqueous nature of glycine-betaine ionic liquids as substitute solvents for the extraction of HMR from Norway spruce tree knots. Fluorescence microscopy pictures (Figure 3C) demonstrated the impact of the HMR-rich extract that precipitated, $[(C_2)_3NC_2]Br$, and the HMR extract in an aqueous solution, [(C2)3NC2]Br (both at 100 μg mL^{-1}), on cellular oxidative stress. Hoechst 33342-stained cell nuclei were blue, whereas the H2DCFDA probe would show green cellular reactive oxygen species (ROS). The absence of green in Figure 3C's second and third lines indicates that all samples' cells at 100 μg mL−1 are free of oxidative damage. The ability of the interacted HMR and the HMR in IL solutions of water lessen oxidative pressure caused by lipopolysaccharide (LPS) is depicted in Figure 3D. The reduction in the intensity of the green color of the cells indicates that HMR does, in fact, have significant cellular antioxidant activity, lowering the ROS generation brought on by LPS. The excellent antioxidant property of the HMR-rich extracts and their biocompatible nature are supported by the cytotoxicity experiments of the [(C2)3NC2]Br aqueous solutions. Compared to the reversed HMR-rich solid extracts, the free radical scavenging and anti-inflammatory properties of HMR-rich aqueous IL solutions were shown to be more promising, (Ferreira et al., 2017). To ensure the accurate identification of analytes in the sample, laboratory analysis usually involves several processes, such as extraction, purification, and pre-concentration. In most situations, sample preparation techniques are the most time-consuming part of the entire analytical procedure.

In analytical chemistry, efficient and successful sample preparation methods are essential. Due to its easy operation, high efficiency, quick speed, and superior kinetic features, dispersive solid-phase extraction has been utilized extensively in the detection of trace pharmaceuticals. The magnetic ILs adsorbent uses a tiny amount of organic solvent and is reusable at least five times. By the use of generated ionic liquid as an extractant, the established detection method was straightforward, highly selective, sensitive, and effective, (Hassan & Tanimu, 2021).

Ionic liquids have drawn a lot of interest in this regard and are frequently employed as additives to enhance target separations, (Hussain et al., 2019). Using 1-hexyl-3-

methylimidazolium trifluoroborate as a mobile phase additive, Ding et al. created a reversed-phase high-performance liquid chromatographic methodology for the continuous detection of jatrorrhizine, palmatine, and berberine in Phellodendron bark.

Figure 3. (A) HMR extraction yield for a solid–liquid (S/L) ratio = 0.10 (blue square) and for a S/L ratio = 0.02 (orange square) using various aqueous solutions of ILs (at 0.5 M), acetone, and water (T = 25 C, t = 180 min). HMR2/HMR1 ratio for a S/L ratio of 0.10 (blue circle) or 0.02 (orange circle). The yield of total HMR extracted using aqueous solutions of [(C2)3NC2]Br at 25 C is displayed in (B) response surface and contour plots, which combine the effects of (i) extraction time (t) and IL concentration; (ii) solid–liquid ratio (S/L ratio) and concentration; and (iii) solid–liquid ratio (S/L ratio) and extraction time (t). (C) The effects of oxidative stress on macrophage cells when various HMR and pure IL samples are presented. 20 μm is the scale bar. (D) LPS-stimulated macrophages treated with precipitated HMR, HMR in IL solution, and pure IL showed cellular oxidative stress. 20 μm is the scale bar

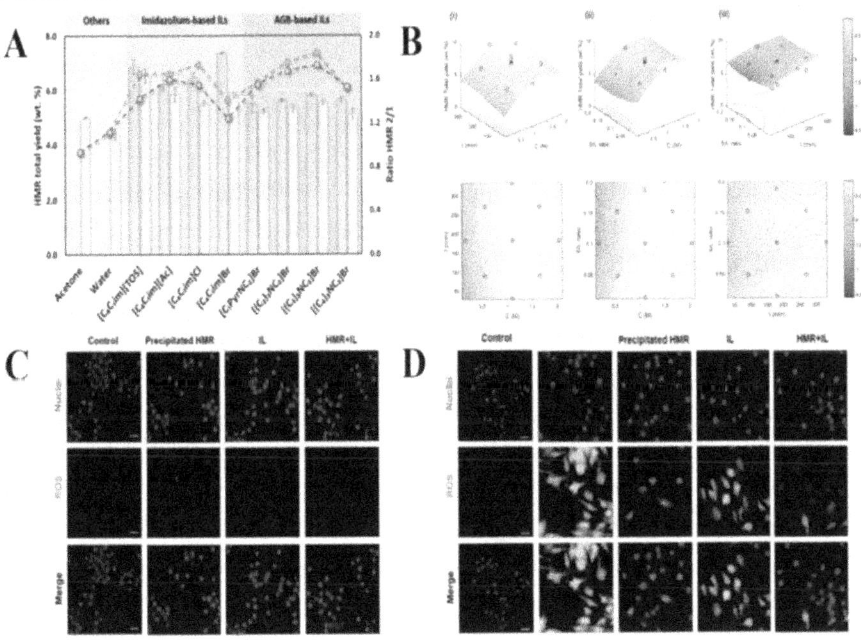

(Ferreira et al., 2017)

For GC and LC, ILs can be utilized to modify or functionalize stationary phases in addition to serving as mobile phase additives, (Zang et al., 2018). Because of its benefits in terms of speed, sensitivity, ease of use, and precise quantification, gas chromatography (GC) is frequently utilized for the separation of easy boiling chemical molecules and enantiomers. Chiral ionic liquids (CILs) can be employed as chiral stationary phases (CSPs) in GC due to their exceptional temperature stability. The chiral ionic liquid ([L-PheEtO][(C2F5SO2)2N]) that Kimaru et al. produced has a good thermal stability and can be employed as a stationary phase in GC. Fluorescence spectroscopy was used to demonstrate its chirality recognition capabilities using 2, 2, 2-trifluoroanthrylethanol (TFAE) as a model chiral analyte, (Mir et al., 2024). Among the most popular methods for enantiomeric separation and analysis has been shown to be high-performance liquid chromatography (HPLC), (Muderawan, Ong, & Ng, 2006). Rahim, Tay, and Mohamad (2018) revealed new chiral stationary phases (β-CD-BIMOTs, β-CD-DIMOTs) created with ionic liquids implemented with β-cyclodextrin to increase the enantiomeric flavonoid and β-blocker separation. They demonstrated that the superposition of the bonding of hydrogen, hydrophobic, and also p–p interactions gives β-CD-BIMOTs stationary phases superior separation capabilities when comparing the enantiomeric separation ability of non-steroidal anti-inflammatory medicines between β-CD-BIMOTs and β-CD-DIMOTs stationary phases with the native β-CD the stationary state, (Rahim, Tay, & Mohamad, 2018).

Drug Solubilization

The drug's solubility in biological or aqueous fluids and its porousness through biological membranes at body temperature are key factors in determining the bioavailability of pharmacological formulations, (Mizuuchi et al., 2008). Poor solubility and permeability cause developmental problems for about 40% of medications. These problems are intended to be resolved by conventional techniques such complexation with cyclodextrins, solid dispersions, lipid formulations, and the use of organic solubilizers. However, because they may solvate a variety of chemicals, ionic liquids (ILs), which are salts that melt at about 99°C, present a possible substitute. Many poorly soluble medications can become more soluble when ILs are used as adaptable solvents with special solvation qualities. By decreasing the interfacial interation between the drug and solvent and interfering with water's ability to self-association, they solubilize compounds. By improving drug penetration via physiological barriers, ILs can increase therapeutic efficacy, (Dobler et al., 2013). The electron distribution, alkyl side chains, and anionic species composition of ILs are among the structural characteristics that affect their solubility effects. Longer alkyl chains in ILs tend to make hydrophobic medicines more soluble, whereas hydrophilic medications are less soluble.

In particular, glycine-based ILs showed the greatest increase in solubility for the chemicals under study, (Mir, Jassal, & Andrews, 2024). Additionally, choline-based ILs have demonstrated potential in enhancing acetaminophen's solubility in aqueous solutions. Both choline lactate and choline bitartrate solutions improved the solubility of acetaminophen. The impact increased with temperature and concentration. The solubility improvement was facilitated by this interaction between the medication and the IL, (Mokhtarpour et al., 2020). ILs can help build medication formulations with better dissolving rates in addition to solubilizing chemicals. To overcome rifampicin's solubility issues, creating ultrafine rifampicin particles using an ionic liquid improved the drug's solubility by 30% and accelerated its rate of dissolution by 83%, (Sarma et al., 2011). Finally, by increasing the solubility and bioavailability of poorly soluble medications, ILs provide a cost-effective and environmentally friendly substitute for conventional solubilization techniques, potentially resulting in more potent drug delivery systems.

Ionic Liquid Engineering in Drug Crystal

Different crystalline forms of the same medicine can have quite different physicochemical characteristics, such as density, melting point, solubility, dissolution rate, and chemical stability. The manufacturing of medicinal formulations and the bioavailability of pharmaceutically active ingredients depend heavily on these variances. Therefore, one of the main goals of drug development is to modify the polymorphism of medications, (Sarma et al., 2011). Ionic liquids (ILs) have been shown in recent research to be efficient crystallization media, allowing for the control of crystal shapes and the improvement of crystal habits. This feature greatly enhances the quality of APIs and makes post-processing of pharmaceutical preparations easier by enabling the creation of crystalline forms or habits that are frequently impossible to achieve with conventional organic solvents, (An et al., 2016). Gabapentin form IV, a very unstable polymorph, was successfully obtained as a result of this study (Figure 4). Results from all five crystallization procedures were satisfactory. The mechanism for the production of pure gabapentin form IV was then clarified using molecular dynamics simulations. Through GBP-RTIL interactions, the simulations showed that the RTILs promoted the creation of form IV. In particular, the H(acidic (C4/C6mim))-O(carboxylate) interaction was a key factor in the formation of this polymorph, (Mir, 2023).

Figure 4. (A) Schematic diagram showing how GBP crystals are obtained. Temperatures denoted by RT, LT, and HT are 24°C, 0°C, and 80°C, respectively. (B) Crystal pictures of GBP polymorphic forms in C4mimBF4 + C6mimBF4 taken four months after crystallization. Form IV gradually transforms into form II, creating a hybrid of forms II and IV, while forms II and III are preserved over time

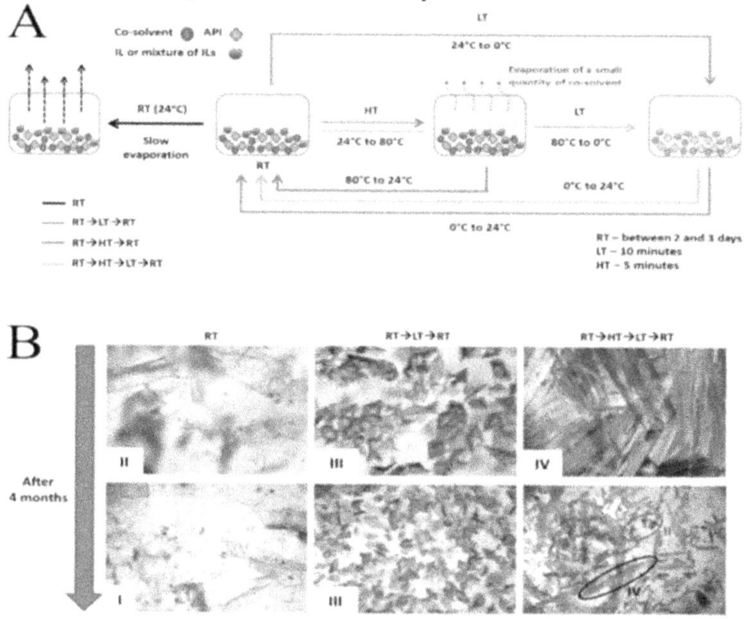

(Mir, 2023)

Drug Crystal Habits and Their Impact on Drug Properties

Drug crystal habits have an impact on flowability, tableting, mechanical strength, and other crucial characteristics in addition to solubility and dissolution rates. Drugs frequently crystallize into needle, rod, flake, or block-like shapes. Among these, crystals that resemble needles or flakes usually possess poor powder flow properties, which can be harmful to the manufacturing of medicinal products. The morphology of the crystals, which in turn affects the drug's dissolving characteristics, is largely determined by the solvents and crystallization methods employed. By improving morphology or decreasing size, ionic liquids (ILs), a novel crystallization medium, provide a way to alter or enhance drug crystal behaviors. Reducing the crystal size were demonstrated as an efficient way to modify the solubility and rate of dissolution of weakly soluble compounds, (Dong et al., 2009). Ibuprofen was crystallized by Karthika, Radhakrishnan, and Kalaichelvi (2017) using imidazole-based ILs with PF6 anions, resulting in needle-like structures with high aspect ratios. The reaction between [bmim][PF6] and the solute's crystal plane were changed by adding different organic solvents to [bmim][PF6], which created hydrogen bonds with the IL. Ibuprofen crystals' aspect ratio decreased as a result of this modification, giving them a more enticing crystal habit. The study mentioned that the reaction of the cation with the particular surface of the solute can drastically change the aspect ratio of the produced crystals when the IL is hydrogen-bonded to the solvent, (Karthika, Radhakrishnan, & Kalaichelvi, 2017).

Table 1. Applications of ionic liquids in pharmaceutical fields

Ionic Liquid	Function	Compound	Activity
[C1OC2MIM][Ms], [C1OC2MIM][TFa], [C4MIM][TFa]	Solvent	Stavudine	Anti-HIV drug
[C1C1MIM][C1SO$_4$]	Dehydrating agent	Hydrazinyl phthalazines	Antimalarial agents
[C1OC2MIM][Ms], [C1OC2MIM][TFa], [C4MIM][TFa]	Solvent	Trifluridine	Anti-HSV drug
[C2Py][BF4], [MClm] (MClm = $AlCl_3$, $FeCl_3$, $ZnCl_2$, $SnCl_2$, SnCl4, CuCl)	Catalyst	1-[(E)-3-methyl -4-benzenesulfonyl -3-methylbut - 2-enyl] -2, 3, 4, 5-tetramethoxy -6-methylbenzene	Coenzyme Q10 intermediate
[C$_2$MIM][Br]	Solvent for enzymatic reactions	(R)-modafinil	Wakefulness-promoting agent
[C$_4$MIM][PF$_6$]	Solvent	Pravadoline	NSAID
[C$_4$C1Py][IC$_{12}$]	Iodination reagent	Clioquinol	Antifungal drug
[C$_4$MIM][PF$_6$]	Ionic liquid phase, whole-cell catalysis	(R)-1-trimethylsilylethanol	Vital synthon for silicon-based drugs
[C$_4$MIM][NTf$_2$]	Ionic liquid phase, whole-cell catalysis	(S)-3-chloro-1-phenyl-1-propanol	Precursor to antidepressant drugs
[C$_4$C$_1$Py][ICl$_2$]	Iodination reagent	Iodoquinol	Antiprotozoal drug
[C$_4$MIM][PF$_6$]	Solvent for enzymatic reactions	11α-hydroxy-16α,17-epoxyprogesterone	Intermediate in steroid drug synthesis
[C$_4$MIM][BF$_4$]	Solvent	Isoxazolines	Precursor to antimicrobial agents
[C$_4$MIM][NTf$_2$]	Solvent	3-phenylglycidol	Precursor for drugs like to moxetine and reboxetine

continued on following page

Table 1. Continued

Ionic Liquid	Function	Compound	Activity
$[C_4MIM][PF_6]$, $[C_8C1Py][BF_4]$	Solvent for enzymatic reactions	(R,S)-1-chloro-3-(3,4-difluorophenoxy)-2-propanol	Intermediate for (S)-lubeluzole synthesis
$[C_4MIM][BF_4]$	Solvent	(S)-naproxen	NSAID
$[C_5MIM][NO_3]$	Catalyst, solvent	α-tocopherol succinate	Vitamin E ester
Many imidazolium and pyridinium ILs	Solvent	Dronedarone intermediates	Anti-arrhythmic drug

BIOMEDICAL APPLICATIONS OF ILS

Drug Delivery Transport

The majority of drug delivery systems (DDSs) are made to make it easier for medicinal substances to enter the body via the proper pathway. DDSs may occasionally participate in diagnostic processes as well. However, the poor pharmacokinetics and distribution of drug molecules in vivo pose serious problems for traditional DDSs, making low bioavailability and adverse systemic side effects. In order to enhance medication bioavailability, reduce drug loss, and avoid adverse reactions, researchers have been working on DDSs consistently, (Azum & Rub, 2021). Without a delivery system, the concentration of medicines in the body swiftly increases to supra-optimal levels and then rapidly decreases to sub-optimal levels as a result of systemic metabolism or circulating degradation. This results in insufficient medication absorption at the intended location, (Chen et al., 2019). Extended-release formulations that can sustain the best therapeutic doses of medications for prolonged periods of time have long been the subject of research. Ionic liquids have attracted a lot of interest as potential remedies for the problems with conventional DDSs because of their chemical stability and biocompatibility, (Liu et al., 2024). Furthermore, structural changes to the anions and cations of ILs can be used to modify their hydrophilicity and lipophilicity, which enhances the bioavailability of medications with poor transdermal delivery efficiency because of solubility problems. Imidazoles, amino acid esters, and choline are examples of ILs that have been found to have transdermal osmosis activity, which makes them viable options for application in DDSs. Choline, amino acid esters, glycine betaine, and protein-derived cations are examples of naturally occurring anions and cations that have been developed and produced as ILs for DDS applications in recent years, (Ali, Moshikur, & Wakabayashi, 2019).

Short alkyl chains are carried by a unique family of ILs known as surfactant ionic liquids (SAILs). With ionic hydrophilic top part and a non-ionic hydrophobic tail part, they structurally resemble cationic surfactants. Researchers have shown a top level of interest in SAILs because of their special qualities, which include high thermal stability and biodegradability, Maneewattanapinyo, Pichayakorn, & Monton, 2023; Banjare et al., 2018). Their ability to create stable drug-carrying vehicles, that covers micelles, vesicles, emulsions, microporous polymers, and active pharmaceutical components, has demonstrated their potential as drug delivery systems (Figure 5), (Tampucci et al., 2020).

Figure 5. An illustration of three common medication carriers

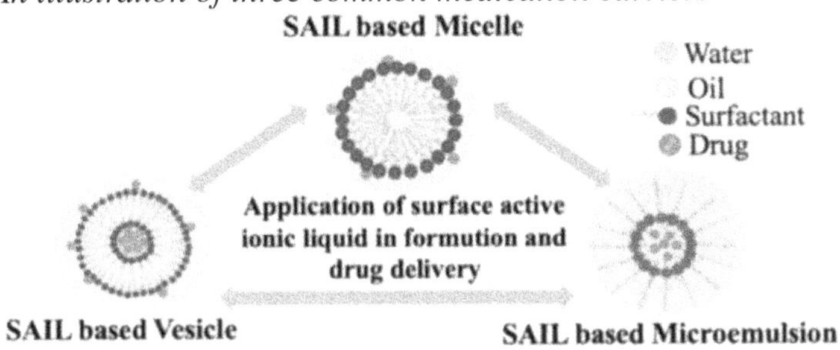

(Mir, 2024)

Antimicrobial Effects of Ionic Liquids

Antibiotic resistance mechanisms are inherent in bacteria, making infection treatment more difficult. Antimicrobial resistance is a major concern in healthcare settings because it increases morbidity and death. Bacteria are more likely to acquire sophisticated resistance to antibiotics the more antibiotics are used worldwide. Antibiotics are rendered ineffective by these resistance mechanisms, which leads to the need for greater dosages or a change to a different type of antibiotic. In order to address the growing problem of antibiotic resistance, scientists have been searching for novel, potent antimicrobial drugs, (Majumder & Rahman, 2020). Ionic liquids (ILs) have become potential agents in the battle against pathogenic bacteria because of their highly adjustable and designable physicochemical features, (Bubalo, Radošević, & Srček, 2014). The electrdynamic interactions between ILs' cationic structures and the anionic bacterial cell wall are what largely cause their antibacterial activity. Furthermore, the bacterial membrane's structure may be upset and bacterial cell death may result from the introduction of hydrophobic side chains into the phospholipid bilayer of the membrane. The length of ILs' alkyl side chains has been linked to both their toxicity and antibacterial activity, according to studies. Side chains with 10–14 carbon atoms often have the strongest antibacterial activity in ILs; side chains with more over 14 carbon atoms have a "cut-off" effect, meaning that toxicity no longer rises, (Gundolf, Weying-Zerrer, & Sommer, 2019; Ranke et al., 2004).

A key component of ILs' antibacterial activity is their electrostatic affinity to bacterial membranes. The electrodynamic interactions and, thus, the antibacterial activity of ILs can be greatly influenced by their charge density. For instance, Zheng et al. produced a number of mono- or bis-imidazolium ILs and discovered that because the imidazole cation in the bis-imidazolium structure had a larger charge density, the bis-imidazolium ILs had stronger antibacterial qualities. An enhanced charge density of the ILs often results in stronger antibacterial action, (Zheng et al., 2016). ILs can damage DNA or bio-membranes, interfere with bacterial membranes, destabilize proteins and enzymes, and interfere with cell metabolism, among other effects on bacterial cells (Figure 6). Hydrophobic interlinkage between cationic side and non-ionic DNA bases further promote their ability to attach to DNA through electrodynamic interlinkages between the positive head groups of ILs and the DNA phosphate groups (Figure 6C). Furthermore, ILs have the ability to disrupt bacterial production of extracellular polymeric substances (EPSs), which are essential for the creation of biofilms. ILs can break down the biofilm and encourage deeper penetration into bacterial aggregation by altering the EPS structure (Figure 6D), (Michalski et al., 2023).

Four steps are usually involved in the mechanism by which ILs produce their antibacterial effects:

1. ILs adhering to the membrane of the bacterial cell.
2. Electrostatic interactions that alter membrane phospholipids and deactivate membrane proteins.
3. Phospholipid bilayer disruption that results in cytoplasmic leaking inside cells.
4. Cell lysis as a result of the bacterial cell wall being destroyed.

Numerous ILs have shown antibacterial qualities, including those based on imidazolium, ammonium, phosphonium, pyridinium, and other substances, (Hajfarajollah et al., 2014; Garcia et al., 2023).

For example, a novel family of room temperature ionic liquids (RTILs) containing alkoxymethyl substituents in the cation was created by Feder-Kubis and Wnuk (2022). It was discovered that the bacterial strain and the length of the alkyl chain had a significant impact on the antibacterial action of these ILs. High selective action against Gram-positive bacteria viz- Staphylococcus species was demonstrated by the IL with the longest alkyl chain in the alkoxymethyl group, (Feder-Kubis & Wnuk, 2022) Likewise, Kumer et al. created ortho-toludinium carboxylate ionic liquids (OTILs) and evaluated its effectiveness against a range of fungus and bacteria. The findings demonstrated encouraging antifungal and antibacterial properties, suggesting that OTILs have the potential to be strong antimicrobial agents, (Kumer & Khan, 2021).

Figure 6. Ionic liquids' antimicrobial properties. (A) Diagram of a Gram-negative cell wall with an ionic liquid-treated membrane. (B) How proteins interact with ionic liquids. (C) How DNA structure interacts with ionic liquids. (D) Disruption of biofilm following contact with ionic solutions

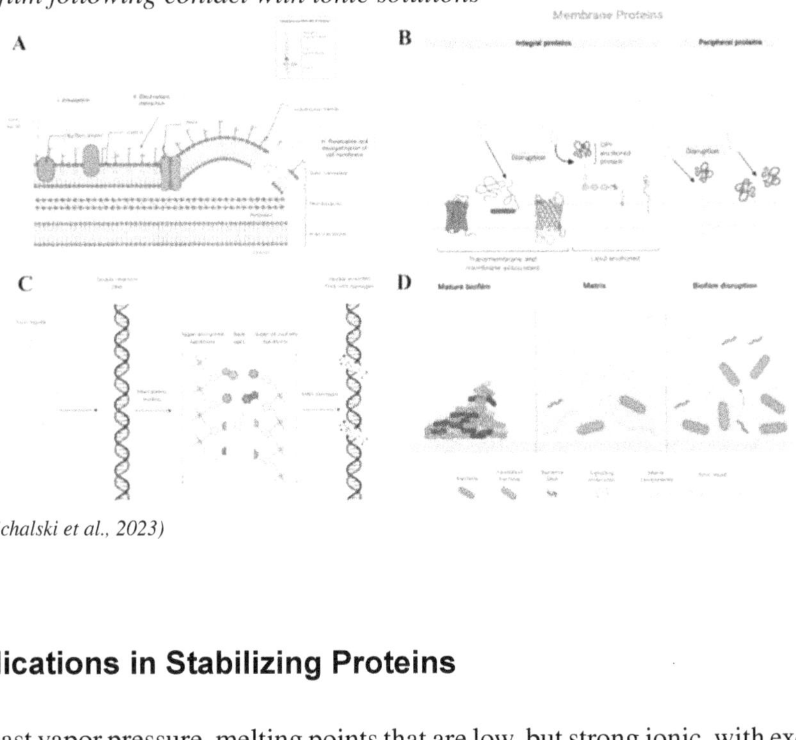

(Michalski et al., 2023)

Applications in Stabilizing Proteins

Least vapor pressure, melting points that are low, but strong ionic, with excellent stability, least toxicity, and great miscibility with other liquids are characteristics of ionic liquids (ILs). Solvents with different order making and order breaking characteristics can be produced by simply changing the cation and anion composition. These characteristics are essential for figuring out a solute's water solubility and protein stability, (Rogers & Seddon, 2003). At the moment, ILs are used extensively in biotechnology, serving as protein stabilizers and solvents and co-solvents for enzymes, (Reslan & Kayser, 2018). As extremely intricate macromolecules that are vital to biological functions, proteins need to maintain their secondary structural components in order to function, (Alves, Melo, & Mertens, 2023). Different covalent (like hydrogen bonds) and non-covalent (like hydrophobic and electrostatic interactions) bonds support proteins in biological systems in maintaining their regular functions, which is essential for maintaining the structural and functional characeyeristics of proteins' bioactive conformations, (Molodenskiy et al., 2017). Folded, partly folded, and unfolded states are all present in the conformational equilibrium of proteins in solution. However, when variables like temperature, pressure, pH, or solvent composition alter, this equilibrium may be upset, (Timasheff, 1993). Protein aggregates may eventually form as a result of proteins' progressive structural unfolding or alteration in response to these changes. The function and activity of proteins and enzymes may be lost as a result of this aggregation, which severely restricts their use in biopharmaceuticals, (Kayser, Chennamsetty, & Voynov, 2011).

Additives are mostly employed to stop proteins from unfolding and aggregating without affecting their secondary and tertiary structures, (Andersson & Hatti-Kaul, 1999). Biocompatible co-solvents can support protein stability because proteins are fixed by a balance by the interactions with the solvent environment and intramolecular interactions, (Alves, Melo, & Mertens, 2023). The hunt for biocompatible ILs as possible protein stabilizers has therefore drawn a lot of attention, (Rawat & Bohidar, 2012). The most promising stabilizers among them have been identified as imidazolium-based and choline-based ILs; choline-based ILs have emerged as being especially well-suited for stabilizing biopharmaceuticals because of their superior protein stability and biocompatibility, (Khachatrian & Mukhametzyanov, 2022).

Finding quantifiable relationships between the characteristics of the ILs and their corresponding effect on protein fixation was the goal of their investigation. The findings demonstrated that the intermolecular interactions of ILs in aqueous solutions are related to their impact on protein stability. The ability of ILs to take protons is frequently linked to their stabilizing effect on lysozyme. By tracking variations in protein fluorescence, the interaction between ionic liquids and hen egg-white lysozyme (HEWL) at standard temperatures was evaluated. Furthermore, circular dichroism (CD) spectroscopy was used to assess the thermal stability of HEWL in water with ILs. According to CD results, the investigated ILs generally had no influence on lysozyme's fluorescence and did not change its native 2^0 and 3^0 structure at room temperature, (Khachatrian & Mukhametzyanov, 2022).

Kumari and Sharma (2022) investigated how ammonium and choline-based ILs, such as tetrabutylammonium chloride, choline chloride, and tetrabutylammonium methane sulfonate (TBAMS), inhibited the amyloid formation of superoxide dismutase 1 (SOD1). They discovered that the nucleation stage of fibrillation was the primary site of action for both TBAMS and TBACl solutions in water, which successfully inhibited SOD1 fibrillation. The inhibition impact increased with amount. At lower doses, CholCl prevented the fibrillar aggregation of SOD1, while at higher amount, the reverse affect was seen. CholCl inhibited both the nucleation and elongation phases of fibrillation. Measurements of thioflavin T (ThT) intensity indicated that these ILs' inhibitory effects occurred in the following order: TBACl > TBAMS > CholCl. Using molecular docking have and molecular dynamics simulation research, the potential binding sites and variable binding of ILs with various forms of SOD1 were identified (Figure 7). Using Ligplot (AutoDock 4.2) (Figure 10(A2–C2)) and Discovery Studio v4.5 (Figure 7(A3–C3)), the various types of interactions that occur when ILs bind to SOD1 were seen, (Kumar & Sharma, 2022).

Figure 7. SOD1-[tetrabutylammonium cation] complex (A1), SOD1-[methane sulfonate anion] complex (B1), and SOD1-[cholinium cation] complex (C1) are the lowest energy-binding models of the SOD1-IL complex. ILs bind to SOD1 by a variety of interactions (A2, B2, C2) and (A3, B3, C3). The colors green and red represent hydrogen bonding and hydrophobic interactions, respectively

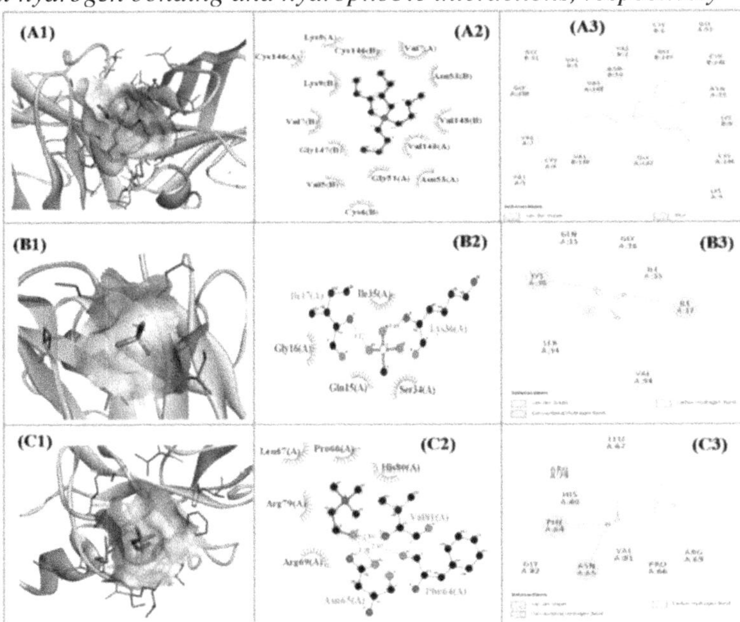

(Kumari & Sharma, 2022)

Ionic Liquids and their Antimicrobial and Antifungal Properties

Ionic liquids (ILs) have drawn maximum attention from researchers due to their antibacterial qualities. When it was found that various types of ionic liquids inhibited the growth of numerous microbial and fungal colonies, the medical and industrial potential of ILs became clear. It has been demonstrated that ammonium, imidazolium, pyrrolidinium, pyridinium, piperidinium, and other ILs inhibit both pathogenic and nonpathogenic bacteria, fungi, and other microbes. These results show both advantages and disadvantages. The high antibacterial activity of ILs can be a beneficial characteristic in medicine, but it may also limit their utility in biotechnology. For example, pyridinium inhibited Clostridium sp., which is important in uranium bio-adsorption, while Propionibacterium freudenreichii subsp. freudenreichii, a probiotic bacterium employed in the dairy industry (Hajfrarajollah et al., 2014), was found to be toxic to imidazolium ILs., (Zhang, Malhotra, & Francis, 2014). In hydrolysates of pretreated lignocellulose biomass, 1-ethyl-3-methylimidazolium acetate ([C2MIM][OAc]) inhibited ethanol formation and growth of the model yeast Saccharomyces cerevisiae, (Ouellet et al., 2011). Although both ILs hindered yeast growth at greater concentrations, 1-ethyl-3-methylimidazolium acetate and 1-ethyl-3-methylimidazolium methyl phosphonate, when present in modest levels (5%), caused a metabolic shift including fermentation and respiration, (Mehmood et al., 2015). It has also been observed that copper and silver compounds exhibit antibacterial qualities against methicillin-resistant Staphylococcus aureus. Silver/polymer core-shell and silver nanoparticle composite materials have attracted a lot of interest for their superior antibacterial properties because of their striking chemical and visual properties. Numerous microorganisms with a range of physiological and respiratory functions were examined for antimicrobial activity using imidazolium and pyridinium ILs with different alkyl chain lengths. On pure cultures of Bacillus subtilis, Escherichia coli, Saccharomyces cerevisiae, and Pseudomonas fluorescens, hexyl- and octyl-chain imidazolium and pyridinium bromides demonstrated exceptional antimicrobial activity, (Pham, Cho, & Yun, 2010).

Cancer Prevention Tools by Ionic Liquids

The biological activity of a chemical, or cytotoxicity, has been extensively researched in connection with ILs and different cell types. The effects of ILs on both tumor and non-tumor tissues including vertebrate and invertebrate organisms, such as rat (C6 glial cells), rat pheochromocytoma, insects (S2, Drosophila melanogaster), squamous carcinoma (A431), immortalized keratinocytes (HaCaT), colon cancer (HCT-116), embryonic kidney (HEK), cervical cancer (HeLa), hepatocyte carcinoma (HepG2), breast cancer (MCF-7), brain cancer (T98G), histiocytic lymphoma (UC937), mouse (3T3 fibroblasts), fish (CCO, channel catfish ovary cell line), and others have been the subject of numerous investigations, (Bubalo, Radošević, & Srček, 2015). The utmost method for analysing a chemical's cytotoxicity is the MTT assay, which uses the reduction of tetrazolium dyes like MTT (3-(4,5-dimethylthiazol-2-yl)-2,5-diphenyltetrazolium bromide), MTS, XTT, and WST dyes to track the capability of mitochondrial enzymes, (Mosmann, 1983). Furthermore, studies on the cytotoxicity mechanism of ILs has demonstrated that they cause apoptosis, DNA damage, and oxidative stress in cells, (Mosmann, 1983). The structure of ILs has a significant impact on their cytotoxicity, which can vary greatly from micromolar to millimolar quantities. When compared to ILs that contain other common cations like imidazolium and pyridinium, cholinium-based ILs often show lesser cytotoxicity. A lot of work has gone into creating models that forecast the cytotoxicity of ILs. However, because cytotoxicity depends on environmental factors, such as cell type, it cannot yet be distinguished based only on structural arrangement. Research on ILs' possible application as anti-cancer medicines has started as their cytotoxic effects have become more understood. This creates the intriguing prospect of creating treatments with little toxicity to the human body and customized chemotherapeutic activity, (Torrecilla et al., 2009).

Cytotoxic Potential of Ionic Liquids

With many exceptions to the general rule, like when the alkyl chain lengthens at locations R1 or R3, which raises the cytotoxicity linked to the imidazolium ring's alkyl chain lengthening. In cell lines such as HeLa carcinoma cells, substituted benzyl, alkyl, and allyl chains have shown enhanced cytotoxicity. As a result, toxicity cannot be accurately predicted from the structure alone. [NTf2] is the most poisonous anion. The lipophilicity of the imidazolium cation has been closely associated with the cytotoxicity of ILs (such as IPC-81 cells), indicating that membrane rupture rather than halide cytotoxicity may be the primary cause of toxicity. Cellular apoptosis, or planned cell death, is usually the result of this cytotoxic effect and is defined by membrane damage. Necrosis, or cell death brought on by injury or trauma, happens less frequently. The cytotoxicity of amino acids to the NIH/3T3 cell line and ionic liquids to the CaCo-2 cell line are shown in Tables 2 and 3, (Egorova & Ananikov, 2014).

Table 2. Cytotoxicity study of ionic liquids on CaCo-2 cell line

S. No	Ionic Liquid	Molecular Mass (g/mol)	24h IC_{50} (mM)
1	[BMIM][Gly]	212.99	15.421 ± 0.22
2	[BMIM][(L)-lac]	227.95	20.01 ± 0.02
3	[EMIM][Cl]	145.75	31.09 ± 0.42

(Egorova & Ananikov, 2014)

Table 3. Cytotoxicity effect of ionic liquids and amino acids on NIH/3T3 cell line

S. No	Ionic Liquid	Molecular Mass (g/mol)	24h IC_{50} (mM)
1	[BMIM][PF6]	283.57	15.55 ± 0.09
2	Gly_HCl	110.34	29.99 ± 0.11
3	Gly	74.12	>79

(Egorova & Ananikov, 2014)

CONCLUSION

In pharmaceutical and biomedical applications, ionic liquids (ILs) have shown enormous promise, especially in drug delivery, production, and bioavailability augmentation. They are a sustainable substitute for conventional organic solvents because of their adjustable qualities, which enable the creation of personalized drug carriers, permeation enhancers, and green solvents. As ILs have evolved from the first to the fourth generation, their utility in transdermal drug administration and therapeutic formulations has increased because to their increased biodegradability, decreased toxicity, and improved biocompatibility. Furthermore, ILs provide innovative answers to persistent problems in drug development and are essential for antibacterial therapies, protein stability, and cancer therapy. Notwithstanding these benefits, a number of issues still exist, such as environmental effect, regulatory obstacles, and toxicity concerns. For ILs to be widely used in pharmaceuticals, these restrictions must be addressed by cutting-edge research on biodegradable and biocompatible ILs. Future research should concentrate on enhancing IL formulations, investigating how they interact with biological systems, and evaluating their long-term efficacy and safety. By offering safer, more efficient, and long-lasting therapeutic treatments, IL-based drug delivery systems have the potential to completely transform contemporary medicine with continued innovation.

REFERENCES

Agatemor, C., Ibsen, K. N., Tanner, E. E., & Mitragotri, S. (2018). Ionic liquids for addressing unmet needs in healthcare. *Bioengineering & Translational Medicine*, *3*(1), 7–25. DOI: 10.1002/btm2.10083 PMID: 29376130

Algieri, V., Algieri, C., Costanzo, P., Fiorani, G., & De Maria, P. (2023). Novel regioselective synthesis of 1,3,4,5-tetrasubstituted pyrazoles and biochemical valuation on F1F0-ATPase and mitochondrial permeability transition pore formation. *Pharmaceutics*, *15*(2), 498. DOI: 10.3390/pharmaceutics15020498 PMID: 36839821

Ali, M. K., Moshikur, R. M., Wakabayashi, R., Tahara, Y., Moniruzzaman, M., Kamiya, N., & Goto, M. (2019). Synthesis and characterization of choline-fatty-acid-based ionic liquids: A new biocompatible surfactant. *Journal of Colloid and Interface Science*, *551*, 72–80. DOI: 10.1016/j.jcis.2019.04.095 PMID: 31075635

Alves, M. M. S., Melo, M. N., Mertens, H. D. T., Pereiro, A. B., & Archer, M. (2023). Studies on the interaction between model proteins and fluorinated ionic liquids. *Pharmaceutics*, *15*(1), 157. DOI: 10.3390/pharmaceutics15010157 PMID: 36678790

Amaral, M., Pereiro, A. B., Gaspar, M. M., & Reis, C. P. (2021). Recent advances in ionic liquids and nanotechnology for drug delivery. *Nanomedicine (London)*, *16*(1), 63–80. DOI: 10.2217/nnm-2020-0340 PMID: 33356551

An, J. H., Jin, F., Kim, H. S., Ryu, H. C., Kim, J. S., Kim, H. M., Kiyonga, A. N., Min, D. S., Youn, W., Kim, K. H., & Jung, K. (2016). Application of ionic liquid to polymorphic transformation of anti-viral/HIV drug adefovir dipivoxil. *Archives of Pharmacal Research*, *39*(5), 646–659. DOI: 10.1007/s12272-016-0721-0 PMID: 26908332

Andersson, M. M., & Hatti-Kaul, R. (1999). Protein stabilising effect of polyethyleneimine. *Journal of Biotechnology*, *72*(1–2), 21–31. DOI: 10.1016/S0168-1656(99)00050-4

Azum, N., Rub, M. A., & Asiri, A. M. (2021). Mixed micellization and spectroscopic studies of anti-allergic drug and non-ionic surfactant in the presence of ionic liquid. *Polymers*, *13*(16), 2756. DOI: 10.3390/polym13162756 PMID: 34451295

Banjare, M. K., Behera, K., Satnami, M. L., Pandey, S., & Ghosh, K. K. (2018). Self-assembly of a short-chain ionic liquid within deep eutectic solvents. *RSC Advances*, *8*(15), 7969–7979. DOI: 10.1039/C7RA13557B PMID: 35542011

Brahma, S., & Gardas, R. L. (2024). In Rajkhowa, S., Singh, P., Sen, A., & Sarma, J. (Eds.), *"History and development of ionic liquids," in Handbook of ionic liquids: fundamentals, applications, and sustainability* (pp. 1–28). John Wiley and Sons.

Bubalo, M. C., Radošević, K., & Gaurina Srček, V. (2015). Cytotoxicity towards CCO cells of imidazolium ionic liquids with functionalized side chains: Preliminary QSTR modeling using regression and classification-based approaches. *Ecotoxicology and Environmental Safety*, *112*, 22–28. DOI: 10.1016/j.ecoenv.2014.10.029 PMID: 25463849

Bubalo, M. C., Radošević, K., & Srček, V. G. (2014). A brief overview of the potential environmental hazards of ionic liquids. *Ecotoxicology and Environmental Safety*, *99*, 1–12. DOI: 10.1016/j.ecoenv.2013.10.019 PMID: 24210364

Chen, S., Hanning, S., Falconer, J., Locke, M., & Wen, J. (2019). Recent advances in non-ionic surfactant vesicles (niosomes): Fabrication, characterization, pharmaceutical and cosmetic applications. *European Journal of Pharmaceutics and Biopharmaceutics*, *144*, 18–39. DOI: 10.1016/j.ejpb.2019.08.015 PMID: 31446046

Das, S., Kashyap, N., Kalita, S., Bora, D. B., & Borah, R. (2020). A brief insight into the physicochemical properties of room-temperature acidic ionic liquids and their catalytic applications in C–C bond formation reactions. *Advances in Physical Organic Chemistry*, *54*, 1–98. DOI: 10.1016/bs.apoc.2020.07.002

Dias, A. R., Costa-Rodrigues, J., Teixeira, C., Prudêncio, C., Gomes, P., & Ferraz, R. (2019). Ionic liquids for topical delivery in cancer. *Current Medicinal Chemistry*, *26*(40), 7520–7532. DOI: 10.2174/0929867325666180914104045 PMID: 30362415

Dobler, D., Schmidts, T., Klingenhöfer, I., & Runkel, F. (2013). Ionic liquids as ingredients in topical drug delivery systems. *International Journal of Pharmaceutics*, *441*(1–2), 620–627. DOI: 10.1016/j.ijpharm.2012.10.035 PMID: 23123180

Dong, Y., Ng, W. K., Shen, S., Kim, S., & Tan, R. B. (2009). Preparation and characterization of spironolactone nanoparticles by antisolvent precipitation. *International Journal of Pharmaceutics*, *375*(1–2), 84–88. DOI: 10.1016/j.ijpharm.2009.03.013 PMID: 19481693

Egorova, K. S., & Ananikov, V. P. (2014). Toxicity of ionic liquids: Eco(cyto) activity as complicated, but unavoidable parameter for task-specific optimization. *ChemSusChem*, *7*(2), 336–360. DOI: 10.1002/cssc.201300459 PMID: 24399804

Egorova, K. S., & Ananikov, V. P. (2018). Fundamental importance of ionic interactions in the liquid phase: A review of recent studies of ionic liquids in biomedical and pharmaceutical applications. *Journal of Molecular Liquids*, *272*, 271–300. DOI: 10.1016/j.molliq.2018.09.025

Egorova, K. S., Gordeev, E. G., & Ananikov, V. P. (2017). Biological activity of ionic liquids and their application in pharmaceutics and medicine. *Chemical Reviews*, *117*(10), 7132–7189. DOI: 10.1021/acs.chemrev.6b00562 PMID: 28125212

Elgharbawy, A. A., Moniruzzaman, M., & Goto, M. (2020). Recent advances of enzymatic reactions in ionic liquids: Part II. *Biochemical Engineering Journal*, *154*, 107426. DOI: 10.1016/j.bej.2019.107426

Feder-Kubis, J., & Wnuk, A. (2022). Imidazolium room-temperature ionic liquids with alkoxymethyl substituent: A quest for improved microbiological selectivity. *Chemical Engineering Journal*, *442*, 136062. DOI: 10.1016/j.cej.2022.136062

Ferraz, R., Branco, L. C., Prudêncio, C., Noronha, J. P., & Petrovski, Ž. (2011). Ionic liquids as active pharmaceutical ingredients. *ChemMedChem*, *6*(6), 975–985. DOI: 10.1002/cmdc.201100082 PMID: 21557480

Ferreira, A. M., Morais, E. S., Leite, A. C., Mohamadou, A., Holmbom, B., Holmbom, T., Neves, B. M., Coutinho, J. A. P., Freire, M. G., & Silvestre, A. J. D. (2017). Enhanced extraction and biological activity of 7-hydroxymatairesinol obtained from Norway spruce knots using aqueous solutions of ionic liquids. *Green Chemistry*, *19*(11), 2626–2635. DOI: 10.1039/C7GC01091E

Gano, M., Klebeko, J., & Peech, R. (2021). Efficient esterification of curcumin in bis(trifluoromethylsulfonyl)imide-based ionic liquids. *Journal of Molecular Liquids*, *337*, 116420. DOI: 10.1016/j.molliq.2021.116420

Garcia, M. T., Bautista, E., de la Fuente, A., & Perez, L. (2023). Cholinium-based ionic liquids as promising antimicrobial agents in pharmaceutical applications: Surface activity, antibacterial activity and ecotoxicological profile. *Pharmaceutics*, *15*(7), 1806. DOI: 10.3390/pharmaceutics15071806 PMID: 37513993

Gundolf, T., Weyhing-Zerrer, N., Sommer, J., Kalb, R., Schoder, D., Rossmanith, P., & Mester, P. (2019). Biological impact of ionic liquids based on sustainable fatty acid anions examined with a tripartite test system. *ACS Sustainable Chemistry & Engineering*, *7*(19), 15865–15873. DOI: 10.1021/acssuschemeng.8b06201

Hajfarajollah, H., Mokhtarani, B., Mirzaei, M., Afaghi, A., & Noghabi, K. A. (2014). Toxicity of various kinds of ionic liquids towards the cell growth and end product formation of the probiotic strain, Propionibacterium freudenreichii. *RSC Advances*, *4*(25), 13153–13160. DOI: 10.1039/C4RA00925H

Hajfarajollah, H., Mokhtarani, B., Noghabi, K. A., Sharifi, A., & Mirzaei, M. (2014). Antibacterial and antiadhesive properties of butylmethylimidazolium ionic liquids toward pathogenic bacteria. *RSC Advances*, *4*(77), 42751–42757. DOI: 10.1039/C4RA07055K

Hassan, A. A., Tanimu, A., & Alhooshani, K. (2021). Iron and cobalt-containing magnetic ionic liquids for dispersive micro-solid phase extraction coupled with HPLC-DAD for the preconcentration and quantification of carbamazepine drug in urine and environmental water samples. *Journal of Molecular Liquids*, *336*, 116370. DOI: 10.1016/j.molliq.2021.116370

Hussain, A., AlAjmi, M. F., Hussain, I., & Ali, I. (2019). Future of ionic liquids for chiral separations in high-performance liquid chromatography and capillary electrophoresis. *Critical Reviews in Analytical Chemistry*, *49*(4), 289–305. DOI: 10.1080/10408347.2018.1523706 PMID: 30582712

Indrayanto, G. (2018). Validation of chromatographic methods of analysis: Application for drugs that derived from herbs. *Profiles of Drug Substances, Excipients, and Related Methodology*, *43*, 359–392. DOI: 10.1016/bs.podrm.2018.01.003 PMID: 29678264

Jaitely, V., Karatas, A., & Florence, A. T. (2008). Water-immiscible room temperature ionic liquids (RTILs) as drug reservoirs for controlled release. *International Journal of Pharmaceutics*, *354*(1–2), 168–173. DOI: 10.1016/j.ijpharm.2008.01.034 PMID: 18342462

Kapre, S., Palakurthi, S. S., Jain, A., & Palakurthi, S. (2024). DES-igning the future of drug delivery: A journey from fundamentals to drug delivery applications. *Journal of Molecular Liquids*, *400*, 124517. DOI: 10.1016/j.molliq.2024.124517

Karthika, S., Radhakrishnan, T. K., & Kalaichelvi, P. (2017). The role of hydrogen bonding propensity in tuning the morphology of crystals obtained from imidazolium based ionic liquids. *Journal of Crystal Growth*, *463*, 168–175. DOI: 10.1016/j.jcrysgro.2017.02.028

Kayser, V., Chennamsetty, N., Voynov, V., Helk, B., Forrer, K., & Trout, B. L. (2011). Evaluation of a non-Arrhenius model for therapeutic monoclonal antibody aggregation. *Journal of Pharmaceutical Sciences*, *100*(7), 2526–2542. DOI: 10.1002/jps.22493 PMID: 21268027

Khachatrian, A. A., Mukhametzyanov, T. A., Yakhvarov, D. G., Sinyashin, O. G., Garifullin, B. F., Rakipov, I. T., Mironova, D. A., Burilov, V. A., & Solomonov, B. N. (2022). Intermolecular interactions between imidazolium- and cholinium-based ionic liquids and lysozyme: Regularities and peculiarities. *Journal of Molecular Liquids*, *348*, 118426. DOI: 10.1016/j.molliq.2021.118426

Kumari, M., Sharma, S., & Deep, S. (2022). Tetrabutylammonium based ionic liquids (ILs) inhibit the amyloid aggregation of superoxide dismutase 1 (SOD1). *Journal of Molecular Liquids*, *353*, 118761. DOI: 10.1016/j.molliq.2022.118761

Kumari, P., Pillai, V. V. S., & Benedetto, A. (2020). Mechanisms of action of ionic liquids on living cells: The state of the art. *Biophysical Reviews*, *12*(5), 1187–1215. DOI: 10.1007/s12551-020-00754-w PMID: 32936423

Kumer, A., & Khan, M. W. (2021). Synthesis, characterization, antimicrobial activity and computational exploration of ortho toludinium carboxylate ionic liquids. *Journal of Molecular Structure*, *1245*, 131087. DOI: 10.1016/j.molstruc.2021.131087

Liu, T., Liu, Y., Zhao, X., Zhang, L., Wang, W., Bai, D., Liao, Y., Wang, Z., Wang, M., & Zhang, J. (2024). Thermodynamically stable ionic liquid microemulsions pioneer pathways for topical delivery and peptide application. *Bioactive Materials*, *32*, 502–513. DOI: 10.1016/j.bioactmat.2023.10.002 PMID: 38026438

Lu, B., Liu, T., Wang, H., Wu, C., Chen, H., Liu, Z., & Zhang, J. (2022). Ionic liquid transdermal delivery system: Progress, prospects, and challenges. *Journal of Molecular Liquids*, *351*, 118643. DOI: 10.1016/j.molliq.2022.118643

MacFarlane, D. R., Chong, A. L., Forsyth, M., Kar, M., Vijayaraghavan, R., Somers, A., & Seddon, K. R. (2018). New dimensions in salt–solvent mixtures: A 4th evolution of ionic liquids. *Faraday Discussions*, *206*, 9–28. DOI: 10.1039/C7FD00189D PMID: 29034392

Magina, S., Barros-Timmons, A., Ventura, S. P. M., & Evtuguin, D. V. (2021). Evaluating the hazardous impact of ionic liquids–challenges and opportunities. *Journal of Hazardous Materials*, *412*, 125215. DOI: 10.1016/j.jhazmat.2021.125215 PMID: 33951860

Majumder, M. A. A., Rahman, S., Cohall, D., Bharatha, A., Singh, K., Haque, M., & Gittens-St Hilaire, M. (2020). Antimicrobial stewardship: Fighting antimicrobial resistance and protecting global public health. *Infection and Drug Resistance*, *13*, 4713–4738. DOI: 10.2147/IDR.S290835 PMID: 33402841

Maneewattanapinyo, P., Pichayakorn, W., Monton, C., Dangmanee, N., Wunnakup, T., & Suksaeree, J. (2023). Effect of ionic liquid on silver-nanoparticle-complexed Ganoderma applanatum and its topical film formulation. *Pharmaceutics*, *15*(4), 1098. DOI: 10.3390/pharmaceutics15041098 PMID: 37111583

Mehmood, N., Husson, E., Jacquard, C., Wewetzer, S., Büchs, J., Sarazin, C., & Gosselin, I. (2015). Impact of two ionic liquids, 1-ethyl-3-methylimidazolium acetate and 1-ethyl-3-methylimidazolium methylphosphonate, on Saccharomyces cerevisiae: Metabolic, physiologic, and morphological investigations. *Biotechnology for Biofuels*, *8*(1), 1–14. DOI: 10.1186/s13068-015-0206-2 PMID: 25688291

Michalski, J., Odrzygozdz, C., Mester, P., Narozna, D., & Clapa, T. (2023). Defeat undefeatable: Ionic liquids as novel antimicrobial agents. *Journal of Molecular Liquids*, *369*, 120782. DOI: 10.1016/j.molliq.2022.120782

Millard, J., Alvarez-Nunez, F., & Yalkowsky, S. (2002). Solubilization by cosolvents: Establishing useful constants for the log-linear model. *International Journal of Pharmaceutics*, *245*(1–2), 153–166. DOI: 10.1016/S0378-5173(02)00334-4 PMID: 12270252

Mir, M. A. (2023). Molecular dynamic, Hirshfeld surface, molecular docking and drug likeness studies of a potent anti-oxidant, anti-malaria and anti-inflammatory medicine: Pyrogallol. *Results in Chemistry*, *5*, 100763. DOI: 10.1016/j.rechem.2023.100763

Mir, M. A. (2023). DFT, Hirshfeld surface, molecular docking and drug likeness studies of medicinally important coumarin molecule. *Arabian Journal for Science and Engineering*, *48*(6), 7445–7462. DOI: 10.1007/s13369-022-07476-z

Mir, M. A. (2024). DFT analysis for structure, vibrational assignments, and molecular properties of medicinally important isoquinoline. *Journal of Dispersion Science and Technology*, *45*(8), 1–11. DOI: 10.1080/01932691.2023.2299360

Mir, M. A., Ahmad, W., Andrews, K., & Chang, S. K. (2024). Chemical synthesis, experimental, molecular docking and drug-likeness studies of salidroside. *Arabian Journal for Science and Engineering*, *49*(7), 9451–9466. DOI: 10.1007/s13369-024-08721-3

Mir, M. A., Ahmad, W., Andrews, K., & Chang, S. K. (2025). Microplastics in food products: Prevalence, artificial intelligence-based detection, and potential health impacts on humans. *Emerging Contaminants*, *11*(1), 100477. DOI: 10.1016/j.emcon.2025.100477

Mir, M. A., & Banik, B. K. (2025). Sustainable healing: Natural compounds facilitating the future cancer treatment. *World Development Sustainability*, *4*, 100215. DOI: 10.1016/j.wds.2025.100215

Mir, M. A., Chang, S. K., & Hefni, D. (2024). A comprehensive review on challenges and choices of food waste in Saudi Arabia: Exploring environmental and economic impacts. *Environmental Systems Research*, *13*(1), 40. DOI: 10.1186/s40068-024-00364-5

Mir, M. A., Jassal, M. M. S., & Andrews, K. (2024). Molecular dynamic, Hirshfeld surface, computational quantum and spectroscopic analysis of 4-hydroxy-1-naphthaldehyde. *Current Organocatalysis*, *11*(1), 1–12. DOI: 10.2174/2213337210666230816091246

Mizuuchi, H., Jaitely, V., Murdan, S., & Florence, A. T. (2008). Room temperature ionic liquids and their mixtures: Potential pharmaceutical solvents. *European Journal of Pharmaceutical Sciences*, *33*(4–5), 326–331. DOI: 10.1016/j.ejps.2008.01.002 PMID: 18291630

Mokhtarpour, M., Basteholia, N., Shekaari, H., & Zafarani-Moattar, M. T. (2020). Effect of choline-based ionic liquids as novel green solvents on the aqueous solubility enhancement and thermodynamic properties of acetaminophen. *Journal of Molecular Liquids*, *306*, 112504. DOI: 10.1016/j.molliq.2020.112504

Molodenskiy, D., Shirshin, E., Tikhonova, T., Gruzinov, A., Peters, G., & Spinozzi, F. (2017). Thermally induced conformational changes and protein–protein interactions of bovine serum albumin in aqueous solution under different pH and ionic strengths as revealed by SAXS measurements. *Physical Chemistry Chemical Physics*, *19*(25), 17143–17155. DOI: 10.1039/C6CP08809K PMID: 28636681

Moshikur, R. M., Carrier, R. L., Moniruzzaman, M., & Goto, M. (2023). Recent advances in biocompatible ionic liquids in drug formulation and delivery. *Pharmaceutics*, *15*(4), 1179. DOI: 10.3390/pharmaceutics15041179 PMID: 37111664

Moshikur, R. M., & Goto, M. (2021). In Goto, M., & Moniruzzaman, M. (Eds.), *"Ionic liquids as active pharmaceutical ingredients (APIs)," in Application of ionic liquids in drug delivery* (pp. 13–33). Springer Singapore.

Mosmann, T. (1983). Rapid colorimetric assay for cellular growth and survival: Application to proliferation and cytotoxicity assays. *Journal of Immunological Methods*, *65*(1–2), 55–63. DOI: 10.1016/0022-1759(83)90303-4 PMID: 6606682

Muderawan, I. W., Ong, T. T., & Ng, S. C. (2006). Urea bonded cyclodextrin derivatives onto silica for chiral HPLC. *Journal of Separation Science*, *29*(12), 1849–1871. DOI: 10.1002/jssc.200500491 PMID: 16970188

Ouellet, M., Datta, S., Dibble, D. C., Tamrakar, P. R., & Simmons, P. I. (2011). Impact of ionic liquid pretreated plant biomass on Saccharomyces cerevisiae growth and biofuel production. *Green Chemistry*, *13*(10), 2743–2749. DOI: 10.1039/c1gc15327g

Pham, T. P., Cho, C. W., & Yun, Y. S. (2010). Environmental fate and toxicity of ionic liquids: A review. *Water Research*, *44*(2), 352–372. DOI: 10.1016/j.watres.2009.09.030 PMID: 19854462

Phukan, P., & Sarma, D. (2021). Synthesis of medicinally relevant scaffolds-triazoles and pyrazoles in green solvent ionic liquids. *Current Organic Chemistry*, *25*(13), 1523–1538. DOI: 10.2174/1385272825666210412160142

Rahim, N. Y., Tay, K. S., & Mohamad, S. (2018). Chromatographic and spectroscopic studies on β-cyclodextrin functionalized ionic liquid as chiral stationary phase: Enantio separation of NSAIDs. *Adsorption Science and Technology*, *36*(1–2), 130–148. DOI: 10.1177/0263617416686798

Ranke, J., Mölter, K., Stock, F., Bottin-Weber, U., & Hoffmann, J. (2004). Biological effects of imidazolium ionic liquids with varying chain lengths in acute Vibrio fischeri and WST-1 cell viability assays. *Ecotoxicology and Environmental Safety*, *58*(3), 396–404. DOI: 10.1016/S0147-6513(03)00105-2 PMID: 15223265

Rawat, K., & Bohidar, H. B. (2012). Universal charge quenching and stability of proteins in 1-methyl-3-alkyl (hexyl/octyl) imidazolium chloride ionic liquid solutions. *The Journal of Physical Chemistry B*, *116*(36), 11065–11074. DOI: 10.1021/jp3049108 PMID: 22891622

Reslan, M., & Kayser, V. (2018). Ionic liquids as biocompatible stabilizers of proteins. *Biophysical Reviews*, *10*(3), 781–793. DOI: 10.1007/s12551-018-0407-6 PMID: 29511969

Rogers, R. D., & Seddon, K. R. (2003). Ionic liquids—Solvents of the future? *Science*, *302*(5646), 792–793. DOI: 10.1126/science.1090313 PMID: 14593156

Sangwan, R., Saini, M., Verma, R., Kumar, S., Banerjee, M., & Jain, S. (2020). Synthesis of 1,8-dioxooctahydroxanthene derivatives using ionic liquids, quantum chemical studies and anticancer activity. *Journal of Molecular Structure*, *1208*, 127786. DOI: 10.1016/j.molstruc.2020.127786

Sarma, B., Chen, J., Hsi, H.-Y., & Myerson, A. S. (2011). Solid forms of pharmaceuticals: Polymorphs, salts and cocrystals. *Korean Journal of Chemical Engineering*, *28*(2), 315–322. DOI: 10.1007/s11814-010-0520-0

Shamshina, J. L., Barber, P. S., & Rogers, R. D. (2013). Ionic liquids in drug delivery. *Expert Opinion on Drug Delivery*, *10*(10), 1367–1381. DOI: 10.1517/17425247.2013.808185 PMID: 23795613

Shukla, M. K., Tiwari, H., Verma, R., Dong, W.-L., Azizov, S., Kumar, B., Pandey, S., & Kumar, D. (2023). Role and recent advancements of ionic liquids in drug delivery systems. *Pharmaceutics*, *15*(2), 702. DOI: 10.3390/pharmaceutics15020702 PMID: 36840024

Shukla, M. K., Tiwari, H., Verma, R., Dong, W.-L., Azizov, S., Kumar, B., Pandey, S., & Kumar, D. (2023). Role and recent advancements of ionic liquids in drug delivery systems. *Pharmaceutics*, *15*(2), 702. DOI: 10.3390/pharmaceutics15020702 PMID: 36840024

Sidat, Z., Marimuthu, T., Kumar, P., du Toit, L. C., Kondiah, P. P. D., Choonara, Y. E., & Pillay, V. (2019). Ionic liquids as potential and synergistic permeation enhancers for transdermal drug delivery. *Pharmaceutics*, *11*(2), 96. DOI: 10.3390/pharmaceutics11020096 PMID: 30813375

Tampucci, S., Guazzelli, L., Burgalassi, S., & Monti, D. (2020). pH-responsive nanostructures based on surface active fatty acid-protic ionic liquids for imiquimod delivery in skin cancer topical therapy. *Pharmaceutics*, *12*(11), 1078. DOI: 10.3390/pharmaceutics12111078 PMID: 33187215

Timasheff, S. N. (1993). The control of protein stability and association by weak interactions with water: How do solvents affect these processes? *Annual Review of Biophysics and Biomolecular Structure*, *22*(1), 67–97. DOI: 10.1146/annurev.bb.22.060193.000435 PMID: 8347999

Torrecilla, J. S., García, J., Rojo, E., & Rodríguez, F. (2009). Estimation of toxicity of ionic liquids in leukemia rat cell line and acetylcholinesterase enzyme by principal component analysis, neural networks and multiple linear regressions. *Journal of Hazardous Materials*, *164*(1), 182–194. DOI: 10.1016/j.jhazmat.2008.08.022 PMID: 18805639

Trujillo-Rodríguez, M. J., Pino, V., & Míró, M. (2020). High-throughput microscale extraction using ionic liquids and derivatives: A review. *Journal of Separation Science*, *43*(9–10), 1890–1907. DOI: 10.1002/jssc.202000045 PMID: 32074395

Welton, T. (2018). Ionic liquids: A brief history. *Biophysical Reviews*, *10*(3), 691–706. DOI: 10.1007/s12551-018-0419-2 PMID: 29700779

Wu, F., Zhao, Y., Wang, Y., Xu, Y., Tang, M., Wang, Z., Han, B., & Liu, Z. (2022). Hydrogen-bonding and acid cooperative catalysis for benzylation of arenes with benzyl alcohols over ionic liquids. *Green Chemistry*, *24*(8), 3137–3142. DOI: 10.1039/D1GC04485K

Wu, X., Zhu, Q., Chen, Z., Wu, W., Lu, Y., & Qi, J. (2021). Ionic liquids as a useful tool for tailoring active pharmaceutical ingredients. *Journal of Controlled Release*, *338*, 268–283. DOI: 10.1016/j.jconrel.2021.08.032 PMID: 34425167

Zang, H., Yao, S., Luo, Y., Tang, D., & Song, H. (2018). Complex-precipitation using functionalized chiral ionic liquids with L-proline anion and chromatographic analysis for enantio separation of racemic amino acids. *Chirality*, *30*(10), 1182–1194. DOI: 10.1002/chir.23011 PMID: 30188593

Zhang, C., Malhotra, S. V., & Francis, A. J. (2014). Toxicity of ionic liquids to Clostridium sp. and effects on uranium biosorption. *Journal of Hazardous Materials*, *264*, 246–253. DOI: 10.1016/j.jhazmat.2013.11.003 PMID: 24316798

Zheng, Z., Xu, Q., Guo, J., Qin, J., Mao, H., Wang, B., & Yan, F. (2016). Structure–antibacterial activity relationships of imidazolium-type ionic liquid monomers, poly(ionic liquids) and poly(ionic liquid) membranes: Effect of alkyl chain length and cations. *ACS Applied Materials & Interfaces*, *8*(19), 12684–12692. DOI: 10.1021/acsami.6b03391 PMID: 27145107

Chapter 9
Nanomaterials for Wastewater Treatment

Himadri Sekhar Das
https://orcid.org/0000-0002-3509-3388
Haldia Institute of Technology, India

Gourisankar Roymahapatra
https://orcid.org/0000-0002-8018-5206
Haldia Institute of Technology,, India

Sudipta Banerjee
https://orcid.org/0000-0003-0150-6794
Symbiosis Institute of Technology, Symbiosis International University, India

Heranmoy Maity
Pailan College of Management and Technology, India

ABSTRACT

Nano-materials have emerged as a promising solution for wastewater treatment, offering enhanced efficiency in pollutant removal through advanced physicochemical mechanisms. Their high surface area, reactivity, and tunable properties enable superior removal of contaminants, including heavy metals, organic pollutants, pharmaceuticals, and microbial pathogens. Key mechanisms such as adsorption, photocatalysis, membrane filtration, electrochemical treatment, and antimicrobial action provide innovative pathways for wastewater purification. This chapter explores the recent developments, applications, and future prospects of nanomaterials in wastewater treatment, highlighting their transformative potential in addressing global water challenges.

DOI: 10.4018/979-8-3373-1245-3.ch009

© 2026, IGI Global Scientific Publishing. Copying or distributing in print or electronic forms without written permission of IGI Global Scientific Publishing is prohibited. Use of this chapter to [gene]rative artificial intelligence (AI) technologies is expressly prohibited. The publisher reserves all rights to license its use for generative AI training and machine learning model development.

INTRODUCTION

Water pollution caused by industrialization, urbanization, and agricultural runoff poses significant environmental and health challenges. Traditional water treatment methods, including chemical coagulation, biological treatment, and membrane filtration, have limitations in efficiency and sustainability. Nanotechnology offers innovative solutions for addressing these challenges through the use of nanomaterials with enhanced adsorption, catalytic, and antimicrobial properties.

Historical Background: Evolution of Nanomaterials in Wastewater Treatment

The use of nanomaterials in wastewater treatment has evolved significantly over the past few decades, driven by advancements in nanotechnology, environmental concerns, and the need for more efficient water purification methods, (Zhang et al., 2016). The application of nanoscale materials for contaminant removal can be traced back to early studies in colloidal science, but rapid developments in material science and engineering have propelled nanomaterials into the forefront of wastewater treatment technologies.

Early Developments and Theoretical Foundations

The foundation of nanotechnology dates back to the 1959 lecture by physicist Richard Feynman, *"There's Plenty of Room at the Bottom,"* which introduced the concept of manipulating materials at the atomic and molecular scale (Pisano, R., Durlo, A. (2023). However, it was not until the 1980s and 1990s that scientists began to explore nanomaterials for environmental applications. During this period, researchers discovered that materials at the nanoscale exhibit unique physicochemical properties, including high surface area, increased reactivity, and enhanced adsorption capacity.

One of the earliest applications of nanomaterials in water treatment involved activated carbon nanoparticles, which were used to enhance adsorption processes. Researchers also explored the potential of zeolites, clays, and metal hydroxides as nanoscale adsorbents for removing heavy metals from water. The high affinity of these materials for pollutants demonstrated their potential for improving conventional treatment methods (Ayesha Naseer, (2024). Figure 1 shows the different way of nanomaterials in wastewater treatment.

Figure 1. Nanomaterials in wastewater treatment

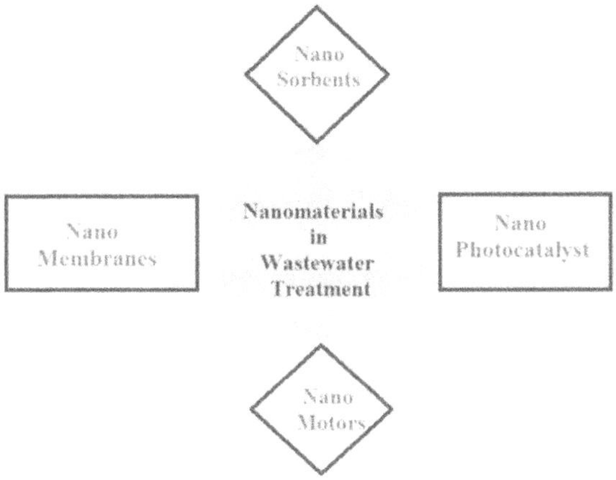

Emergence of Engineered Nanomaterials (2000s)

The early 2000s witnessed a significant shift with the development of engineered nanomaterials **(ENMs)** specifically designed for water purification. Breakthroughs in carbon-based nanomaterials, including carbon nanotubes (CNTs) and graphene oxide (GO), offered superior adsorption capacities and mechanical stability, making them ideal candidates for removing organic pollutants, heavy metals, and pathogens (Elhenawy et al., 2024).

During this period, metal and metal oxide nanoparticles, such **as** titanium dioxide (TiO_2), zinc oxide (ZnO), and silver (Ag) nanoparticles, gained attention for their photocatalytic and antimicrobial properties. TiO_2-based photocatalysis became a promising technique for degrading organic contaminants under UV light, while silver nanoparticles were extensively studied for their antimicrobial activity against bacteria and viruses.

Additionally, researchers developed nanocomposite membranes, incorporating nanoparticles into polymeric matrices to enhance filtration performance, permeability, and fouling resistance. These membranes exhibited improved selectivity, enabling the removal of emerging contaminants such as pharmaceuticals, pesticides, and microplastics.

Advancements in the 2010s: Towards Sustainable and Smart Nanomaterials

The 2010s saw an increasing focus on the sustainability and environmental safety of nanomaterials. Concerns over the toxicity and long-term stability of nanoparticles led to the development of biodegradable, eco-friendly nanomaterials, such as cellulose nanocrystals, biochar-based nanomaterials, and functionalized biopolymers, (Saraswat, 2023).

Simultaneously, the integration of nanotechnology with smart sensing systems enabled real-time monitoring of water quality. Advances in nano-enabled sensors allowed for the rapid detection of heavy metals, toxins, and microbial contaminants, improving wastewater management strategies.

CURRENT TRENDS AND FUTURE DIRECTIONS

Recent research focuses on hybrid nanomaterials combining multiple functions, such as adsorption, photocatalysis, and antimicrobial action, into a single system. Additionally, efforts are being made to scale up production while ensuring cost-effectiveness and regulatory compliance. The integration of artificial intelligence (AI) and nanotechnology in wastewater treatment is also an emerging area, optimizing pollutant detection and removal processes, (Godja, 2024).

As the field progresses, interdisciplinary collaboration between material scientists, environmental engineers, and policymakers will be essential in translating nanotechnology advancements into real-world wastewater treatment solutions.

RECENT ADVANCES IN NANOTECHNOLOGY FOR WASTEWATER TREATMENT

The past decade has witnessed significant advancements in nanotechnology for wastewater treatment, with breakthroughs in novel nanomaterials, hybrid systems, and scalable applications. Recent research has focused on improving the efficiency, selectivity, sustainability, and real-world applicability of nanomaterials. This section highlights key developments in nanotechnology applications, including advanced adsorbents, photocatalytic systems, membrane technologies, and smart nanomaterials (Satyam, 2024).

Advanced Nanomaterial-Based Adsorbents

Adsorption remains one of the most efficient wastewater treatment methods, and recent studies have focused on developing high-capacity and selective nanoadsorbents.

- **Metal-Organic Frameworks (MOFs):** These porous materials exhibit extremely high surface areas and tunable functional groups, making them highly effective in adsorbing heavy metals (e.g., Pb^{2+}, Cr^{6+}) and organic contaminants. Recent studies have introduced **bio-derived MOFs**, reducing environmental concerns associated with synthetic MOFs (Hanshuo Zhang, (2022).
- **Functionalized Carbon-Based Nanomaterials: Graphene oxide (GO), carbon nanotubes (CNTs), and biochar-derived nanocarbons** have been modified with functional groups (e.g., amine, thiol, carboxyl) to enhance selectivity toward pollutants such as PFAS, dyes, and pharmaceuticals (A. Kumar et al, (2022).
- **MXenes:** A new class of **2D transition metal carbides and nitrides**, MXenes have demonstrated exceptional adsorption capacity for heavy metals, anionic contaminants, and organic pollutants due to their hydrophilic surface and tunable chemistry.

Breakthroughs in Photocatalytic Nanomaterials

Photocatalysis has gained momentum due to its ability to **degrade persistent organic pollutants and microbial pathogens** using light energy. Recent breakthroughs include:

- **Visible-Light-Activated Photocatalysts:** Traditional TiO_2 photocatalysts require UV light, but **doped and composite-based materials** (e.g., Ag-TiO_2, ZnO-GO hybrids, black phosphorus nanosheets) now enable efficient pollutant degradation under visible light, making them more practical for real-world applications (N, M, Chauke et al, (2025).
- **Z-Scheme Photocatalysts:** Recent studies have demonstrated **Z-scheme heterojunctions** (e.g., **g-C_3N_4/TiO_2, $BiVO_4$/ZnO**) that mimic photosynthesis, significantly enhancing charge separation and photocatalytic efficiency (Praveen Kumar Basivi et al, (2025).
- **Biodegradable Photocatalysts:** Research has focused on eco-friendly photocatalysts, such as **biochar-supported TiO_2 and cellulose-TiO_2 hybrids,** which offer high photocatalytic activity with minimal environmental impact (Hanane Chakhtouna et al, (2023).

Nano-Enabled Membrane Technologies

Membrane-based separation technologies have been improved using nanomaterials, enhancing their **selectivity, fouling resistance, and permeability**.

- **Nanocomposite Membranes:** Incorporating carbon nanomaterials (GO, CNTs), metal nanoparticles (Ag, Cu), and MOFs has led to membranes with improved rejection rates for microplastics, heavy metals, and pharmaceuticals (Bo Han et al., 2024).
- **Electrospun Nanofiber Membranes:** The use of electrospun nanofibers (e.g., polymeric nanofibers embedded with TiO_2, ZnO, or Ag NPs) has resulted in membranes with high surface area, tunable pore size, and antimicrobial properties (Maliszewska et al., 2022).
- **Smart Self-Cleaning Membranes:** Advanced membranes now integrate photocatalytic and electrochemical functionalities to prevent biofouling, enhancing long-term stability in wastewater treatment plants.

Electrochemical and Nanozyme-Based Technologies

Recent advancements have explored nanomaterial-enhanced electrochemical treatment and nanozyme-based catalysis.

- **Nanozyme Catalysis:** Engineered nanomaterials with enzyme-mimicking properties, such as Fe_3O_4, CeO_2, and MOF-based nanozymes, have been developed for highly efficient oxidation-reduction reactions, breaking down complex pollutants.
- **Graphene-Based Electrodes:** The use of graphene-modified electrodes in electrochemical oxidation has demonstrated enhanced efficiency in degrading emerging contaminants like antibiotics, pesticides, and endocrine disruptors.
- **Hybrid Electro-Fenton Processes:** Integrating Fe-based nanocatalysts with electrochemical Fenton reactions has significantly improved the removal of recalcitrant organic pollutants and heavy metals (Wong et al., 2021).

Smart and AI-Integrated Nanomaterials

The latest frontier in nanotechnology for wastewater treatment involves the integration of artificial intelligence (AI), sensor technology, and smart nanomaterials.

- **AI-Driven Optimization:** Machine learning models are being employed to optimize nanomaterial synthesis, predict adsorption capacity, and enhance real-time monitoring of contaminant removal.
- **Nano-Enabled Sensors:** Graphene-based and plasmonic nanosensors are being developed for real-time detection of contaminants such as heavy metals, PFAS, and microbial pathogens in wastewater streams.
- **Stimuli-Responsive Nanomaterials:** Research has explored pH-responsive, light-activated, and magnetically separable nanomaterials that allow for controlled pollutant capture and release (C. L. Arellano Vidal et al, (2024).

Recent advancements in nanotechnology have significantly improved wastewater treatment efficiency, with breakthroughs in adsorption, photocatalysis, membrane filtration, electrochemical degradation, and smart sensing technologies. The development of biodegradable and eco-friendly nanomaterials is addressing environmental concerns, while AI-driven optimization and real-time monitoring are enhancing practical applications. As research continues, the focus is on scaling up these innovations to ensure cost-effectiveness, regulatory compliance, and large-scale applicability for global wastewater management.

COMPARATIVE ANALYSIS OF NANOTECHNOLOGY FOR WASTEWATER TREATMENT

Nanotechnology offers multiple approaches for wastewater treatment, each with unique advantages and limitations. This comparative analysis evaluates five major nanotechnology-based treatment methods—adsorption, Photocatalysis, membrane filtration, electrochemical treatment, and antimicrobial action—based on key parameters such as efficiency, scalability, cost-effectiveness, environmental impact, and application scope (Nishu, 2023). Table 1 shows the performance comparison between different purification process.

Table 1. Key performance comparison

Treatment Method	Efficiency	Cost	Scalability	Environmental Impact	Pollutant Types	Limitations
Adsorption	High (removes heavy metals, dyes, pharmaceuticals)	Moderate	High (easily scalable)	Low (some adsorbents are reusable)	Heavy metals, dyes, organic contaminants	Adsorbents require regeneration or disposal
Photocatalysis	Very High (degrades organic pollutants and microbes)	Moderate to High	Moderate (requires catalyst recovery)	Low to Moderate (dependent on catalyst materials)	Organic pollutants, pharmaceuticals, pathogens	Requires UV or visible light, catalyst deactivation over time
Membrane Filtration	High (removes microplastics, heavy metals, and microbes)	High	Moderate to High	Moderate (membrane fouling is a concern)	Microplastics, bacteria, heavy metals, organics	Fouling reduces efficiency, membrane replacement needed
Electrochemical Treatment	Very High (removes emerging contaminants and recalcitrant pollutants)	High	Moderate	Moderate (electrode materials may degrade)	Persistent organic pollutants, pharmaceuticals, heavy metals	Energy-intensive, electrode corrosion
Antimicrobial Action	High (kills bacteria, viruses, fungi)	Moderate	High	Moderate (metal-based nanomaterials may be toxic)	Microbial contaminants	Potential development of microbial resistance, toxicity concerns

Strengths and Weaknesses Analysis

Adsorption-Based Treatment

Strengths:
Simple and cost-effective
Efficient for heavy metal removal
Scalable and applicable in diverse wastewater settings
Can be regenerated for multiple uses
Weaknesses:
Adsorbent materials require disposal or regeneration
Not effective for complete organic pollutant degradation
Selectivity depends on surface modifications

Photocatalytic Treatment

Strengths:
Highly effective in degrading organic pollutants and pathogens
Produces fewer secondary pollutants
Can operate under solar light (if optimized)
Weaknesses:
Requires UV or visible light for activation
Photocatalyst deactivation over time
Catalyst recovery and reusability remain challenges

Membrane-Based Filtration

Strengths:
Highly efficient for removing particulates, bacteria, and microplastics
Provides clean water with minimal chemical addition
Can be integrated into existing treatment plants
Weaknesses:
High initial cost and energy consumption
Prone to biofouling, reducing efficiency over time
Requires periodic maintenance and replacement

Electrochemical Treatment

Strengths:
Effective for degrading persistent pollutants and emerging contaminants
Can be combined with advanced oxidation processes (e.g., Electro-Fenton)
No need for chemical additives
Weaknesses:
High energy consumption
Electrode stability and durability issues
Not always suitable for large-scale applications

Antimicrobial Nanomaterials

Strengths:
Effective in killing a wide range of pathogens
Can be integrated into membranes and coatings
Long-lasting disinfection effect
Weaknesses:

Concerns about environmental toxicity of metal-based nanomaterials (e.g., Ag, Cu)
Possible microbial resistance over time
Regulatory concerns for safe deployment

Selection Criteria for Wastewater Treatment Technologies

The choice of a nanotechnology-based wastewater treatment method depends on **application needs, pollutant type, regulatory compliance, and cost considerations**.

Table 2.

Criteria	Best Performing Method
Best for Heavy Metal Removal	Adsorption (e.g., MOFs, graphene oxide)
Best for Organic Pollutant Degradation	Photocatalysis (e.g., TiO_2, ZnO-based systems)
Best for Microbial Disinfection	Antimicrobial Nanoparticles (e.g., Ag, Cu, ZnO)
Best for Energy Efficiency	Adsorption (low energy requirement)
Best for Large-Scale Application	Membrane Filtration (e.g., nanocomposite membranes)
Best for Emerging Contaminants	Electrochemical Treatment (e.g., graphene-based electrodes)

Each nanotechnology-based wastewater treatment method has distinct advantages and challenges. While adsorption and membrane filtration offer scalability and efficiency, photocatalysis and electrochemical methods provide high-performance degradation of emerging contaminants. Hybrid approaches, combining adsorption with photocatalysis **or** membrane filtration with antimicrobial coatings, are emerging as the most promising solutions for enhanced wastewater treatment.

Future research should focus on:

- Developing hybrid nanomaterial systems that integrate multiple treatment mechanisms.
- Enhancing the sustainability of nanomaterials by designing biodegradable and non-toxic options.
- Improving cost-effectiveness by optimizing scalable synthesis methods.
- Addressing regulatory challenges to ensure safe and responsible deployment in real-world applications.

TYPES OF NANOMATERIALS FOR WASTEWATER TREATMENT

Metal and Metal Oxide Nanoparticles

The use of metal and metal oxide nanoparticles in wastewater treatment has gained significant attention due to their unique physicochemical properties, including high surface area, catalytic activity, and strong adsorption capabilities. These nanoparticles play a crucial role in removing organic pollutants, pathogens, and heavy metals from contaminated water. Below are some of the most widely researched and applied nanoparticles in wastewater treatment:

Titanium Dioxide (TiO_2)

- **Role in Wastewater Treatment:** TiO_2 nanoparticles are widely used as photocatalysts for the degradation of organic pollutants, including dyes, pharmaceuticals, and pesticides.
- **Mechanism:** Under UV or visible light irradiation, TiO_2 generates reactive oxygen species (ROS), such as hydroxyl radicals (•OH), which break down organic contaminants into harmless byproducts like CO_2 and H_2O.
- **Advantages:** High stability, non-toxicity, and excellent photocatalytic efficiency.
- **Challenges:** Limited activity under visible light; requires modification (e.g., doping with metals or nonmetals) for enhanced efficiency (A.M. Gatou, (2024).

Zinc Oxide (ZnO)

- Role in Wastewater Treatment: ZnO is another semiconductor material with strong photocatalytic and antimicrobial properties.
- Mechanism: Similar to TiO_2, ZnO generates ROS under light exposure, aiding in the breakdown of organic pollutants and deactivating bacteria and viruses.
- Advantages: Broad-spectrum antimicrobial activity, high chemical stability, and lower toxicity compared to some other metal oxides.
- Challenges: Susceptibility to photocorrosion, which can reduce long-term efficiency (Aqsa Nawaz,(2024).

Silver (Ag) Nanoparticles

- **Role in Wastewater Treatment:** Silver nanoparticles are highly effective in microbial disinfection due to their antimicrobial properties.
- **Mechanism:** Ag nanoparticles release Ag^+ ions, which interact with bacterial cell membranes, disrupting cellular processes and leading to cell death. They can also generate ROS, further enhancing microbial inactivation.
- **Advantages:** Effective against a wide range of microorganisms, including bacteria, viruses, and fungi.
- **Challenges:** Potential toxicity to aquatic life and humans; requires careful concentration control in wastewater applications (T. Frippiat, (2025).

Iron Oxide (Fe_2O_3) Nanoparticles

- **Role in Wastewater Treatment:** Iron oxide nanoparticles are used primarily for magnetic separation, heavy metal adsorption, and catalytic degradation of contaminants.
- **Mechanism:** Fe_3O_4 nanoparticles possess high surface reactivity, allowing them to adsorb heavy metal ions (e.g., Pb^{2+}, Cd^{2+}, As^{3+}). Their magnetic properties enable easy separation from treated water using external magnetic fields.
- **Advantages:** Recyclability, cost-effectiveness, and high adsorption capacity.
- **Challenges:** Potential aggregation of nanoparticles, which can reduce effectiveness; requires surface modification for enhanced stability (Fatemeh Bahmanzadegan, (2025).

Carbon-Based Nanomaterials

Carbon-Based Nanomaterials for Wastewater Treatment

Carbon-based nanomaterials have emerged as highly effective materials in wastewater treatment due to their large surface area, tunable surface chemistry, and remarkable adsorption and catalytic properties. These materials offer promising solutions for removing organic pollutants, heavy metals, and pathogens from water. The key carbon-based nanomaterials used in wastewater treatment include graphene, carbon nanotubes (CNTs), and fullerenes.

Graphene and Graphene Oxide (GO)

- **Role in Wastewater Treatment:** Gaphene and its oxidized derivative, graphene oxide (GO), are widely studied for their exceptional adsorption capacity and ability to remove organic pollutants, dyes, and heavy metals (e.g., Pb^{2+}, Hg^{2+}, As^{3+}).
- **Mechanism:**
    - o The high surface area and π–π interactions facilitate adsorption of organic pollutants.
    - o Functional groups (e.g., hydroxyl, epoxy, and carboxyl) in GO enhance its affinity for metal ions and contaminants.
- **Advantages:** High adsorption efficiency, mechanical strength, chemical stability, and reusability.
- **Challenges:** Prone to aggregation, which reduces efficiency; requires surface modification or nanocomposite integration for enhanced performance (A. Khalajiolyaie, (2025).

Carbon Nanotubes (CNTs)

- **Role in Wastewater Treatment:** CNTs exhibit high adsorption capabilities for organic pollutants, pharmaceuticals, and heavy metals due to their large surface area, hollow structure, and functionalizability.
- **Mechanism:**
    - o Hydrophobic interactions and electrostatic attraction enable the adsorption of organic and inorganic contaminants.
    - o Functionalized CNTs enhance the removal of heavy metals and organic pollutants via surface modification.
    - o CNTs can also act as catalysts in advanced oxidation processes (AOPs), breaking down complex contaminants.
- **Advantages:** High adsorption efficiency, strong mechanical stability, and potential for multifunctional applications (adsorption + catalysis).
- **Challenges:** High production costs, toxicity concerns, and difficulty in large-scale deployment (S. Rezania, (2024).

Fullerenes

- **Role in Wastewater Treatment:** Fullerenes (C_{60} and its derivatives) are increasingly explored in advanced oxidation processes (AOPs) for degrading organic pollutants.
- **Mechanism:**

- Fullerenes generate reactive oxygen species (ROS), such as singlet oxygen ($^1O_2$) and superoxide anions ($O_2^{-}\bullet$), which facilitate the breakdown of organic contaminants.
- Their unique electron transfer properties contribute to catalytic degradation of pollutants.

- **Advantages:** High oxidative potential, stability, and efficient degradation of persistent organic pollutants (POPs).
- **Challenges:** Limited water solubility, potential environmental toxicity, and high production costs.

Carbon-based nanomaterials, including graphene, CNTs, and fullerenes, offer excellent capabilities in wastewater treatment through adsorption, catalysis, and oxidation mechanisms. However, challenges such as cost, toxicity, and scalability must be addressed to enable their widespread application in real-world water treatment systems. Future research should focus on developing eco-friendly, cost-effective, and functionalized carbon nanomaterials for sustainable wastewater remediation.

Polymeric Nanomaterials

Polymeric Nanomaterials for Wastewater Treatment

Polymeric nanomaterials have gained significant attention in wastewater treatment due to their high selectivity, tunable surface properties, and reusability. These nanomaterials, including **molecularly imprinted polymers (MIPs)** and **dendrimers**, offer promising solutions for the selective removal of heavy metals, organic pollutants, and emerging contaminants (A. A. Yaqoob, (2020).

Molecularly Imprinted Polymers (MIPs)

- **Role in Wastewater Treatment:** MIPs are synthetic polymeric materials engineered with molecular recognition sites that selectively bind to target pollutants, such as pharmaceuticals, pesticides, and heavy metals.
- **Mechanism:**
  - MIPs are synthesized by polymerizing functional monomers around a template molecule (e.g., a pollutant or metal ion).
  - After polymerization, the template is removed, leaving behind selective binding cavities that recognize and adsorb specific contaminants.
- **Advantages:**
  - High selectivity for specific pollutants.
  - Reusable and stable under harsh environmental conditions.

- Effective in removing pharmaceuticals, pesticides, and endocrine-disrupting chemicals.
- **Challenges:**
  - Synthesis complexity and cost.
  - Limited adsorption capacity compared to some inorganic nanomaterials.
  - Regeneration and recycling require optimization.

Dendrimers

- **Role in Wastewater Treatment:** Dendrimers are highly branched, nanoscale polymers with a well-defined structure, offering high surface functionality and excellent binding capacity for pollutants.
- **Mechanism:**
  - The numerous functional groups on dendrimer surfaces enable strong electrostatic and chemical interactions with heavy metals (e.g., Pb^{2+}, Hg^{2+}, Cd^{2+}) and organic contaminants.
  - Dendrimers can also be functionalized with chelating agents to enhance their affinity for specific pollutants.
- **Advantages:**
  - High adsorption capacity due to their branched architecture.
  - Tunable surface chemistry for selective pollutant removal.
  - Potential for multifunctional applications, including catalysis and filtration.
- **Challenges:**
  - High synthesis cost and complex fabrication processes.
  - Potential toxicity concerns if not properly designed for environmental applications.
  - Requires further studies for large-scale implementation in wastewater treatment plants.

Polymeric nanomaterials, including MIPs and dendrimers, offer unique advantages in wastewater treatment due to their high selectivity, adsorption efficiency, and potential for reuse. However, challenges related to cost, large-scale synthesis, and recyclability need to be addressed. Future research should focus on developing biodegradable, cost-effective, and environmentally safe polymeric nanomaterials to improve their sustainability and commercial viability.

Nanocomposites

- **Hybrid Materials:** Combining metal oxides with carbon-based nanomaterials for enhanced performance.
- **Polymer-Based Nanocomposites:** Used in membrane filtration and adsorption systems.

MECHANISMS OF POLLUTANT REMOVAL USING NANOMATERIALS

Adsorption

Nanomaterials, due to their high surface area and functionalized surfaces, effectively capture pollutants through electrostatic interactions, van der Waals forces, and chemical bonding. This mechanism is widely used for removing heavy metals, dyes, and organic contaminants. Nanomaterials, owing to their high surface area, tunable porosity, and functionalized surfaces, exhibit exceptional adsorption capabilities. They effectively capture pollutants through electrostatic interactions, van der Waals forces, hydrogen bonding, and chemical complexation. This mechanism is widely applied in the removal of heavy metals (e.g., Pb^{2+}, Cd^{2+}, Hg^{2+}), organic dyes, pharmaceuticals, and other persistent organic pollutants. Advanced nanomaterials, such as carbon-based nanostructures (graphene oxide, carbon nanotubes), metal-organic frameworks (MOFs), and functionalized nanoparticles, further enhance adsorption efficiency, offering promising solutions for wastewater remediation (Abhinay Thakur, (2024).

Photocatalysis

Certain nanomaterials, such as titanium dioxide (TiO_2) and zinc oxide (ZnO), utilize light energy to generate reactive oxygen species (ROS), which break down organic pollutants, pharmaceuticals, and pathogens into non-toxic byproducts. Certain nanomaterials, such as titanium dioxide (TiO_2), zinc oxide (ZnO), and graphene-based composites, harness light energy to generate reactive oxygen species (ROS), including hydroxyl radicals (•OH) and superoxide anions (O_2^-•). These highly reactive species degrade organic pollutants, pharmaceuticals, pesticides, and microbial contaminants into non-toxic byproducts, making photocatalysis an efficient and eco-friendly wastewater treatment method. Recent advancements focus on bandgap engineering, doping with noble metals (e.g., Ag, Au), and coupling with carbon nanomaterials (e.g., graphene, carbon quantum dots) to enhance visible-light absorption and improve photocatalytic efficiency under solar irradiation. The integration of photocatalysts into membranes and hybrid treatment systems further enhances their practical applicability in large-scale wastewater remediation (Puri, 2023).

Membrane Filtration

Nanomaterial-integrated membranes enhance the selective removal of contaminants, providing superior permeability, fouling resistance, and improved separation of heavy metals, microplastics, and organic pollutants. Nanomaterial-integrated membranes offer enhanced selectivity, permeability, and fouling resistance, making them highly effective for wastewater treatment. By incorporating nanoparticles, carbon-based nanomaterials (e.g., graphene oxide, carbon nanotubes), and metal-organic frameworks (MOFs) into traditional membranes, these systems achieve improved mechanical strength, chemical stability, and pollutant rejection efficiency (R. Kamalesh, (2024).

Key advantages include:

- Selective removal of heavy metals (e.g., Pb^{2+}, Hg^{2+}), microplastics, and organic contaminants
- Antifouling properties, reducing biofilm formation and extending membrane lifespan
- Higher water flux and permeability, improving energy efficiency in filtration processes.

Emerging developments focus on smart membranes with stimuli-responsive properties and self-cleaning capabilities, further optimizing their long-term performance in wastewater treatment applications.

Electrochemical Treatment

Nanomaterial-coated electrodes facilitate advanced oxidation and reduction processes, improving the degradation of persistent organic pollutants, heavy metals, and emerging contaminants through electrochemical reactions. Nanomaterial-coated electrodes play a crucial role in wastewater treatment by enhancing oxidation and reduction processes for the degradation of persistent organic pollutants, heavy metals, and emerging contaminants. The integration of nanomaterials such as graphene, carbon nanotubes (CNTs), metal oxides (e.g., TiO_2, MnO_2), and noble metal nanoparticles (e.g., Pt, Ag) significantly improves the electrocatalytic activity, conductivity, and stability of electrodes (Long, 2025).

Key mechanisms include:

- **Electrochemical oxidation** – Generates hydroxyl radicals (•OH) and other reactive species to degrade complex organic pollutants.
- **Electrocoagulation** – Uses nanostructured electrodes to destabilize and remove suspended contaminants, including heavy metals.
- **Electro-Fenton processes** – Enhances pollutant breakdown via in-situ production of hydrogen peroxide (H_2O_2) and hydroxyl radicals.

Future advancements focus on scalable electrode designs, energy-efficient electrocatalysts, and hybrid electrochemical-membrane systems to enhance real-world applications.

Antimicrobial Action

Nanomaterials such as silver (Ag), copper (Cu), and carbon-based nanostructures (e.g., graphene oxide) exhibit strong antimicrobial properties by generating reactive oxygen species or directly interacting with microbial cell membranes, leading to effective disinfection. Nanomaterials exhibit potent antimicrobial properties, making them effective in disinfecting wastewater by targeting bacteria, viruses, and fungi. Silver (Ag), copper (Cu), zinc oxide (ZnO), and carbon-based nanostructures (e.g., graphene oxide, carbon nanotubes) are among the most widely studied antimicrobial nanomaterials (L Wang, (2017).

Mechanisms of Antimicrobial Action

I). **Generation of Reactive Oxygen Species (ROS):** Nanomaterials induce oxidative stress in microbial cells by producing hydroxyl radicals (•OH) and superoxide anions (O_2^-•), leading to cellular damage and eventual cell death.

II). **Direct Interaction with Microbial Membranes:** Nanoparticles can physically disrupt bacterial cell walls, causing leakage of intracellular contents and cell lysis.

III). **Metal Ion Release:** Silver and copper nanoparticles release ions (Ag^+, Cu^{2+}) that interfere with microbial DNA replication and protein function, leading to antimicrobial effects (A Abdal Dayem, (2017).

IV). **Photocatalytic Disinfection:** Certain nanomaterials, such as TiO_2 and ZnO, leverage light energy to enhance their antimicrobial activity through ROS generation.

Future developments focus on biodegradable antimicrobial nanomaterials, surface-functionalized nanoparticles for targeted action, and hybrid nanomaterial systems to improve disinfection efficiency while minimizing environmental toxicity.

Nanoparticles remove pollutants through various mechanisms such as oxidation, adsorption, precipitation, coagulation, filtration, and ion exchange. Oxidation, especially with metal oxide nanoparticles, degrades organic contaminants into less harmful substances. Adsorption allows pollutants to adhere to the nanoparticle surface due to their large surface area and reactive sites. Precipitation transforms dissolved pollutants into solid forms for removal. Coagulation helps in aggregating fine particles, enhancing sedimentation and separation. Filtration involves the physical trapping of pollutants within nanoporous structures. Ion exchange enables the replacement of toxic ions with non-toxic ones on the nanoparticle surface. These synergistic mechanisms ensure efficient and sustainable pollutant removal (Figure 2).

Figure 2. Mechanism of pollutants removal with nanoparticles

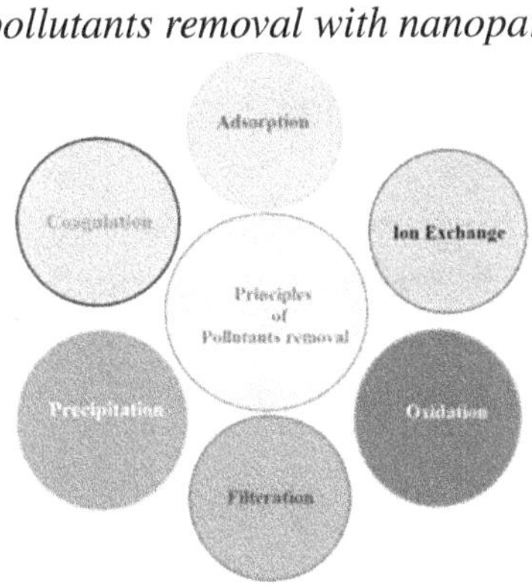

ADVANTAGES AND CHALLENGES OF NANOMATERIALS IN WASTEWATER TREATMENT

Nanomaterials have demonstrated significant potential in wastewater treatment due to their **high efficiency, selectivity, and adaptability** to various treatment processes. However, challenges such as **cost, environmental impact, and scalability** must be addressed for widespread adoption. This section outlines the key advantages and challenges associated with the use of nanomaterials in wastewater treatment.

Advantages

High Efficiency in Pollutant Removal

Nanomaterials exhibit exceptional efficiency in removing a wide range of pollutants, including heavy metals, organic pollutants, pharmaceuticals, and microbial contaminants. Their high surface area, catalytic activity, and adsorption capacity enable rapid and effective water purification.

Selective Adsorption and Degradation

Many nanomaterials, such as molecularly imprinted polymers (MIPs) and functionalized nanoparticles, provide high selectivity for specific contaminants. This allows for targeted removal of hazardous substances without affecting beneficial water components.

Potential for Reuse and Regeneration

Certain nanomaterials, such as iron oxide nanoparticles (Fe_3O_4) and polymer-based nanocomposites, can be regenerated and reused multiple times, reducing operational costs and waste generation. Magnetic separation and chemical regeneration techniques further enhance their reusability.

Compatibility with Existing Water Treatment Systems

Nanomaterials can be integrated into membrane filtration, adsorption, photocatalysis, and advanced oxidation processes (AOPs), making them compatible with conventional wastewater treatment systems. This facilitates their scalability and practical implementation in real-world applications.

Challenges

High Production Costs of Nanomaterials

The synthesis of nanomaterials often requires specialized equipment, high-purity precursors, and complex fabrication processes, leading to high costs. Cost-effective and scalable production methods, such as green synthesis and waste-derived nanomaterials, need further development.

Potential Toxicity and Environmental Impact

The long-term environmental and health impacts of nanomaterials remain a major concern. Some nanoparticles, particularly metal-based nanomaterials (e.g., Ag, ZnO, TiO_2), may leach into the environment, causing bioaccumulation and toxicity to aquatic organisms. More research is needed to assess their biodegradability and safe disposal methods.

Scalability and Long-Term Stability Issues

While nanomaterials perform well in laboratory settings, challenges arise in scaling up production and maintaining stability in real-world applications. Issues such as nanoparticle aggregation, membrane fouling, and reduced catalytic activity over time must be addressed through surface modifications and hybrid nanomaterial strategies.

Need for Regulatory Guidelines and Standardization

The lack of standardized protocols for nanomaterial synthesis, characterization, and environmental impact assessment hinders their widespread adoption. Regulatory bodies need to establish clear safety guidelines for nanomaterial production, usage, and disposal to ensure their safe and sustainable application in wastewater treatment.

Nanomaterials offer unparalleled advantages in wastewater treatment, including high efficiency, selectivity, and reusability. However, challenges such as high costs, environmental risks, and scalability must be addressed through sustainable synthesis methods, advanced material modifications, and regulatory frameworks. Future research should focus on eco-friendly and cost-effective nanomaterials to ensure their safe and practical implementation in large-scale wastewater treatment systems.

FUTURE PROSPECTS

The future of nanomaterials in wastewater treatment lies in improving cost-effectiveness, sustainability, and environmental safety. Research is focused on developing biodegradable and non-toxic nanomaterials, optimizing their synthesis, and integrating them with smart monitoring systems for real-time water quality assessment. The future of nanomaterials in wastewater treatment hinges on enhancing their cost-effectiveness, sustainability, and environmental safety. Ongoing research is dedicated to developing biodegradable, non-toxic, and recyclable nanomaterials to minimize environmental impact while maintaining high treatment efficiency.

Key advancements include the optimization of synthesis techniques to reduce production costs and improve scalability. Additionally, the integration of nanomaterials with smart monitoring systems and AI-driven analytics is gaining traction, enabling real-time water quality assessment and adaptive treatment processes.

Future innovations will also explore hybrid nanomaterial systems that combine multiple functionalities for enhanced contaminant removal. As regulatory frameworks evolve, efforts will focus on ensuring the safe deployment of nanomaterials in real-world applications. With continued progress, nanomaterials hold immense potential to revolutionize wastewater treatment, paving the way for sustainable and efficient water management solutions.

CONCLUSION

Nanomaterials offer revolutionary approaches to wastewater treatment, addressing critical challenges in pollutant removal. However, further advancements are necessary to overcome technical and regulatory hurdles. Continued interdisciplinary research will pave the way for scalable and eco-friendly solutions. Nanomaterials have emerged as a transformative approach in wastewater treatment, offering exceptional efficiency in removing contaminants, including heavy metals, organic pollutants, and pathogens. Their unique physicochemical properties-such as high surface area, reactivity, and tunability-enable advanced treatment solutions beyond conventional methods.

Despite their promise, challenges remain in terms of large-scale implementation, environmental safety, cost-effectiveness, and regulatory compliance. Addressing these concerns requires interdisciplinary collaboration among material scientists, engineers, environmentalists, and policymakers. Future research should focus on developing sustainable, recyclable, and non-toxic nanomaterials while optimizing their performance in real-world applications. With continued advancements, nanomaterials have the potential to revolutionize wastewater treatment, contributing significantly to global water sustainability and environmental protection.

REFERENCES

Abdal Dayem, A., Hossain, M. K., Lee, S. B., Kim, K., Saha, S. K., Yang, G.-M., Choi, H. Y., & Cho, S.-G. (2017). The role of reactive oxygen species (ROS) in the biological activities of metallic nanoparticles. *International Journal of Molecular Sciences*, *18*(1), 120. Advance online publication. DOI: 10.3390/ijms18010120 PMID: 28075405

Arellano Vidal, C. L., Sanahuja, A., & Ramos, V. (2024). Machine learning techniques for improving nanosensors in agroenvironmental applications. *Agronomy (Basel)*, *14*(2), 341. Advance online publication. DOI: 10.3390/agronomy14020341

Bahmanzadegan, F., & Aghaei, A. (2025). A comprehensive review on novel zeolite-based adsorbents for environmental pollutant. *Journal of Hazardous Materials Advances*, *17*, 100617. Advance online publication. DOI: 10.1016/j.hazadv.2025.100617

Basivi, P. K., Giri, A., & Kumar, V. (2025). Graphitic carbon nitride (g–C3N4)–based Z-scheme photocatalysts: Innovations for energy and environmental applications. *Materials Today Sustainability*, *29*, 101069. Advance online publication. DOI: 10.1016/j.mtsust.2024.101069

Chakhtouna, H., Benzeid, H., Zari, N., Qaiss, A. E. K., & Bouhfid, R. (2023). Photocatalytic and bactericidal behaviors of Ag/TiO2 doped biochar through ball–milling approach. *Journal of Photochemistry and Photobiology A Chemistry*, *444*, 114971. Advance online publication. DOI: 10.1016/j.jphotochem.2023.114971

Chauke, N. M., Mukwevho, E., & Mamba, B. B. (2025). Transformative advancements in visible-light-activated titanium dioxide for industrial wastewater remediation. *International Journal of Environmental Science and Technology*, *22*(9), 8521–8552. Advance online publication. DOI: 10.1007/s13762-025-06397-2

Elhenawy, S., Khraisheh, M., AlMomani, F., Al-Ghouti, M., & Hassan, M. K. (2024). Emerging nanomaterials for drinking water purification: A new era of water treatment technology. *Nanomaterials (Basel, Switzerland)*, *14*(21), 1707. Advance online publication. DOI: 10.3390/nano14211707 PMID: 39513787

Frippiat, T., Art, T., & Delguste, C. (2025). Silver nanoparticles as antimicrobial agents in veterinary medicine: Current applications and future perspectives. *Nanomaterials (Basel, Switzerland)*, *15*(3), 202. Advance online publication. DOI: 10.3390/nano15030202 PMID: 39940178

Gatou, A. M., Lagopati, N., & Pavlatou, E. A. (2024). Photocatalytic TiO2-based nanostructures as a promising material for diverse environmental applications: A review. *Reactions*, *5*(1), 135–194. DOI: 10.3390/reactions5010007

Godja, N.-C., & Munteanu, F.-D. (2024). Hybrid nanomaterials: A brief overview of versatile solutions for sensor technology in healthcare and environmental applications. *Biosensors (Basel)*, *14*(2), 67. Advance online publication. DOI: 10.3390/bios14020067 PMID: 38391986

Han, B., Zhang, Y., & Wang, L. (2024). Thin-film nanocomposite (TFN) membrane technologies for the removal of emerging contaminants from wastewater. *Journal of Cleaner Production*, *480*, 144043. Advance online publication. DOI: 10.1016/j.jclepro.2024.144043

Kamalesh, R., & Selvakumar, P. (2024). Emerging breakthroughs in membrane filtration techniques and their application in agricultural wastewater treatment: Reusability aspects. *Sustainable Chemistry for the Environment*, *8*, 100183. Advance online publication. DOI: 10.1016/j.scenv.2024.100183

Khalajiolyaie, A., & Ghaffari, M. (2025). Advances in graphene-based materials for metal ion sensing and wastewater treatment: A review. *Environments (Basel, Switzerland)*, *12*(2), 43. Advance online publication. DOI: 10.3390/environments12020043

Kumar, A., Barik, B., Jablonski, P. G., Sonkaria, S., & Khare, V. (2022). Functionalized and biomimicked carbon-based materials and their impact for improving surface coatings for protection and functionality: Insights and technological trends. *Coatings*, *12*(11), 1674. Advance online publication. DOI: 10.3390/coatings12111674

Long, F., Wang, L., & Zhang, X. (2025). Versatile electrode materials applied in the electrochemical advanced oxidation processes for wastewater treatment: A systematic review. *Separation and Purification Technology*, *354*(Part 1), 128725. Advance online publication. DOI: 10.1016/j.seppur.2024.128725

Maliszewska, I., Czapka, T., & Grząbka-Zasadzińska, A. (2022). Electrospun polymer nanofibers with antimicrobial activity. *Polymers*, *14*(9), 1661. Advance online publication. DOI: 10.3390/polym14091661 PMID: 35566830

Naseer, A., & Ahmad, F. (2024). Role of nanocomposites and nano adsorbents for heavy metals removal and dyes: An overview. *Desalination and Water Treatment*, *320*, 100662. Advance online publication. DOI: 10.1016/j.dwt.2024.100662

Nawaz, A., & Ahmad, M. (2024). Zinc oxide nanoparticles: Pathways to micropollutant adsorption, dye removal, and antibacterial actions—A study of mechanisms, challenges, and future prospects. *Journal of Molecular Structure*, *1312*(Part 1), 138545. Advance online publication. DOI: 10.1016/j.molstruc.2024.138545

Nishu, S. K., & Gaur, R. (2023). Smart and innovative nanotechnology applications for water purification. *Hybrid Advances, 3*, 100044. Advance online publication. DOI: 10.1016/j.hybadv.2023.100044

Pisano, R., & Durlo, A. (2023). Feynman's frameworks on nanotechnology in historiographical debate. In Condé, M. L., & Salomon, M. (Eds.), *Handbook for the historiography of science* (pp. 465–484). Springer., DOI: 10.1007/978-3-031-27510-4_26

Puri, N., & Gupta, A. (2023). Water remediation using titanium and zinc oxide nanomaterials through disinfection and photocatalysis process: A review. *Environmental Research, 227*, 115786. Advance online publication. DOI: 10.1016/j.envres.2023.115786 PMID: 37004858

Rezania, S., & Kordkandi, S. A. (2024). Recent advances in the adsorption of different pollutants from wastewater using carbon-based and metal-oxide nanoparticles. *Applied Sciences (Basel, Switzerland), 14*(24), 11492. Advance online publication. DOI: 10.3390/app142411492

Saraswat, P., Singh, R. B., Prasad, M., Misra, R., Rajput, V. D., & Ranjan, R. (2023). Applications of bio-based nanomaterials in environment and agriculture: A review on recent progresses. *Hybrid Advances, 4*, 100097. Advance online publication. DOI: 10.1016/j.hybadv.2023.100097

Satyam, S., & Kumar, V. (2024). Innovations and challenges in adsorption-based wastewater remediation: A comprehensive review. *Heliyon, 10*(9), e29573. Advance online publication. DOI: 10.1016/j.heliyon.2024.e29573 PMID: 38699034

Thakur, A., Kumar, A., & Singh, A. (2024). Adsorptive removal of heavy metals, dyes, and pharmaceuticals: Carbon-based nanomaterials in focus. *Carbon, 217*, 118621. Advance online publication. DOI: 10.1016/j.carbon.2023.118621

Wang, L., & Zhang, X. (2017). The antimicrobial activity of nanoparticles: Present situation and prospects for the future. *International Journal of Nanomedicine, 12*, 1227–1249. DOI: 10.2147/IJN.S121956 PMID: 28243086

Wong, E. L. S., Vuong, K. Q., & Chow, E. (2021). Nanozymes for environmental pollutant monitoring and remediation. *Sensors (Basel), 21*(2), 408. Advance online publication. DOI: 10.3390/s21020408 PMID: 33430087

Yaqoob, A. A., Parveen, T., Umar, K., & Mohamad Ibrahim, M. N. (2020). Role of nanomaterials in the treatment of wastewater: A review. *Water (Basel), 12*(2), 495. Advance online publication. DOI: 10.3390/w12020495

Zhang, H., & Zhang, X. (2022). MIL series of metal organic frameworks (MOFs) as novel adsorbents for heavy metals in water: A review. *Journal of Hazardous Materials*, *429*, 128271. Advance online publication. DOI: 10.1016/j.jhazmat.2022.128271 PMID: 35093745

Zhang, Y., Wu, B., Xu, H., Liu, H., Wang, M., He, Y., & Pan, B. (2016). Nanomaterials-enabled water and wastewater treatment. *NanoImpact*, *3–4*, 22–39. DOI: 10.1016/j.impact.2016.09.004

Compilation of References

Abbott, A. P., Cullis, P. M., Gibson, M. J., Harris, R. C., & Raven, E. (2007). Extraction of glycerol from biodiesel into a eutectic based ionic liquid. *Green Chemistry*, *9*(8), 868–872. DOI: 10.1039/b702833d

Abdal Dayem, A., Hossain, M. K., Lee, S. B., Kim, K., Saha, S. K., Yang, G.-M., Choi, H. Y., & Cho, S.-G. (2017). The role of reactive oxygen species (ROS) in the biological activities of metallic nanoparticles. *International Journal of Molecular Sciences*, *18*(1), 120. Advance online publication. DOI: 10.3390/ijms18010120 PMID: 28075405

Abhijeet Kr Shrivastava. Dr. (2024). Nanocatalysis: new dimensions in catalytic industry. In *Futuristic Trends in Chemical Material Sciences & Nano Technology Volume 3 Book 14* (pp. 137–148). Iterative International Publishers, Selfypage Developers Pvt Ltd. DOI: 10.58532/V3BECS14P4CH1

Afreen, U., Fahelelbom, K. M., Shah, S. N. H., Ashames, A., Almas, U., Khan, S. A., Yameen, M. A., Nisar, N., Asad, M. H. H. B., & Murtaza, G. (2022). Formulation and evaluation of niosomes-based chlorpheniramine gel for the treatment of mild to moderate skin allergy. *Journal of Experimental Nanoscience*, *17*(1), 467–495. DOI: 10.1080/17458080.2022.2094915

Agatemor, C., Ibsen, K. N., Tanner, E. E., & Mitragotri, S. (2018). Ionic liquids for addressing unmet needs in healthcare. *Bioengineering & Translational Medicine*, *3*(1), 7–25. DOI: 10.1002/btm2.10083 PMID: 29376130

Ahmad, A. L., Yasin, N. H. M., Derek, C. J. C., & Lim, J. K. (2011). Microalgae as a sustainable energy source for biodiesel production: A review. *Renewable & Sustainable Energy Reviews*, *15*(1), 584–593. DOI: 10.1016/j.rser.2010.09.018

Akhtar, M. S., Ali, S., & Zaman, W. (2024). Recent Advancements in Catalysts for Petroleum Refining. *Catalysts*, *14*(12), 841. DOI: 10.3390/catal14120841

Akoh, C. C., Chang, S.-W., Lee, G.-C., & Shaw, J.-F. (2007). Enzymatic approach to biodiesel production. *Journal of Agricultural and Food Chemistry*, *55*(22), 8995–9005. DOI: 10.1021/jf071724y PMID: 17902621

Algieri, V., Algieri, C., Costanzo, P., Fiorani, G., & De Maria, P. (2023). Novel regioselective synthesis of 1,3,4,5-tetrasubstituted pyrazoles and biochemical valuation on F1F0-ATPase and mitochondrial permeability transition pore formation. *Pharmaceutics*, *15*(2), 498. DOI: 10.3390/pharmaceutics15020498 PMID: 36839821

Alguacil, F. J., & Robla, J. I. (2024). Using ionic liquids to improve CO2 capture. *Molecules (Basel, Switzerland)*, *29*(22), 5388. DOI: 10.3390/molecules29225388 PMID: 39598774

Al-Harbi, N. & Abd-Elrahman, N. K. (2024). Physical methods for preparation of nanomaterials, their characterization and applications: A review. *Journal of Umm Al-Qura University for Applied Sciences*, *209*. Advance online publication. DOI: 10.1007/s43994-024-00165-7

Ali, A., Zafar, H., Zia, M., ul Haq, I., Phull, A. R., Ali, J. S., & Hussain, A. (2016). Synthesis, characterization, applications, and challenges of iron oxide nanoparticles. *Nanotechnology, Science and Applications*, *9*, 49–67. DOI: 10.2147/NSA.S99986 PMID: 27578966

Ali, M. K., Moshikur, R. M., Wakabayashi, R., Tahara, Y., Moniruzzaman, M., Kamiya, N., & Goto, M. (2019). Synthesis and characterization of choline-fatty-acid-based ionic liquids: A new biocompatible surfactant. *Journal of Colloid and Interface Science*, *551*, 72–80. DOI: 10.1016/j.jcis.2019.04.095 PMID: 31075635

Alkhatib, I. I. I., Bahamon, D., al Hajaj, A., & Vega, L. F. (2022). Molecular Thermodynamic Modeling of Hybrid Ionic Liquids for Biogas Upgrading. *Industrial & Engineering Chemistry Research*, *61*(33), 12190–12207. DOI: 10.1021/acs.iecr.2c00710

Almutairi, R. (2023). *Ionanofluids: A review on its properties and thermal applications*. 16–37. DOI: 10.21741/9781644902790-3

Alqahtani, A. S. (2024). Indisputable roles of different ionic liquids, deep eutectic solvents and nanomaterials in green chemistry for sustainable organic synthesis. *Journal of Molecular Liquids*, *399*, 124469. DOI: 10.1016/j.molliq.2024.124469

Altammar, K. A. (2023). A review on nanoparticles: Characteristics, synthesis, applications, and challenges. *Frontiers in Microbiology*, *14*, 1155622. DOI: 10.3389/fmicb.2023.1155622 PMID: 37180257

Alves, M. M. S., Melo, M. N., Mertens, H. D. T., Pereiro, A. B., & Archer, M. (2023). Studies on the interaction between model proteins and fluorinated ionic liquids. *Pharmaceutics, 15*(1), 157. DOI: 10.3390/pharmaceutics15010157 PMID: 36678790

Amaral, M., Pereiro, A. B., Gaspar, M. M., & Reis, C. P. (2021). Recent advances in ionic liquids and nanotechnology for drug delivery. *Nanomedicine (London), 16*(1), 63–80. DOI: 10.2217/nnm-2020-0340 PMID: 33356551

Amin, F., Fozia, , Khattak, B., Alotaibi, A., Qasim, M., Ahmad, I., Ullah, R., Bourhia, M., Gul, A., Zahoor, S., & Ahmad, R. (2021). Fozia, Khattak, B., Alotaibi, A., Qasim, M., Ahmad, I., Ullah, R., Bourhia, M., Gul, A., Zahoor, S., & Ahmad, R. (2021). Green Synthesis of Copper Oxide Nanoparticles Using Aerva javanica Leaf Extract and Their Characterization and Investigation of In Vitro Antimicrobial Potential and Cytotoxic Activities. *Evidence-Based Complementary and Alternative Medicine : eCAM, 2021*(1), 1–12. DOI: 10.1155/2021/5589703 PMID: 34239581

Anastas, P. T., & Eghbali, N. (2010). Green chemistry: Principles and practice. *Chemical Society Reviews, 39*(1), 301–312. DOI: 10.1039/B918763B PMID: 20023854

Anastas, P. T., & Warner, J. C. (2000). Green Chemistry. *Green Chemistry*. Advance online publication. DOI: 10.1093/oso/9780198506980.001.0001

Andersson, M. M., & Hatti-Kaul, R. (1999). Protein stabilising effect of polyethyleneimine. *Journal of Biotechnology, 72*(1–2), 21–31. DOI: 10.1016/S0168-1656(99)00050-4

Andrade, Â. L., Valente, M. A., Ferreira, J. M., & Fabris, J. D. (2012). Preparation of size-controlled nanoparticles of magnetite. *Journal of Magnetism and Magnetic Materials, 324*(10), 1753–1757. DOI: 10.1016/j.jmmm.2011.12.033

Andreani, L., & Rocha, J. D. (2012). Use of ionic liquids in biodiesel production. A review. *Brazilian Journal of Chemical Engineering, 29*(1), 1–13. DOI: 10.1590/S0104-66322012000100001

Aniya, M., & Ikeda, M. (2019). A model for Non-Arrhenius ionic conductivity. *Nanomaterials (Basel, Switzerland), 9*(6), 911. DOI: 10.3390/nano9060911 PMID: 31238516

An, J. H., Jin, F., Kim, H. S., Ryu, H. C., Kim, J. S., Kim, H. M., Kiyonga, A. N., Min, D. S., Youn, W., Kim, K. H., & Jung, K. (2016). Application of ionic liquid to polymorphic transformation of anti-viral/HIV drug adefovir dipivoxil. *Archives of Pharmacal Research, 39*(5), 646–659. DOI: 10.1007/s12272-016-0721-0 PMID: 26908332

Anthony, J. L., Maginn, E. J., & Brennecke, J. F. (2001). Solution thermodynamics of imidazolium-based Ionic liquids and water. *The Journal of Physical Chemistry B, 105*(44), 10942–10949. DOI: 10.1021/jp0112368

Arai, S., Nakashima, K., Tanino, T., Ogino, C., Kondo, A., & Fukuda, H. (2010). Production of biodiesel fuel from soybean oil catalyzed by fungus whole-cell biocatalysts in ionic liquids. *Enzyme and Microbial Technology, 46*(1), 51–55. DOI: 10.1016/j.enzmictec.2009.08.008

Arellano Vidal, C. L., Sanahuja, A., & Ramos, V. (2024). Machine learning techniques for improving nanosensors in agroenvironmental applications. *Agronomy (Basel), 14*(2), 341. Advance online publication. DOI: 10.3390/agronomy14020341

Atadashi, I. M., Aroua, M. K., & Aziz, A. A. (2011). Biodiesel separation and purification: A review. *Renewable Energy, 36*(2), 437–443. DOI: 10.1016/j.renene.2010.07.019

Aziz, T., Li, W., Zhu, J., & Chen, B. (2024). Developing multifunctional cellulose derivatives for environmental and biomedical applications: Insights into modification processes and advanced material properties. *International Journal of Biological Macromolecules, 278*, 134695. DOI: 10.1016/j.ijbiomac.2024.134695 PMID: 39151861

Azum, N., Rub, M. A., & Asiri, A. M. (2021). Mixed micellization and spectroscopic studies of anti-allergic drug and non-ionic surfactant in the presence of ionic liquid. *Polymers, 13*(16), 2756. DOI: 10.3390/polym13162756 PMID: 34451295

Bablee, A., Amarasekara, A., Gabitto, J., Bhuiyan, A., & Shamim, N. (2025). A Comprehensive Review of the Thermophysical Properties of Energetic Ionic Liquids. *Energies, 18*(2), 267. DOI: 10.3390/en18020267

Bahmanzadegan, F., & Aghaei, A. (2025). A comprehensive review on novel zeolite-based adsorbents for environmental pollutant. *Journal of Hazardous Materials Advances, 17*, 100617. Advance online publication. DOI: 10.1016/j.hazadv.2025.100617

Bakar, R., Darvishi, S., Aydemir, U., Yahsi, U., Tav, C., Menceloglu, Y. Z., & Senses, E. (2023). Decoding Polymer Architecture Effect on Ion Clustering, Chain Dynamics, and Ionic Conductivity in Polymer Electrolytes. *ACS Applied Energy Materials, 6*(7), 4053–4064. DOI: 10.1021/acsaem.3c00310 PMID: 37064412

Bakthavatchalam, B., Habib, K., Saidur, R., Saha, B. B., & Irshad, K. (2020a). Comprehensive study on nanofluid and ionanofluid for heat transfer enhancement: A review on current and future perspective. *Journal of Molecular Liquids, 305*, 112787. DOI: 10.1016/j.molliq.2020.112787

Banjare, M. K., Behera, K., Satnami, M. L., Pandey, S., & Ghosh, K. K. (2018). Self-assembly of a short-chain ionic liquid within deep eutectic solvents. *RSC Advances*, *8*(15), 7969–7979. DOI: 10.1039/C7RA13557B PMID: 35542011

Bartlewicz, O., Dąbek, I., Szymańska, A., & Maciejewski, H. (2020). Heterogeneous Catalysis with the Participation of Ionic Liquids. *Catalysts*, *10*(11), 1227. DOI: 10.3390/catal10111227

Basivi, P. K., Giri, A., & Kumar, V. (2025). Graphitic carbon nitride (g–C3N4)–based Z-scheme photocatalysts: Innovations for energy and environmental applications. *Materials Today Sustainability*, *29*, 101069. Advance online publication. DOI: 10.1016/j.mtsust.2024.101069

Basu, S., & Banik, B. K. (2024). Nanoparticles as Catalysts: Exploring Potential Applications. *Current Organocatalysis*, *11*(4), 265–272. DOI: 10.2174/01221333 722856102312270949 59

Bauer, G., Lima, S., Chenevard, J., Sugnaux, M., & Fischer, F. (2017). Biodiesel via in situ wet microalgae biotransformation: Zwitter-type ionic liquid supported extraction and transesterification. *ACS Sustainable Chemistry & Engineering*, *5*(2), 1931–1937. DOI: 10.1021/acssuschemeng.6b02665

Beil, S., Markiewicz, M., Pereira, C. S., Stepnowski, P., Thöming, J., & Stolte, S. (2021). Toward the Proactive Design of Sustainable Chemicals: Ionic Liquids as a Prime Example. *Chemical Reviews*, *121*(21), 13132–13173. DOI: 10.1021/acs.chemrev.0c01265 PMID: 34523909

Berton, P., Abidi, N., & Shamshina, J. L. (2022). Ionic liquids: Implementing objectives of sustainability for the next generation chemical processes and industrial applications. *Current Opinion in Green and Sustainable Chemistry*, *35*, 100625. DOI: 10.1016/j.cogsc.2022.100625

Bharathi, C., Rajeswari, R., Janaki, P., Sivamurugan, A. P., Senthil, G. K., Radhamani, S., Sathya Priya, R., Sangeetha, M., Thenmozhi, S., Chiranjeevirajan, N., Sharmila, R., Ramya, B., & Balamurugan, R. (2025). Bio-mediated synthesis of nanoparticles: A new paradigm for environmental sustainability. *Plant Science Today*, *12*(sp1). Advance online publication. DOI: 10.14719/pst.5192

Binnig, G., Quate, C. F., & Gerber, Ch. (1986). Atomic Force Microscope. *Physical Review Letters*, *56*(9), 930–933. DOI: 10.1103/PhysRevLett.56.930 PMID: 10033323

Bligh, E. G., & Dyer, W. J. (1959). A rapid method of total lipid extraction and purification. *Canadian Journal of Biochemistry and Physiology*, *37*(8), 911–917. DOI: 10.1139/o59-099 PMID: 13671378

Boamah, P. O., Wang, L., Shen, W., Tang, S., & Lee, H. K. (2023). Applications of ionic liquids in the microextraction of pesticides: A mini-review. *Journal of Chromatography Open, 4*, 100090. Advance online publication. DOI: 10.1016/j.jcoa.2023.100090

Bonhôte, P., Dias, A., Papageorgiou, N., Kalyanasundaram, K., & Grätzel, M. (1996). Hydrophobic, highly conductive ambient-temperature molten salts. *Inorganic Chemistry, 35*(5), 1168–1178. DOI: 10.1021/ic951325x PMID: 11666305

Boon, J. A., Levisky, J. A., Pflug, J. L., & Wilkes, J. S. (1986). Friedel-crafts reactions in ambient-temperature molten salts. *The Journal of Organic Chemistry, 51*(4), 480–483. DOI: 10.1021/jo00354a013

Boualem, A. D., Argoub, K., Benkouider, A. M., Yahiaoui, A., & Toubal, K. (2022). Viscosity prediction of ionic liquids using NLR and SVM approaches. *Journal of Molecular Liquids, 368*(Part A). *Journal of Molecular Liquids, 120610*, 120610. Advance online publication. DOI: 10.1016/j.molliq.2022.120610

Bragg, W. L. (1913). The Structure of Some Crystals as Indicated by Their Diffraction of X-rays. *Proceedings of the Royal Society of London. Series A, Containing Papers of a Mathematical and Physical Character, 89*(610), 248–277. DOI: 10.1098/rspa.1913.0083

Brahma, S., & Gardas, R. L. (2024). In Rajkhowa, S., Singh, P., Sen, A., & Sarma, J. (Eds.), *"History and development of ionic liquids," in Handbook of ionic liquids: fundamentals, applications, and sustainability* (pp. 1–28). John Wiley and Sons.

Brennecke, J. F., & Maginn, E. J. (2001). Ionic liquids: Innovative fluids for chemical processing. *AIChE Journal. American Institute of Chemical Engineers, 47*(11), 2384–2389. DOI: 10.1002/aic.690471102

Brigger, I., Dubernet, C., & Couvreur, P. (2012). Nanoparticles in cancer therapy and diagnosis. *Advanced Drug Delivery Reviews, 64*, 24–36. DOI: 10.1016/j.addr.2012.09.006 PMID: 12204596

Bubalo, M. C., Radošević, K., & Gaurina Srček, V. (2015). Cytotoxicity towards CCO cells of imidazolium ionic liquids with functionalized side chains: Preliminary QSTR modeling using regression and classification-based approaches. *Ecotoxicology and Environmental Safety, 112*, 22–28. DOI: 10.1016/j.ecoenv.2014.10.029 PMID: 25463849

Bubalo, M. C., Radošević, K., & Srček, V. G. (2014). A brief overview of the potential environmental hazards of ionic liquids. *Ecotoxicology and Environmental Safety, 99*, 1–12. DOI: 10.1016/j.ecoenv.2013.10.019 PMID: 24210364

Cañadas, R., González-Miquel, M., González, E. J., Díaz, I., & Rodríguez, M. (2020). Overview of neoteric solvents as extractants in food industry: A focus on phenolic compounds separation from liquid streams. *Food Research International*, *136*, 109558. DOI: 10.1016/j.foodres.2020.109558 PMID: 32846603

Canton, P., Perosa, A., & Selva, M. (2017). Metal Nanoparticles Stabilized in Ionic Liquids for Catalytic Multiphase Reactions. *Current Organic Chemistry*, *21*(24). Advance online publication. DOI: 10.2174/1385272820666161009222123

Chacón, G., & Dupont, J. (2019). Arene Hydrogenation by Metal Nanoparticles in Ionic Liquids. *ChemCatChem*, *11*(1), 333–341. DOI: 10.1002/cctc.201801363

Chakhtouna, H., Benzeid, H., Zari, N., Qaiss, A. E. K., & Bouhfid, R. (2023). Photocatalytic and bactericidal behaviors of Ag/TiO2 doped biochar through ball–milling approach. *Journal of Photochemistry and Photobiology A Chemistry*, *444*, 114971. Advance online publication. DOI: 10.1016/j.jphotochem.2023.114971

Chauke, N. M., Mukwevho, E., & Mamba, B. B. (2025). Transformative advancements in visible-light-activated titanium dioxide for industrial wastewater remediation. *International Journal of Environmental Science and Technology*, *22*(9), 8521–8552. Advance online publication. DOI: 10.1007/s13762-025-06397-2

Chauvin, Y., Mußmann, L., & Olivier, H. (1995). Flüssige 1,3-Dialkylimidazoliumsalze als Lösungsmittel fur die Katalyse in Zweiphasensystemen: Durch Rhodiumkomplexe katalysierte Hydrierung, Isomerisierung und Hydroformylierung von Alkenen. *Angewandte Chemie*, *107*(23-24), 2941–2943. DOI: 10.1002/ange.19951072341

Chen, M., Wu, T., Niu, L., Ye, T., Dai, W., Zeng, L., Kornyshev, A. A., Wang, Z., Liu, Z., & Feng, G. (2024). Organic Solvent Boosts Charge Storage and Charging Dynamics of Conductive MOF Supercapacitors. *Advanced Materials*, *36*(30), 2403202. Advance online publication. DOI: 10.1002/adma.202403202 PMID: 38751336

Chen, S., Hanning, S., Falconer, J., Locke, M., & Wen, J. (2019). Recent advances in non-ionic surfactant vesicles (niosomes): Fabrication, characterization, pharmaceutical and cosmetic applications. *European Journal of Pharmaceutics and Biopharmaceutics*, *144*, 18–39. DOI: 10.1016/j.ejpb.2019.08.015 PMID: 31446046

Cheong, L.-Z., Tan, C.-P., & Cheah, Y.-K. (2011). Extraction and enrichment of n-3 polyunsaturated fatty acids and ethyl esters through reversible π–π complexation with aromatic rings containing ionic liquids. *Journal of Agricultural and Food Chemistry*, *59*(16), 8961–8967. DOI: 10.1021/jf202043w PMID: 21790198

Cheville, N. F., & Stasko, J. (2013). Techniques in Electron Microscopy of Animal Tissue. *Veterinary Pathology*, *51*(1), 28–41. DOI: 10.1177/0300985813505114 PMID: 24114311

Chiou, Y.-J., Fu, C., & Liao, Y.-H. (2023). A study on rose-window instability in a dielectric droplet exposed to corona discharge. *Journal of Mechanisms*, *39*, 491–499. DOI: 10.1093/jom/ufad041

Chiozzi, V., Agriopoulou, S., & Varzakas, T. (2022). Advances, Applications, and Comparison of Thermal (Pasteurization, Sterilization, and Aseptic Packaging) against Non-Thermal (Ultrasounds, UV Radiation, Ozonation, High Hydrostatic Pressure) Technologies in Food Processing. *Applied Sciences (Basel, Switzerland)*, *12*(4), 2202. DOI: 10.3390/app12042202

Choi, O. K., Lim, H. C., Cho, Y., Cho, A. E., Al Abdulgader, H., Lithoxoos, G. P., Aljama, H. A., Choi, S.-H., Choi, Y. C., & Lee, J. W. (2023). Room-temperature ionic liquids as candidate materials for produced water desalination: Experiments and molecular dynamic analysis. *Desalination*, *557*, 116608. Advance online publication. DOI: 10.1016/j.desal.2023.116608

Choi, S.-A., Oh, Y.-K., Lee, J., Sim, S. J., & Park, J. Y. (2014). Effects of ionic liquid mixtures on lipid extraction from *Chlorella vulgaris*. *Renewable Energy*, *65*, 169–174. DOI: 10.1016/j.renene.2013.08.015

Chun, S., Dzyuba, S. V., & Bartsch, R. A. (2001). Influence of structural variation in room-temperature ionic liquids on the selectivity and efficiency of competitive alkali metal salt extraction by a crown ether. *Analytical Chemistry*, *73*(15), 3737–3741. DOI: 10.1021/ac010061v PMID: 11510842

Clarke, C. J., Tu, W.-C., Levers, O., Bröhl, A., & Hallett, J. P. (2018). Green and sustainable solvents in chemical processes. *Chemical Reviews*, *118*(2), 747–800. DOI: 10.1021/acs.chemrev.7b00571 PMID: 29300087

Clark, J. H., Farmer, T. J., Hunt, A. J., & Sherwood, J. (2015). Opportunities for bio-based solvents created as petrochemical and fuel products transition towards renewable resources. *International Journal of Molecular Sciences*, *16*(8), 17101–17159. DOI: 10.3390/ijms160817101 PMID: 26225963

Compagno, N., Profeta, R., & Scarso, A. (2023). Recent advances in the synthesis of active pharmaceutical and agrochemical ingredients in micellar media. *Current Opinion in Green and Sustainable Chemistry*, *39*, 100729. DOI: 10.1016/j.cogsc.2022.100729

Dada, A. O., Adekola, F. A., Adeyemi, O. S., Bello, O. M., Oluwaseun, A. C., Awakan, O. J., & Grace, F.-A. A. (2018). *Exploring the Effect of Operational Factors and Characterization Imperative to the Synthesis of Silver Nanoparticles*. Silver Nanoparticles - Fabrication, Characterization and Applications., DOI: 10.5772/intechopen.76947

Dai, S., Ju, Y. H., & Barnes, C. E. (1999). Solvent extraction of strontium nitrate by a crown ether using room-temperature ionic liquids. *Journal of the Chemical Society, Dalton Transactions: Inorganic Chemistry*, (8), 1201–1202. DOI: 10.1039/a809672d

Das, S., Kashyap, N., Kalita, S., Bora, D. B., & Borah, R. (2020). A brief insight into the physicochemical properties of room-temperature acidic ionic liquids and their catalytic applications in C–C bond formation reactions. *Advances in Physical Organic Chemistry*, *54*, 1–98. DOI: 10.1016/bs.apoc.2020.07.002

Das, S., Rahman, M., Kundu, D., Majee, A., & Hajra, A. (2010). Task-specific ionic-liquid-catalyzed efficient synthesis of indole derivatives under solvent-free conditions. *Canadian Journal of Chemistry*, *88*(2), 150–154. DOI: 10.1139/V09-154

Davis, J. H.Jr. (2004). Task-specific ionic liquids. *Chemistry Letters*, *33*(9), 1072–1077. DOI: 10.1246/cl.2004.1072

Davis, J. H.Jr, & Forrester, K. J. (1999). Thiazolium-ion based organic Ionic liquids (OILs).1,2 novel OILs which promote the benzoin condensation. *Tetrahedron Letters*, *40*(9), 1621–1622. DOI: 10.1016/S0040-4039(99)00025-8

de Castro, C. N., & Ribeiro, C. A. P., I.C., S., P. Franca, J. M., J.V., M., V., F., M.S, S., Goodrich, P., & Hardacre, C. (2013). Synthesis, Properties and Physical Applications of IoNanofluids. In *Ionic Liquids - New Aspects for the Future*. InTech. DOI: 10.5772/52596

De Diego, T., Manjón, A., Lozano, P., & Iborra, J. L. (2011). A recyclable enzymatic biodiesel production process in ionic liquids. *Bioresource Technology*, *102*(10), 6336–6339. DOI: 10.1016/j.biortech.2011.02.071 PMID: 21392972

De Jong, W. H., & Borm, P. J. (2008). Drug delivery and nanoparticles: Applications and hazards. *International Journal of Nanomedicine*, *3*(2), 133–149. DOI: 10.2147/IJN.S596 PMID: 18686775

De Los Ríos, A. P., Hernández Fernández, F. J., Gómez, D., Rubio, M., & Víllora, G. (2011). Biocatalytic transesterification of sunflower and waste cooking oils in ionic liquid media. *Process Biochemistry (Barking, London, England)*, *46*(7), 1475–1480. DOI: 10.1016/j.procbio.2011.03.021

Dehe, S., & Hardt, S. (2021). Deformation modes of an oil-water interface under a local electric field: From Taylor cones to surface dimples. *Physical Review Fluids, 6*(12), 123702. Advance online publication. DOI: 10.1103/PhysRevFluids.6.123702

Di Pietro, M. E., Castiglione, F., & Mele, A. (2020). Anions as dynamic probes for ionic liquid mixtures. *The Journal of Physical Chemistry B, 124*(14), 2879–2891. DOI: 10.1021/acs.jpcb.0c00026 PMID: 32186377

Dias, A. R., Costa-Rodrigues, J., Teixeira, C., Prudêncio, C., Gomes, P., & Ferraz, R. (2019). Ionic liquids for topical delivery in cancer. *Current Medicinal Chemistry, 26*(40), 7520–7532. DOI: 10.2174/0929867325666180914104045 PMID: 30362415

Dobler, D., Schmidts, T., Klingenhöfer, I., & Runkel, F. (2013). Ionic liquids as ingredients in topical drug delivery systems. *International Journal of Pharmaceutics, 441*(1–2), 620–627. DOI: 10.1016/j.ijpharm.2012.10.035 PMID: 23123180

Dong, Y., Ng, W. K., Shen, S., Kim, S., & Tan, R. B. (2009). Preparation and characterization of spironolactone nanoparticles by antisolvent precipitation. *International Journal of Pharmaceutics, 375*(1–2), 84–88. DOI: 10.1016/j.ijpharm.2009.03.013 PMID: 19481693

Do-Thanh, C., Luo, H., & Dai, S. (2023). A perspective on task-specific ionic liquids for the separation of rare earth elements. *RSC Sustainability, 1*(5), 1168–1176. DOI: 10.1039/D3SU00007A

Dulog, J. C., Unabia, R., Delicana, J. D. P., Lalem, A., Sayson, N. L. B., Capangpangan, R. Y., Lubguban, A. A., & Alguno, A. C. (2024). Synthesis and Characterization of AuNP/TiO_2 Hybrid Nanoparticles for Possible Photocatalytic Application. *Materials Science Forum, 1113*, 21–26. DOI: 10.4028/p-ogEu80

Dupont, J., Consorti, C. S., & Spencer, J. (2000). Room temperature molten salts: Neoteric "green" solvents for chemical reactions and processes. *Journal of the Brazilian Chemical Society, 11*(4), 337–344. DOI: 10.1590/S0103-50532000000400002

Dupont, J., Fonseca, G. S., Umpierre, A. P., Fichtner, P. F., & Teixeira, S. R. (2002). Transition-metal nanoparticles in Imidazolium Ionic Liquids: Recycable catalysts for biphasic hydrogenation reactions. *Journal of the American Chemical Society, 124*(16), 4228–4229. DOI: 10.1021/ja025818u PMID: 11960449

Dupont, J., Leal, B. C., Lozano, P., Monteiro, A. L., Migowski, P., & Scholten, J. D. (2024). Ionic Liquids in Metal, Photo-, Electro-, and (Bio) Catalysis. *Chemical Reviews, 124*(9), 5227–5420. DOI: 10.1021/acs.chemrev.3c00379 PMID: 38661578

Dupont, J., & Scholten, J. D. (2010). On the structural and surface properties of transition-metal nanoparticles in ionic liquids. *Chemical Society Reviews, 39*(5), 1780–1804. DOI: 10.1039/b822551f PMID: 20419219

Durga, G., & Mishra, A. (2016). Ionic Liquids: Industrial Applications. In *Encyclopedia of Inorganic and Bioinorganic Chemistry* (pp. 1–13). Wiley. DOI: 10.1002/9781119951438.eibc2434

Dutta, D., & Sit, N. (2024). A comprehensive review on types and properties of biopolymers as sustainable bio-based alternatives for packaging. *Food Biomacromolecules, 1*(2), 58–87. DOI: 10.1002/fob2.12019

Earle, M. J., Esperança, J., Gilea, M. A., Canongia Lopes, J. N., Rebelo, L. P., Magee, J. W., Seddon, K. R., & Widegren, J. A. (2006). The distillation and volatility of Ionic liquids. *Nature, 439*(7078), 831–834. DOI: 10.1038/nature04451 PMID: 16482154

Egorova, K. S., & Ananikov, V. P. (2014). Toxicity of ionic liquids: Eco(cyto) activity as complicated, but unavoidable parameter for task-specific optimization. *ChemSusChem, 7*(2), 336–360. DOI: 10.1002/cssc.201300459 PMID: 24399804

Egorova, K. S., & Ananikov, V. P. (2018). Fundamental importance of ionic interactions in the liquid phase: A review of recent studies of ionic liquids in biomedical and pharmaceutical applications. *Journal of Molecular Liquids, 272*, 271–300. DOI: 10.1016/j.molliq.2018.09.025

Egorova, K. S., Gordeev, E. G., & Ananikov, V. P. (2017). Biological activity of ionic liquids and their application in pharmaceutics and medicine. *Chemical Reviews, 117*(10), 7132–7189. DOI: 10.1021/acs.chemrev.6b00562 PMID: 28125212

Elango, G., Kumaran, S. M., Kumar, S. S., Muthuraja, S., & Roopan, S. M. (2015). Green synthesis of SnO2 nanoparticles and its photocatalytic activity of phenolsulfonphthalein dye. *Spectrochimica Acta. Part A: Molecular and Biomolecular Spectroscopy, 145*, 176–180. DOI: 10.1016/j.saa.2015.03.033 PMID: 25770967

Elgharbawy, A. A., Moniruzzaman, M., & Goto, M. (2020). Recent advances of enzymatic reactions in ionic liquids: Part II. *Biochemical Engineering Journal, 154*, 107426. DOI: 10.1016/j.bej.2019.107426

Elhenawy, S., Khraisheh, M., AlMomani, F., Al-Ghouti, M., & Hassan, M. K. (2024). Emerging nanomaterials for drinking water purification: A new era of water treatment technology. *Nanomaterials (Basel, Switzerland), 14*(21), 1707. Advance online publication. DOI: 10.3390/nano14211707 PMID: 39513787

Elmobarak, W. F., & Almomani, F. (2021). Application of Fe3O4 magnetite nanoparticles grafted in silica (SiO2) for oil recovery from oil in water emulsions. *Chemosphere, 265*, 129054. DOI: 10.1016/j.chemosphere.2020.129054 PMID: 33280845

Erfirt, K., Wandzik, I., Walczak, K., Matuszek, K., & Chrobok, A. (2014). Hydrogen-bond-rich ionic liquids as effective organocatalysts for Diels–Alder reactions. *Green Chemistry, 16*(7), 3508–3514. DOI: 10.1039/C4GC00380B

European Committee for Standardization. (2003). *Automotive fuels—Fatty acid methyl esters (FAME) for diesel engines—Requirements and test methods* (EN 14214).

Fang, D., Yang, J., & Jiao, C. (2011). Dicationic ionic liquids as environmentally benign catalysts for biodiesel synthesis. *ACS Catalysis, 1*(1), 42–47. DOI: 10.1021/cs100026q

Farra, A., Fratini, E., & Baglioni, P. (2015). Green solvents in carbohydrate chemistry: From raw materials to fine chemicals. *Chemical Reviews, 115*(14), 6811–6853. DOI: 10.1021/cr500719h PMID: 26121409

Feder-Kubis, J., & Wnuk, A. (2022). Imidazolium room-temperature ionic liquids with alkoxymethyl substituent: A quest for improved microbiological selectivity. *Chemical Engineering Journal, 442*, 136062. DOI: 10.1016/j.cej.2022.136062

Ferdous, A. R., Shah, S. N. A., Shah, S. S., & Aziz, M. (2024). Advancements in nanotechnology applications: Transforming catalysts, sensors, and coatings in petrochemical industries. *Fuel, 371*, 132020. DOI: 10.1016/j.fuel.2024.132020

Ferraz, R., Branco, L. C., Prudêncio, C., Noronha, J. P., & Petrovski, Ž. (2011). Ionic liquids as active pharmaceutical ingredients. *ChemMedChem, 6*(6), 975–985. DOI: 10.1002/cmdc.201100082 PMID: 21557480

Ferreira, A. M., Morais, E. S., Leite, A. C., Mohamadou, A., Holmbom, B., Holmbom, T., Neves, B. M., Coutinho, J. A. P., Freire, M. G., & Silvestre, A. J. D. (2017). Enhanced extraction and biological activity of 7-hydroxymatairesinol obtained from Norway spruce knots using aqueous solutions of ionic liquids. *Green Chemistry, 19*(11), 2626–2635. DOI: 10.1039/C7GC01091E

Figueiredo, N. M., Voroshylova, I. V., Ferreira, E. S. C., Marques, J. M. C., & Cordeiro, M. N. D. S. (2024). Magnetic Ionic Liquids: Current Achievements and Future Perspectives with a Focus on Computational Approaches. *Chemical Reviews, 124*(6), 3392–3415. DOI: 10.1021/acs.chemrev.3c00678 PMID: 38466339

Flieger, J., Feder-Kubis, J., & Tatarczak-Michalewska, M. (2020). Chiral ionic liquids: Structural diversity, properties and applications in selected separation techniques. *International Journal of Molecular Sciences*, *21*(12), 4253. DOI: 10.3390/ijms21124253 PMID: 32549300

Flieger, J., & Flieger, M. (2020). Ionic Liquids Toxicity—Benefits and Threats. *International Journal of Molecular Sciences*, *21*(17), 6267. DOI: 10.3390/ijms21176267 PMID: 32872533

Flora, F. C., Relvas, S. B., Silva, F. A. E., Freire, M. G., Chu, V., & Conde, J. P. (2023). Combined Use of Ionic Liquid-Based Aqueous Biphasic Systems and Microfluidic Devices for the Detection of Prostate-Specific Antigen. *Biosensors (Basel)*, *13*(3), 334. Advance online publication. DOI: 10.3390/bios13030334 PMID: 36979546

Freemantle, M., Welton, T., & Rogers, R. D. (2009). An introduction to Ionic liquids. DOI: 10.1039/9781839168604

Freire, M. G., Neves, C. M. S. S., Carvalho, P. J., Gardas, R. L., Fernandes, A. M., Marrucho, I. M., Santos, L. M. N. B. F., & Coutinho, J. A. P. (2007). Mutual solubilities of water and hydrophobic ionic liquids. *The Journal of Physical Chemistry B*, *111*(45), 13082–13089. DOI: 10.1021/jp076271e PMID: 17958353

Frippiat, T., Art, T., & Delguste, C. (2025). Silver nanoparticles as antimicrobial agents in veterinary medicine: Current applications and future perspectives. *Nanomaterials (Basel, Switzerland)*, *15*(3), 202. Advance online publication. DOI: 10.3390/nano15030202 PMID: 39940178

Gano, M., Klebeko, J., & Peech, R. (2021). Efficient esterification of curcumin in bis(trifluoromethylsulfonyl)imide-based ionic liquids. *Journal of Molecular Liquids*, *337*, 116420. DOI: 10.1016/j.molliq.2021.116420

Garcia, M. T., Bautista, E., de la Fuente, A., & Perez, L. (2023). Cholinium-based ionic liquids as promising antimicrobial agents in pharmaceutical applications: Surface activity, antibacterial activity and ecotoxicological profile. *Pharmaceutics*, *15*(7), 1806. DOI: 10.3390/pharmaceutics15071806 PMID: 37513993

Gatou, A. M., Lagopati, N., & Pavlatou, E. A. (2024). Photocatalytic TiO2-based nanostructures as a promising material for diverse environmental applications: A review. *Reactions*, *5*(1), 135–194. DOI: 10.3390/reactions5010007

Gazal, U., Khan, I., Bhat, A. H., & Pathak, V. (2021). Ionic liquids as green solvents in the pharmaceutical industry. In *Green Sustainable Process for Chemical and Environmental Engineering and Science* (pp. 41–54). Elsevier., DOI: 10.1016/B978-0-12-821885-3.00002-5

Ginot, L., el Bakkouche, A., Giusti, F., Dourdain, S., & Pellet-Rostaing, S. (2024). Hydrophobic Porous Liquids with Controlled Cavity Size and Physico-Chemical Properties. *Advancement of Science, 11*(4), 2305906. Advance online publication. DOI: 10.1002/advs.202305906 PMID: 38036426

Godja, N.-C., & Munteanu, F.-D. (2024). Hybrid nanomaterials: A brief overview of versatile solutions for sensor technology in healthcare and environmental applications. *Biosensors (Basel), 14*(2), 67. Advance online publication. DOI: 10.3390/bios14020067 PMID: 38391986

Gonçalves, H. M. R., Pereira, R. F. P., Lepleux, E., Pacheco, L., Valente, A. J. M., Duarte, A. J., & de Zea Bermudez, V. (2020). Non-Newtonian Thermosensitive Nanofluid Based on Carbon Dots Functionalized with Ionic Liquids. *Small, 16*(28), 1907661. Advance online publication. DOI: 10.1002/smll.201907661 PMID: 32462808

Gong, K., Fang, D., Wang, H. L., Zhou, X. L., & Liu, Z. L. (2009). The one-pot synthesis of 14-alkyl- or aryl-14H-dibenzo[a,j]xanthenes catalyzed by task-specific ionic liquid. *Dyes and Pigments, 80*(1), 30–33. DOI: 10.1016/j.dyepig.2008.02.011

Granger, P., & Parvulescu, V. I. (2020). Application of Ionic Liquids for Sustainable Catalysis. In *Heterogeneous Catalysis for Energy Applications* (pp. 304–360). The Royal Society of Chemistry., DOI: 10.1039/9781788019576-00304

Grasso, A., Ferrante, M., Arena, G., Salemi, R., Zuccarello, P., Fiore, M., & Copat, C. (2021). Chemical characterization and quantification of silver nanoparticles (Ag-NPs) and dissolved ag in seafood by single particle icp-ms: Assessment of dietary exposure. *International Journal of Environmental Research and Public Health, 18*(8), 4076. DOI: 10.3390/ijerph18084076 PMID: 33924319

Greaves, T. L., & Drummond, C. J. (2007). Protic Ionic Liquids: Properties and applications. *Chemical Reviews, 108*(1), 206–237. DOI: 10.1021/cr068040u PMID: 18095716

Green, L., Hemeon, I., & Singer, R. D. (2000). 1-Ethyl-3-methylimidazolium halogenoaluminate Ionic liquids as reaction media for the acylative cleavage of ethers. *Tetrahedron Letters, 41*(9), 1343–1346. DOI: 10.1016/S0040-4039(99)02289-3

Greer, A. J., Jacquemin, J., & Hardacre, C. (2020). Industrial Applications of Ionic Liquids. *Molecules (Basel, Switzerland), 25*(21), 5207. DOI: 10.3390/molecules25215207 PMID: 33182328

Guan, Y., Li, L., & Piao, H. (2025). Factors Influencing Nanoscale Catalysts Prepared Impregnation Method. *Applied and Computational Engineering*, *122*(1), 35–39. DOI: 10.54254/2755-2721/2025.19593

Gundolf, T., Weyhing-Zerrer, N., Sommer, J., Kalb, R., Schoder, D., Rossmanith, P., & Mester, P. (2019). Biological impact of ionic liquids based on sustainable fatty acid anions examined with a tripartite test system. *ACS Sustainable Chemistry & Engineering*, *7*(19), 15865–15873. DOI: 10.1021/acssuschemeng.8b06201

Guo, F., Fang, Z., Tian, X.-F., Long, Y.-D., & Jiang, L.-Q. (2011). One-step production of biodiesel from Jatropha oil with high-acid value in ionic liquids. *Bioresource Technology*, *102*(11), 6469–6472. DOI: 10.1016/j.biortech.2011.02.079 PMID: 21420854

Guzmán-Lucero, D., Guzmán-Pantoja, J., Velázquez, H. D., Likhanova, N. V., Bazaldua-Domínguez, M., Vega-Paz, A., & Martínez-Palou, R. (2021). Isobutane/butene alkylation reaction using ionic liquids as catalysts. Toward a sustainable industry. *Molecular Catalysis*, *515*, 111892. DOI: 10.1016/j.mcat.2021.111892

Hajfarajollah, H., Mokhtarani, B., Mirzaei, M., Afaghi, A., & Noghabi, K. A. (2014). Toxicity of various kinds of ionic liquids towards the cell growth and end product formation of the probiotic strain, Propionibacterium freudenreichii. *RSC Advances*, *4*(25), 13153–13160. DOI: 10.1039/C4RA00925H

Hajfarajollah, H., Mokhtarani, B., Noghabi, K. A., Sharifi, A., & Mirzaei, M. (2014). Antibacterial and antiadhesive properties of butylmethylimidazolium ionic liquids toward pathogenic bacteria. *RSC Advances*, *4*(77), 42751–42757. DOI: 10.1039/C4RA07055K

Hama, S., & Kondo, A. (2013). Enzymatic biodiesel production: An overview of potential feedstocks and process development. *Bioresource Technology*, *135*, 386–395. DOI: 10.1016/j.biortech.2012.08.014 PMID: 22985827

Han, B., Zhang, Y., & Wang, L. (2024). Thin-film nanocomposite (TFN) membrane technologies for the removal of emerging contaminants from wastewater. *Journal of Cleaner Production*, *480*, 144043. Advance online publication. DOI: 10.1016/j.jclepro.2024.144043

Handy, S. T. (2003). Greener solvents: Room temperature ionic liquids from biorenewable sources. *Chemistry (Weinheim an der Bergstrasse, Germany)*, *9*(13), 2938–2944. DOI: 10.1002/chem.200304799

Handy, S. T., & Zhang, X. (2001). Organic synthesis in ionic liquids: The Stille coupling. *Organic Letters*, *3*(2), 233–236. DOI: 10.1021/ol0068849 PMID: 11430042

Han, H., Geng, X., Zhang, B., Meng, J., Liu, X., He, X., Liu, Z., Gao, Y., Liu, D., & Hu, X. (2019). Synthesis of novel functional Ionic liquids and their application in biomass. *RSC Advances, 9*(51), 29652–29658. DOI: 10.1039/C9RA06243B PMID: 35531511

Han, H., Jang, J. U., Oh, D., Na, K.-H., Choi, W.-Y., Jayakrishnan, N., & Nayak, A. K. (2023). Advances and Perspectives of Titanium-Based Nanocomposites for Energy Generation and Environmental Remediation Applications: A Review. *Energy & Fuels, 37*(23), 17708–17735. DOI: 10.1021/acs.energyfuels.3c02382

Han, M., Yi, W., Wu, Q., Liu, Y., Hong, Y., & Wang, D. (2009). Preparation of biodiesel from waste oils catalyzed by a Brønsted acidic ionic liquid. *Bioresource Technology, 100*(7), 2308–2310. DOI: 10.1016/j.biortech.2008.10.046 PMID: 19071017

Harris, J., Viner, K., & McCabe, B. K. (2018). Advances in microalgal lipid extraction for biofuel production: A review. *Biofuels, Bioproducts & Biorefining, 12*(6), 1118–1135. DOI: 10.1002/bbb.1923

Ha, S. H., Lan, M. N., Lee, S. H., Hwang, S. M., & Koo, Y.-M. (2007). Lipase-catalyzed biodiesel production from soybean oil in ionic liquids. *Enzyme and Microbial Technology, 41*(4), 480–483. DOI: 10.1016/j.enzmictec.2007.03.017

Hasan, S. (2015). A review on nanoparticles: Their synthesis and types. *Research Journal of Recent Sciences, 2277*, 2502.

Hassan, A. A., Tanimu, A., & Alhooshani, K. (2021). Iron and cobalt-containing magnetic ionic liquids for dispersive micro-solid phase extraction coupled with HPLC-DAD for the preconcentration and quantification of carbamazepine drug in urine and environmental water samples. *Journal of Molecular Liquids, 336*, 116370. DOI: 10.1016/j.molliq.2021.116370

Hassan, R., Nazir, F., Roosh, M., Qaisar, A., Habib, U., Sajini, A. A., & Iqbal, M. (2022). Synthesis, characterization, biological evaluation, and in Silico studies of imidazolium-, pyridinium-, and ammonium-based Ionic liquids containing N-butyl side chains. *Molecules (Basel, Switzerland), 27*(19), 6650. DOI: 10.3390/molecules27196650 PMID: 36235187

Hawker, R. R., Haines, R. S., & Harper, J. B. (2018). Rational selection of the cation of an ionic liquid to control the reaction outcome of a substitution reaction. *Chemical Communications, 54*(18), 2296–2299. DOI: 10.1039/C8CC00241J PMID: 29443346

Hayyan, A., Hashim, M. A., Mjalli, F. S., & AlNashef, I. M. (2013b). A novel phosphonium-based deep eutectic catalyst for biodiesel production from industrial low-grade crude palm oil. *Chemical Engineering Science*, *92*, 81–88. DOI: 10.1016/j.ces.2012.12.024

Hayyan, M., Mjalli, F. S., Hashim, M. A., & AlNashef, I. M. (2010). A novel technique for separating glycerine from palm oil-based biodiesel using ionic liquids. *Fuel Processing Technology*, *91*(1), 116–120. DOI: 10.1016/j.fuproc.2009.09.002

Heilgeist, S., Sekine, R., Sahin, O., & Stewart, R. A. (2021). Finding nano: Challenges involved in monitoring the presence and fate of engineered titanium dioxide nanoparticles in aquatic environments. *Water (Basel)*, *13*(5), 734. DOI: 10.3390/w13050734

He, Z., & Alexandridis, P. (2017). Ionic liquid and nanoparticle hybrid systems: Emerging applications. *Advances in Colloid and Interface Science*, *244*, 54–70. DOI: 10.1016/j.cis.2016.08.004 PMID: 27567031

Holbrey, J. D. (2005). Industrial Applications of Ionic Liquids. *ChemInform*, *36*(27), chin.200527275. Advance online publication. DOI: 10.1002/chin.200527275

Holbrey, J. D., & Rogers, R. D. (2002). Green chemistry and ionic liquids: Synergies and ironies. In Rogers, R. D., & Seddon, K. R. (Eds.), *Ionic liquids: Industrial applications for green chemistry* (pp. 2–14). American Chemical Society., DOI: 10.1021/bk-2002-0818.ch001

Hortolà, P. (2005). SEM examination of human erythrocytes in uncoated bloodstains on stone: Use of conventional as environmental-like SEM in a soft biological tissue (and hard inorganic material). *Journal of Microscopy*, *218*(2), 94–103. DOI: 10.1111/j.1365-2818.2005.01477.x PMID: 15857371

Hua, J., Björling, M., Larsson, R., & Shi, Y. (2021). Controlling friction in Ionic Liquid/Glycerol Aqueous Solution lubricated contacts by adjusting CO<inf>2</inf> and water content. *Tribology International*, *161*, 107070. Advance online publication. DOI: 10.1016/j.triboint.2021.107070

Huang, W., Tang, S., Zhao, H., & Tian, S. (2013). Activation of commercial CaO for biodiesel production from rapeseed oil using a novel deep eutectic solvent. *Industrial & Engineering Chemistry Research*, *52*(34), 11943–11947. DOI: 10.1021/ie401292w

Hu, E. H., Sidler, D. R., & Dolling, U. (1998). Unprecedented catalytic three component one-pot condensation Reaction: An efficient synthesis of 5-Alkoxycarbonyl- 4-aryl-3,4-dihydropyrimidin-2(1H)-ones. *The Journal of Organic Chemistry*, *63*(10), 3454–3457. DOI: 10.1021/jo970846u

Hulsbosch, J., De Vos, D. E., Binnemans, K., & Ameloot, R. (2016). Biobased Ionic Liquids: Solvents for a Green Processing Industry? *ACS Sustainable Chemistry & Engineering*, *4*(6), 2917–2931. DOI: 10.1021/acssuschemeng.6b00553

Hussain, A., AlAjmi, M. F., Hussain, I., & Ali, I. (2019). Future of ionic liquids for chiral separations in high-performance liquid chromatography and capillary electrophoresis. *Critical Reviews in Analytical Chemistry*, *49*(4), 289–305. DOI: 10.1080/10408347.2018.1523706 PMID: 30582712

Hynes, N. R. J., Kumar, J. S., Kamyab, H., Sujana, J. A. J., Al-Khashman, O. A., Kuslu, Y., Ene, A., & Suresh Kumar, B. (2020). Modern enabling techniques and adsorbents based dye removal with sustainability concerns in textile industrial sector -A comprehensive review. *Journal of Cleaner Production*, *272*, 122636. DOI: 10.1016/j.jclepro.2020.122636

Indrayanto, G. (2018). Validation of chromatographic methods of analysis: Application for drugs that derived from herbs. *Profiles of Drug Substances, Excipients, and Related Methodology*, *43*, 359–392. DOI: 10.1016/bs.podrm.2018.01.003 PMID: 29678264

Isahak, W. N. R. W., Ismail, M., Yarmo, M. A., Jahangirian, H., & Eshraghi, M. J. (2011). Transesterification of palm oil by using ionic liquids as a new potential catalyst. *Trends in Applied Sciences Research*, *6*(9), 1055–1062. DOI: 10.3923/tasr.2011.1055.1062

Iyengar, S. J., Joy, M., Ghosh, C. K., Dey, S., Kotnala, R. K., & Ghosh, S. (2014). Magnetic, X-ray and Mössbauer studies on magnetite/maghemite core–shell nanostructures fabricated through an aqueous route. *RSC Advances*, *4*(110), 64919–64929. DOI: 10.1039/C4RA11283K

Jaeger, D. A., & Tucker, C. E. (1989). Diels-Alder reactions in ethylammonium nitrate, a low-melting fused salt. *Tetrahedron Letters*, *30*(14), 1785–1788. DOI: 10.1016/S0040-4039(00)99579-0

Jain, A., Shakya, A. K., Prajapati, S. K., Eldesoqui, M., Mody, N., Jain, S. K., Naik, R. R., & Patil, U. K. (2024). An insight into pharmaceutical challenges with ionic liquids: Where do we stand in transdermal delivery? *Frontiers in Bioengineering and Biotechnology*, *12*, 1454247. Advance online publication. DOI: 10.3389/fbioe.2024.1454247 PMID: 39165403

Jaitely, V., Karatas, A., & Florence, A. T. (2008). Water-immiscible room temperature ionic liquids (RTILs) as drug reservoirs for controlled release. *International Journal of Pharmaceutics*, *354*(1–2), 168–173. DOI: 10.1016/j.ijpharm.2008.01.034 PMID: 18342462

Jastorff, B., Mölter, K., Behrend, P., Bottin-Weber, U., Filser, J., Heimers, A., Ondruschka, B., Ranke, J., Schaefer, M., Schröder, H., Stark, A., Stepnowski, P., Stock, F., Störmann, R., Stolte, S., Welz-Biermann, U., Ziegert, S., & Thöming, J. (2005). Progress in evaluation of risk potential of ionic liquids—Basis for an eco-design of sustainable products. *Green Chemistry*, *7*(5), 362. DOI: 10.1039/b418518h

Javan bakht Dalir, S., Djahaniani, H., Nabati, F., & Hekmati, M. (2020). Characterization and the evaluation of antimicrobial activities of silver nanoparticles biosynthesized from Carya illinoinensis leaf extract. *Heliyon, 6*(3), e03624. DOI: 10.1016/j.heliyon.2020.e03624

Jayaseelan, C., Rahuman, A. A., Roopan, S. M., Kirthi, A. V., Venkatesan, J., Kim, S. K., Siva, C., & Iyappan, M. (2013). Biological approach to synthesize TiO2 nanoparticles using Aeromonas hydrophila and its antibacterial activity. *Spectrochimica Acta. Part A: Molecular and Biomolecular Spectroscopy*, *107*, 82–89. DOI: 10.1016/j.saa.2012.12.083 PMID: 23416912

Jhanji, I. (2023). The Role of Ionic Liquids in Green Solvent Chemistry: Properties and Applications. *International Research Journal of Engineering & Applied Sciences*, *11*(4), 57–67. DOI: 10.55083/irjeas.2023.v11i04009

Jiang, T., Ma, X., Zhou, Y., Liang, S., Zhang, J., & Han, B. (2008). Solvent-free synthesis of substituted ureas from CO2 and amines with a functional ionic liquid as the catalyst. *Green Chemistry*, *10*(4), 465–469. DOI: 10.1039/b717868a

Jia, Z., Li, J., Gao, L., Yang, D., & Kanaev, A. (2023). Dynamic light scattering: A powerful tool for in situ nanoparticle sizing. *Colloids and Interfaces*, *7*(1), 15. DOI: 10.3390/colloids7010015

Jongprasitkul, H., Turunen, S., Parihar, V. S., & Kellomäki, M. (2022). Two-step crosslinking to enhance the printability of methacrylated gellan gum biomaterial ink for extrusion-based 3D bioprinting. *Bioprinting (Amsterdam, Netherlands)*, *25*, e00185. Advance online publication. DOI: 10.1016/j.bprint.2021.e00185

Jóźwiak, B., Dziadosz, J., Golba, A., Cwynar, K., Dzido, G., Zorębski, E., Kolanowska, A., Jędrysiak, R., Gancarz, P., Scheller, Ł., Boncel, S., & Dzida, M. (2020). Thermophysical properties of ionanofluids composed of 1-ethyl-3-methylimidazolium thiocyanate and carboxyl-functionalized long multi-walled carbon nanotubes. *Fluids (Basel, Switzerland)*, *5*(4), 214. Advance online publication. DOI: 10.3390/fluids5040214

Kamalesh, R., & Selvakumar, P. (2024). Emerging breakthroughs in membrane filtration techniques and their application in agricultural wastewater treatment: Reusability aspects. *Sustainable Chemistry for the Environment*, *8*, 100183. Advance online publication. DOI: 10.1016/j.scenv.2024.100183

Kapre, S., Palakurthi, S. S., Jain, A., & Palakurthi, S. (2024). DES-igning the future of drug delivery: A journey from fundamentals to drug delivery applications. *Journal of Molecular Liquids*, *400*, 124517. DOI: 10.1016/j.molliq.2024.124517

Karadaghi, L. R., Pan, B., Baddour, F. G., Malmstadt, N., & Brutchey, R. L. (2023). A techno-economic approach to guide the selection of flow recyclable ionic liquids for nanoparticle synthesis. *RSC Sustainability*, *1*(7), 1861–1873. DOI: 10.1039/D3SU00182B

Karadas, F., Atilhan, M., & Aparicio, S. (2010). Review on the use of Ionic liquids (ILs) as alternative fluids for CO_2Capture and natural gas sweetening. *Energy & Fuels*, *24*(11), 5817–5828. DOI: 10.1021/ef1011337

Karthika, S., Radhakrishnan, T. K., & Kalaichelvi, P. (2017). The role of hydrogen bonding propensity in tuning the morphology of crystals obtained from imidazolium based ionic liquids. *Journal of Crystal Growth*, *463*, 168–175. DOI: 10.1016/j.jcrysgro.2017.02.028

Karthikeyan, A., Raphael, W., & Tavares, J. R. (2022). Nanofluids as heat transfer fluids: Hurdles to industrial application and economic considerations. *Canadian Journal of Chemical Engineering*, *100*(11), 3311–3320. DOI: 10.1002/cjce.24338

Kasar, A. K., Rahman, M. H., D'Souza, B., & Menezes, P. L. (2023). Tribological performance of ionic liquid impregnated porous aluminum borate ceramic. *Tribology International*, *180*, 108219. Advance online publication. DOI: 10.1016/j.triboint.2023.108219

Kaur, G., Kumar, H., & Singla, M. (2022). Diverse applications of ionic liquids: A comprehensive review. *Journal of Molecular Liquids*, *351*, 118556. Advance online publication. DOI: 10.1016/j.molliq.2022.118556

Kayser, V., Chennamsetty, N., Voynov, V., Helk, B., Forrer, K., & Trout, B. L. (2011). Evaluation of a non-Arrhenius model for therapeutic monoclonal antibody aggregation. *Journal of Pharmaceutical Sciences*, *100*(7), 2526–2542. DOI: 10.1002/jps.22493 PMID: 21268027

Kazhiyev, S., Kaumbekova, S., & Shah, D. (2023). Malic acid-based deep eutectic solvent and its application in Insulin's structural stability. *Results in Engineering*, *20*, 101529. Advance online publication. DOI: 10.1016/j.rineng.2023.101529

Khachatrian, A. A., Mukhametzyanov, T. A., Yakhvarov, D. G., Sinyashin, O. G., Garifullin, B. F., Rakipov, I. T., Mironova, D. A., Burilov, V. A., & Solomonov, B. N. (2022). Intermolecular interactions between imidazolium- and cholinium-based ionic liquids and lysozyme: Regularities and peculiarities. *Journal of Molecular Liquids*, *348*, 118426. DOI: 10.1016/j.molliq.2021.118426

Khalajiolyaie, A., & Ghaffari, M. (2025). Advances in graphene-based materials for metal ion sensing and wastewater treatment: A review. *Environments (Basel, Switzerland)*, *12*(2), 43. Advance online publication. DOI: 10.3390/environments12020043

Khan, I., Bhat, A. H., Sharma, D. K., Usmani, M. A., & Khan, F. (2019). *Overview of Nanofluids to Ionanofluids: Applications and Challenges* (pp. 199–227). DOI: 10.1007/978-981-13-9833-9_10

Khan, R. A., Mohammed, H. A., Sulaiman, G. M., Subaiyel, A. A., Karuppaiah, A., Rahman, H., Makhathini, S., Ramburrun, P., & Choonara, Y. E. (2022). Molecule(s) of Interest: I. Ionic Liquids–Gateway to Newer Nanotechnology Applications: Advanced Nanobiotechnical Uses', Current Status, Emerging Trends, Challenges, and Prospects. *International Journal of Molecular Sciences*, *23*(22), 14346. DOI: 10.3390/ijms232214346 PMID: 36430823

Kim, K. S., Choi, J. H., Yeon, S. H., & Lee, H. (2006). Facile one-pot synthesis of gold nanoparticles using alcohol ionic liquids. *Journal of Materials Chemistry*, *16*(14), 1315–1317. DOI: 10.1039/b601478j

Kubisa, P. (2005). Ionic liquids in the synthesis and modification of polymers. *Journal of Polymer Science. Part A, Polymer Chemistry*, *43*(20), 4675–4683. DOI: 10.1002/pola.20971

Kumar, A., Barik, B., Jablonski, P. G., Sonkaria, S., & Khare, V. (2022). Functionalized and biomimicked carbon-based materials and their impact for improving surface coatings for protection and functionality: Insights and technological trends. *Coatings*, *12*(11), 1674. Advance online publication. DOI: 10.3390/coatings12111674

Kumar, A., Kumar, V., Singh, P., Tittal, R. K., & Lal, K. (2024). Ionic liquids for the green synthesis of 1,2,3-triazoles: A systematic review. *Green Chemistry*, *26*(7), 3565–3594. DOI: 10.1039/D3GC04898E

Kumar, D. A., Palanichamy, V., & Roopan, S. M. (2014). Green synthesis of silver nanoparticles using Alternanthera dentata leaf extract at room temperature and their antimicrobial activity. *Spectrochimica Acta. Part A: Molecular and Biomolecular Spectroscopy*, *127*, 168–171. DOI: 10.1016/j.saa.2014.02.058 PMID: 24632169

Kumari, M., Sharma, S., & Deep, S. (2022). Tetrabutylammonium based ionic liquids (ILs) inhibit the amyloid aggregation of superoxide dismutase 1 (SOD1). *Journal of Molecular Liquids*, *353*, 118761. DOI: 10.1016/j.molliq.2022.118761

Kumari, M., Widdrat, M., Tompa, É., Uebe, R., Schüler, D., Pósfai, M., & Hirt, A. M. (2014). Distinguishing magnetic particle size of iron oxide nanoparticles with first-order reversal curves. *Journal of Applied Physics*, *116*(12), 124304. DOI: 10.1063/1.4896481

Kumari, P., Pillai, V. V. S., & Benedetto, A. (2020). Mechanisms of action of ionic liquids on living cells: The state of the art. *Biophysical Reviews*, *12*(5), 1187–1215. DOI: 10.1007/s12551-020-00754-w PMID: 32936423

Kumar, S., Yadav, D. V., & Nehra, D. A. (2023). A Review Of Characterization Tools And Techniques For Nanomaterials And Nanocomposites. *International Journal of Psychosocial Rehabilitation*, *27*(2), 42–53. DOI: 10.61841/rpe30a86

Kumer, A., & Khan, M. W. (2021). Synthesis, characterization, antimicrobial activity and computational exploration of ortho toludinium carboxylate ionic liquids. *Journal of Molecular Structure*, *1245*, 131087. DOI: 10.1016/j.molstruc.2021.131087

Laali, K. K., & Gettwert, V. J. (2000). Electrophilic nitration of aromatics in Ionic liquid solvents. *The Journal of Organic Chemistry*, *66*(1), 35–40. DOI: 10.1021/jo000523p PMID: 11429927

Lai, J. Q., Hu, Z. L., Wang, P. W., & Yang, Z. (2012). Enzymatic production of microalgal biodiesel in ionic liquid [BMIm] [PF6]. *Fuel*, *95*, 329–333. DOI: 10.1016/j.fuel.2011.11.001

Lantos, J., Kumar, N., & Saha, B. (2024). A Comprehensive Review of Fine Chemical Production Using Metal-Modified and Acidic Microporous and Mesoporous Catalytic Materials. *Catalysts*, *14*(5), 317. DOI: 10.3390/catal14050317

Lee, B., Yoon, S., Lee, J. W., Kim, Y., Chang, J., Yun, J., Ro, J. C., Lee, J.-S., & Lee, J. H. (2020). Statistical Characterization of the Morphologies of Nanoparticles through Machine Learning Based Electron Microscopy Image Analysis. *ACS Nano*, *14*(12), 17125–17133. DOI: 10.1021/acsnano.0c06809 PMID: 33231065

Liang, J., Ren, X., Wang, J., Jiang, M., & Li, Z. (2010). Preparation of biodiesel by transesterification from cottonseed oil using the basic dication ionic liquids as catalysts. *Journal of Fuel Chemistry & Technology*, *38*(3), 275–280. DOI: 10.1016/S1872-5813(10)60033-3

Liao, X., Zheng, K., Wang, G., Yang, Y., Li, Y., & Coppens, M.-O. (2022). Solubility of CO<inf>2</inf>in Ionic Liquids with Additional Water and Methanol: Modeling with PC-SAFT Equation of State. *Industrial & Engineering Chemistry Research*, *61*(38), 14364–14373. DOI: 10.1021/acs.iecr.2c02778

Li, B., Wang, J., Gui, Q., & Yang, H. (2020). Drug-loaded chitosan film prepared via facile solution casting and air-drying of plain water-based chitosan solution for ocular drug delivery. *Bioactive Materials*, *5*(3), 577–583. DOI: 10.1016/j.bioactmat.2020.04.013 PMID: 32405573

Li, F., Zuo, Z., Cao, B., & Ji, X. (2024). Study on the Thermodynamic Properties of Ionic Liquids 1-Hexyl-3-methylimidazolium Halide with Isopropanol Mixtures. *Journal of Chemical & Engineering Data*, *69*(11), 3794–3804. DOI: 10.1021/acs.jced.4c00343

Li, H., Wang, J., Warr, G. G., & Atkin, R. (2024). Nanostructure and Dynamics of the Locally Concentrated Ionic Liquid 2:1 (wt:wt) HMIM FAP:TFTFE and HMIM FAP on Graphite and Gold Electrodes as a Function of Potential. *Small*, *20*(46), 2403109. Advance online publication. DOI: 10.1002/smll.202403109 PMID: 39105361

Li, K.-X., Chen, L., Yan, Z.-C., & Wang, H.-L. (2010). Application of pyridinium ionic liquid as a recyclable catalyst for acid-catalyzed transesterification of Jatropha oil. *Catalysis Letters*, *139*(3–4), 151–156. DOI: 10.1007/s10562-010-0409-x

Liu, C.-Z., Wang, F., & Ou-Yang, F. (2015). An overview of the role of ionic liquids in biodiesel reactions. *Journal of Industrial and Engineering Chemistry*, *21*, 1–10. DOI: 10.1016/j.jiec.2014.01.046

Liu, Q., Janssen, M. H. A., van Rantwijk, F., & Sheldon, R. A. (2005). Room-temperature ionic liquids that dissolve carbohydrates in high concentrations. *Green Chemistry*, *7*(1), 39–42. DOI: 10.1039/b412848f

Liu, Q., Xu, J., Jiang, W., Gim, J., Tornheim, A. P., Pathak, R., Zhu, Q., Zuo, P., Yang, Z., Pupek, K. Z., Xu, K., & Zhang, Z. (2024). High-Energy LiNiO<inf>2</inf> Li Metal Batteries Enabled by Hybrid Electrolyte Consisting of Ionic Liquid and Weakly Solvating Fluorinated Ether. *Advancement of Science*, *11*(46), 2409662. Advance online publication. DOI: 10.1002/advs.202409662 PMID: 39418091

Liu, T., Liu, Y., Zhao, X., Zhang, L., Wang, W., Bai, D., Liao, Y., Wang, Z., Wang, M., & Zhang, J. (2024). Thermodynamically stable ionic liquid microemulsions pioneer pathways for topical delivery and peptide application. *Bioactive Materials*, *32*, 502–513. DOI: 10.1016/j.bioactmat.2023.10.002 PMID: 38026438

Liu, Y., Chen, X., Chen, X., Chen, J., Zhang, H., Xu, H., Jin, L., Wang, Q., & Tang, Z. (2024). Preparation and in vivo and ex vivo studies of sirolimus nano-in-situ gel ophthalmic formulation. *Journal of Nanobiotechnology*, *22*(1), 417. Advance online publication. DOI: 10.1186/s12951-024-02668-1 PMID: 39014353

Loeschner, K., Johnson, M. E., & Montoro Bustos, A. R. (2023). Application of single particle ICP-MS for the determination of inorganic nanoparticles in food additives and food: A short review. *Nanomaterials (Basel, Switzerland)*, *13*(18), 2547. DOI: 10.3390/nano13182547 PMID: 37764576

Long, F., Wang, L., & Zhang, X. (2025). Versatile electrode materials applied in the electrochemical advanced oxidation processes for wastewater treatment: A systematic review. *Separation and Purification Technology*, *354*(Part 1), 128725. Advance online publication. DOI: 10.1016/j.seppur.2024.128725

Long, T., Deng, Y., Gan, S., & Chen, J. (2010). Application of choline chloride·x-ZnCl2 ionic liquids for preparation of biodiesel. *Chinese Journal of Chemical Engineering*, *18*(2), 322–327. DOI: 10.1016/S1004-9541(08)60359-6

Lu, B., Liu, T., Wang, H., Wu, C., Chen, H., Liu, Z., & Zhang, J. (2022). Ionic liquid transdermal delivery system: Progress, prospects, and challenges. *Journal of Molecular Liquids*, *351*, 118643. DOI: 10.1016/j.molliq.2022.118643

Luo, H., Fan, W., Li, Y., & Nan, G. (2013). Biodiesel production using alkaline ionic liquid and adopted as lubricity additive for low-sulfur diesel fuel. *Bioresource Technology*, *140*, 337–341. DOI: 10.1016/j.biortech.2012.11.112 PMID: 23708846

Luska, K. L., Migowski, P., & Leitner, W. (2015). Ionic liquid-stabilized nanoparticles as catalysts for the conversion of biomass. *Green Chemistry*, *17*(6), 3195–3206. DOI: 10.1039/C5GC00231A

Ma, C., Wang, N., Ye, N., & Ji, X. (2021). CO<inf>2</inf> capture using ionic liquid-based hybrid solvents from experiment to process evaluation. *Applied Energy*, *304*, 117767. Advance online publication. DOI: 10.1016/j.apenergy.2021.117767

MacFarlane, D. R., Chong, A. L., Forsyth, M., Kar, M., Vijayaraghavan, R., Somers, A., & Seddon, K. R. (2018). New dimensions in salt–solvent mixtures: A 4th evolution of ionic liquids. *Faraday Discussions*, *206*, 9–28. DOI: 10.1039/C7FD00189D PMID: 29034392

Magina, S., Barros-Timmons, A., Ventura, S. P. M., & Evtuguin, D. V. (2021). Evaluating the hazardous impact of ionic liquids–challenges and opportunities. *Journal of Hazardous Materials*, *412*, 125215. DOI: 10.1016/j.jhazmat.2021.125215 PMID: 33951860

Majhi, S., Bhattacharyya, S., & Gopmandal, P. P. (2024). Impact of laterally mobile surface charge on diffusiophoresis of hydrophobic rigid colloids. *Journal of Fluid Mechanics*, *997*, A8. Advance online publication. DOI: 10.1017/jfm.2024.778

Majumder, M. A. A., Rahman, S., Cohall, D., Bharatha, A., Singh, K., Haque, M., & Gittens-St Hilaire, M. (2020). Antimicrobial stewardship: Fighting antimicrobial resistance and protecting global public health. *Infection and Drug Resistance*, *13*, 4713–4738. DOI: 10.2147/IDR.S290835 PMID: 33402841

Maliszewska, I., Czapka, T., & Grząbka-Zasadzińska, A. (2022). Electrospun polymer nanofibers with antimicrobial activity. *Polymers*, *14*(9), 1661. Advance online publication. DOI: 10.3390/polym14091661 PMID: 35566830

Maneewattanapinyo, P., Pichayakorn, W., Monton, C., Dangmanee, N., Wunnakup, T., & Suksaeree, J. (2023). Effect of ionic liquid on silver-nanoparticle-complexed Ganoderma applanatum and its topical film formulation. *Pharmaceutics*, *15*(4), 1098. DOI: 10.3390/pharmaceutics15041098 PMID: 37111583

Manic, M. S., Najdanovic-Visak, V., da Ponte, M. N., & Visak, Z. P. (2011). Extraction of free fatty acids from soybean oil using ionic liquids or poly(ethylene glycol)s. *AIChE Journal. American Institute of Chemical Engineers*, *57*(5), 1344–1355. DOI: 10.1002/aic.12349

Mansfield, E., Tyner, K. M., Poling, C. M., & Blacklock, J. L. (2014). Determination of nanoparticle surface coatings and nanoparticle purity using microscale thermogravimetric analysis. *Analytical Chemistry*, *86*(3), 1478–1484. DOI: 10.1021/ac402888v PMID: 24400715

Markiewicz, R., Klimaszyk, A., Jarek, M., Taube, M., Florczak, P., Kempka, M., Fojud, Z., & Jurga, S. (2021). Influence of alkyl chain length on thermal properties, structure, and Self-Diffusion coefficients of Alkyltriethylammonium-Based ionic liquids. *International Journal of Molecular Sciences*, *22*(11), 5935. DOI: 10.3390/ijms22115935 PMID: 34073046

Marr, P. C., & Marr, A. C. (2016). Ionic liquid gel materials: Applications in green and sustainable chemistry. *Green Chemistry*, *18*(1), 105–128. DOI: 10.1039/C5GC02277K

Marrucho, I., Branco, L., & Rebelo, L. (2014). Ionic liquids in pharmaceutical applications. *Annual Review of Chemical and Biomolecular Engineering*, *5*(1), 527–546. DOI: 10.1146/annurev-chembioeng-060713-040024 PMID: 24910920

Marullo, S., D'Anna, F., Rizzo, C., & Billeci, F. (2021). Ionic liquids: "normal" solvents or nanostructured fluids? *Organic & Biomolecular Chemistry*, *19*(10), 2076–2095. DOI: 10.1039/D0OB02214D PMID: 33606870

Mata, T. M., Martins, A. A., & Caetano, N. S. (2010). Microalgae for biodiesel production and other applications: A review. *Renewable & Sustainable Energy Reviews*, *14*(1), 217–232. DOI: 10.1016/j.rser.2009.07.020

Mathews, C. J., Smith, P. J., & Welton, T. (2000). Palladium catalysed Suzuki cross-coupling reactions in ambient temperature Ionic liquids. *Chemical Communications (Cambridge)*, (14), 1249–1250. DOI: 10.1039/b002755n

Ma, Y., Qian, C., Wang, L., & Yang, M. (2000). Lanthanide Triflate catalyzed Biginelli reaction. one-pot synthesis of Dihydropyrimidinones under solvent-free conditions. *The Journal of Organic Chemistry*, *65*(12), 3864–3868. DOI: 10.1021/jo9919052 PMID: 10864778

Md. Javvad Ali, Praveen M., Ghouse A. and Fatima F. (2021) A brief review of ionic liquids: synthesis and applications. International J. of Biomed Research. 1(2); DOI: . (2021). *Application of Ionic Liquids in Drug Delivery*, 165-186. https://doi.org/DOI: 10.31579/IJBR-2021/011

Mehmood, N., Husson, E., Jacquard, C., Wewetzer, S., Büchs, J., Sarazin, C., & Gosselin, I. (2015). Impact of two ionic liquids, 1-ethyl-3-methylimidazolium acetate and 1-ethyl-3-methylimidazolium methylphosphonate, on Saccharomyces cerevisiae: Metabolic, physiologic, and morphological investigations. *Biotechnology for Biofuels*, *8*(1), 1–14. DOI: 10.1186/s13068-015-0206-2 PMID: 25688291

Mehrin, S., Tanisa, N. Y., Awal, R., Alam Khan, M. K., Shaqur, A., Ahmed, S., Islam, S., & Mia, A. (2024). Efficient Bioelectrochemical Cell Generation and Green Synthesis of Silver Nanoparticles Using Pomegranate and Pineapple Peel Extracts: A Comprehensive Characterization Study. *Advances in Materials Science and Engineering*, *2024*(1), 6681559. DOI: 10.1155/2024/6681559

Michalski, J., Odrzygozdz, C., Mester, P., Narozna, D., & Clapa, T. (2023). Defeat undefeatable: Ionic liquids as novel antimicrobial agents. *Journal of Molecular Liquids*, *369*, 120782. DOI: 10.1016/j.molliq.2022.120782

Millard, J., Alvarez-Nunez, F., & Yalkowsky, S. (2002). Solubilization by cosolvents: Establishing useful constants for the log-linear model. *International Journal of Pharmaceutics*, *245*(1–2), 153–166. DOI: 10.1016/S0378-5173(02)00334-4 PMID: 12270252

Mir, M. A. (2023). DFT, Hirshfeld surface, molecular docking and drug likeness studies of medicinally important coumarin molecule. *Arabian Journal for Science and Engineering*, *48*(6), 7445–7462. DOI: 10.1007/s13369-022-07476-z

Mir, M. A. (2023). Molecular dynamic, Hirshfeld surface, molecular docking and drug likeness studies of a potent anti-oxidant, anti-malaria and anti-inflammatory medicine: Pyrogallol. *Results in Chemistry*, *5*, 100763. DOI: 10.1016/j.rechem.2023.100763

Mir, M. A. (2024). Biogenic amines: Catalysis, quality, and safety aspects of food items consumed in Saudi Arabia. *Current Organocatalysis*, *11*(4), 310–320. DOI: 10.2174/0122133372285097240220062811

Mir, M. A. (2024). DFT analysis for structure, vibrational assignments, and molecular properties of medicinally important isoquinoline. *Journal of Dispersion Science and Technology*, *45*(8), 1–11. DOI: 10.1080/01932691.2023.2299360

Mir, M. A. (2024). Utilization of medicinally important plants as antidiabetic medicines in a sustainable manner. In Banik, B. K. (Ed.), *Advances in green and sustainable chemistry: Green approaches in medicinal chemistry for sustainable drug design* (2nd ed., Vol. 1, pp. 397–410). Elsevier., DOI: 10.1016/B978-0-443-16166-7.00017-7

Mir, M. A., Ahmad, W., Andrews, K., & Chang, S. K. (2024). Chemical synthesis, experimental, molecular docking and drug-likeness studies of salidroside. *Arabian Journal for Science and Engineering*, *49*(7), 9451–9466. DOI: 10.1007/s13369-024-08721-3

Mir, M. A., & Banik, B. K. (2025). Sustainable healing: Natural compounds facilitating the future cancer treatment. *World Development Sustainability*, *6*, 100215. Advance online publication. DOI: 10.1016/j.wds.2025.100215

Mir, M. A., & Banik, B. K. (2025). Synthesis of Schiff base ligands under environment-friendly conditions: A systematic review. *Inorganic Chemistry Communications*, *174*(Part 1), 113987. Advance online publication. DOI: 10.1016/j.inoche.2025.113987

Mir, M. A., Chang, S. K., & Hefni, D. (2024). A comprehensive review on challenges and choices of food waste in Saudi Arabia: Exploring environmental and economic impacts. *Environmental Systems Research*, *13*(1), 40. Advance online publication. DOI: 10.1186/s40068-024-00364-5

Mir, M. A., Chang, S. K., & Hefni, D. (2024). Saudi Arabia e-waste management strategies, challenges and opportunities, effect on health and environment: A strategic review. *Emerging Contaminants*, *10*(4), 100357. Advance online publication. DOI: 10.1016/j.emcon.2024.100357

Mir, M. A., Hefni, D., & Chauhan, N. (2025). Heritage to health: The socio-cultural and health impacts of traditional ethnic foods in Saudi Arabian society. *Food Chemistry Advances*, *6*, 100927. Advance online publication. DOI: 10.1016/j.focha.2025.100927

Mir, M. A., Jassal, M. M. S., & Andrews, K. (2024). Molecular dynamic, Hirshfeld surface, computational quantum and spectroscopic analysis of 4-hydroxy-1-naphthaldehyde. *Current Organocatalysis*, *11*(1), 1–12. DOI: 10.2174/221333721 0666230816091246

Mir, M. A., Khan, M. A. A., Banik, B. K., Hasnain, S. M., Alzayer, L., Andrews, K., & Abba, S. I. (2025). Microplastics in food products: Prevalence, artificial intelligence-based detection, and potential health impacts on humans. *Emerging Contaminants*, *11*(4), 100477. Advance online publication. DOI: 10.1016/j.emcon.2025.100477

Mir, M. A., Kumar, A., Madwal, S. P., Chauhan, R., & Kumar, T. (2022). Synthesis and spectral analysis of the Cr (II) lapachol chelate crystal complexes using DMC, H2O mixture. *Inorganic and Nano-Metal Chemistry*, *54*(7), 621–626. DOI: 10.1080/24701556.2022.2081202

Mizuuchi, H., Jaitely, V., Murdan, S., & Florence, A. T. (2008). Room temperature ionic liquids and their mixtures: Potential pharmaceutical solvents. *European Journal of Pharmaceutical Sciences*, *33*(4–5), 326–331. DOI: 10.1016/j.ejps.2008.01.002 PMID: 18291630

Mohite, D. D., Goyal, A., Singh, A. S., Ansari, M. I., Patil, K. A., Yadav, P. D., Patil, M. J., & Londhe, P. V. (2024). Improvement of thermal performance through nanofluids in industrial applications: A review on technical aspects. *Materials Today: Proceedings*. Advance online publication. DOI: 10.1016/j.matpr.2024.04.083

Mokhtarpour, M., Basteholia, N., Shekaari, H., & Zafarani-Moattar, M. T. (2020). Effect of choline-based ionic liquids as novel green solvents on the aqueous solubility enhancement and thermodynamic properties of acetaminophen. *Journal of Molecular Liquids*, *306*, 112504. DOI: 10.1016/j.molliq.2020.112504

Molodenskiy, D., Shirshin, E., Tikhonova, T., Gruzinov, A., Peters, G., & Spinozzi, F. (2017). Thermally induced conformational changes and protein–protein interactions of bovine serum albumin in aqueous solution under different pH and ionic strengths as revealed by SAXS measurements. *Physical Chemistry Chemical Physics*, *19*(25), 17143–17155. DOI: 10.1039/C6CP08809K PMID: 28636681

Moniruzzaman, M., Nakashima, K., Kamiya, N., & Goto, M. (2010). Recent advances of enzymatic reactions in ionic liquids. *Biochemical Engineering Journal*, *48*(3), 295–314. DOI: 10.1016/j.bej.2009.10.002

Moosavi, M., Torkzadeh, M., & Akbarinezhad, Z. (2024). Molecular dynamics investigation of ionanofluids (INFs): Towards a deeper understanding of their thermophysical, structural and dynamical properties. *Journal of Molecular Liquids*, *399*, 124355. DOI: 10.1016/j.molliq.2024.124355

Moradi, P., & Hajjami, M. (2021). Magnetization of graphene oxide nanosheets using nickel magnetic nanoparticles as a novel support for the fabrication of copper as a practical, selective, and reusable nanocatalyst in C–C and C–O coupling reactions. *RSC Advances*, *11*(42), 25867–25879. DOI: 10.1039/D1RA03578A PMID: 35479448

Moshikur, R. M., Carrier, R. L., Moniruzzaman, M., & Goto, M. (2023). Recent advances in biocompatible ionic liquids in drug formulation and delivery. *Pharmaceutics*, *15*(4), 1179. DOI: 10.3390/pharmaceutics15041179 PMID: 37111664

Moshikur, R. M., & Goto, M. (2021). In Goto, M., & Moniruzzaman, M. (Eds.), *"Ionic liquids as active pharmaceutical ingredients (APIs)," in Application of ionic liquids in drug delivery* (pp. 13–33). Springer Singapore.

Mosmann, T. (1983). Rapid colorimetric assay for cellular growth and survival: Application to proliferation and cytotoxicity assays. *Journal of Immunological Methods*, *65*(1–2), 55–63. DOI: 10.1016/0022-1759(83)90303-4 PMID: 6606682

Mota-Martinez, M. T., Brandl, P., Hallett, J. P., & Mac Dowell, N. (2018). Challenges and opportunities for the utilisation of ionic liquids as solvents for CO2capture. *Molecular Systems Design & Engineering*, *3*(3), 560–571. DOI: 10.1039/C8ME00009C

Muderawan, I. W., Ong, T. T., & Ng, S. C. (2006). Urea bonded cyclodextrin derivatives onto silica for chiral HPLC. *Journal of Separation Science*, *29*(12), 1849–1871. DOI: 10.1002/jssc.200500491 PMID: 16970188

Mudhaffar, B., & Salimon, J. (2010). Optimization of process variables using D-optimal design for separating linoleic acid in *Jatropha curcas* seed oil by urea complex fractionation. *Biotechnology (Faisalabad)*, *9*(3), 362–367. DOI: 10.3923/biotech.2010.362.367

Mukherjee, S., Wciślik, S., Chandra Mishra, P., & Chaudhuri, P. (2024). Nanofluids: Critical issues, economics and sustainability perspectives. *Particuology*, *87*, 147–172. DOI: 10.1016/j.partic.2023.06.021

Murshed, S. M. S., de Castro, C. A. N., Lourenço, M. J. V., Lopes, M. L. M., & Santos, F. J. V. (2012). Current research and future applications of nano- and ionanofluids. *Journal of Physics: Conference Series*, *395*, 012117. DOI: 10.1088/1742-6596/395/1/012117

Musiał, M., Malarz, K., Mrozek-Wilczkiewicz, A., Musiol, R., Zorębski, E., & Dzida, M. (2017). Pyrrolidinium-Based ionic liquids as sustainable media in Heat-Transfer processes. *ACS Sustainable Chemistry & Engineering*, *5*(11), 11024–11033. DOI: 10.1021/acssuschemeng.7b02918

Mutelet, F., Butet, V., & Jaubert, J. (2005). Application of inverse gas chromatography and regular solution theory for characterization of Ionic liquids. *Industrial & Engineering Chemistry Research*, *44*(11), 4120–4127. DOI: 10.1021/ie0488061

Naseer, A., & Ahmad, F. (2024). Role of nanocomposites and nano adsorbents for heavy metals removal and dyes: An overview. *Desalination and Water Treatment*, *320*, 100662. Advance online publication. DOI: 10.1016/j.dwt.2024.100662

Nasirpouri, F., Fallah, S., Ahmadpour, G., Moslehifard, E., Samardak, A. Y., Samardak, V. Y., & Samardak, A. S. (2023). Microstructure, ion adsorption and magnetic behavior of mesoporous γ-Fe 2 O 3 ferrite nanoparticles. *RSC Advances*, *13*(36), 25140–25158. DOI: 10.1039/D3RA01663C PMID: 37622013

Nawaz, A., & Ahmad, M. (2024). Zinc oxide nanoparticles: Pathways to micropollutant adsorption, dye removal, and antibacterial actions—A study of mechanisms, challenges, and future prospects. *Journal of Molecular Structure*, *1312*(Part 1), 138545. Advance online publication. DOI: 10.1016/j.molstruc.2024.138545

Nazari, S., Cameron, S., Johnson, M. B., & Ghandi, K. (2013). Physicochemical properties of imidazo-pyridine protic Ionic liquids. *Journal of Materials Chemistry. A, Materials for Energy and Sustainability*, *1*(38), 11570. DOI: 10.1039/c3ta12022h

Ngo, H. L., LeCompte, K., Hargens, L., & McEwen, A. B. (2000). Thermal properties of imidazolium Ionic liquids. *Thermochimica Acta*, *357-358*, 97–102. DOI: 10.1016/S0040-6031(00)00373-7

Niedermeyer, H., Hallett, J. P., Villar-Garcia, I. J., Hunt, P. A., & Welton, T. (2013). A step towards the a priori design of ionic liquids. *Physical Chemistry Chemical Physics*, *15*(27), 11566–11577. DOI: 10.1039/c3cp50521a PMID: 23752557

Nishu, S. K., & Gaur, R. (2023). Smart and innovative nanotechnology applications for water purification. *Hybrid Advances*, *3*, 100044. Advance online publication. DOI: 10.1016/j.hybadv.2023.100044

Nixon, W. C. (1971). The general principles of scanning electron microscopy. *Philosophical Transactions of the Royal Society of London. Series B, Biological Sciences*, *261*(837), 45–50. DOI: 10.1098/rstb.1971.0035

Numpilai, T., Pham, L. K. H., & Wiston, T. (2024). Advances in Ionic Liquid Technologies for CO_2 Capture and Conversion: A Comprehensive review. *Industrial & Engineering Chemistry Research*, *63*(46), 19865–19915. DOI: 10.1021/acs.iecr.4c02072

Ogodo, U. P., & Abosede, O. O. (2025). The Role of Chemistry in Achieving Sustainable Development Goals: Green Chemistry Perspective. *International Research Journal of Pure and Applied Chemistry*, *26*(1), 1–8. DOI: 10.9734/irjpac/2025/v26i1893

Oliver Kappe, C. (1993). 100 years of the biginelli dihydropyrimidine synthesis. *Tetrahedron*, *49*(32), 6937–6963. DOI: 10.1016/S0040-4020(01)87971-0

Omori, N., Bobitan, A. D., Vamvakeros, A., Beale, A. M., & Jacques, S. D. M. (2023). Recent developments in X-ray diffraction/scattering computed tomography for materials science. *Philosophical Transactions. Series A, Mathematical, Physical, and Engineering Sciences*, *381*(2259), 20220350. Advance online publication. DOI: 10.1098/rsta.2022.0350 PMID: 37691470

Orha, L., Akien, G. R., & Horvath, I. T. (2012). Ionic liquids as green solvents for nucleophilic substitution reactions. In P. T. Anastas (Ed.), *Handbook of green chemistry: Vol. 7. Green synthesis* (pp. 93–120). Wiley-VCH.

Oskarsson, A., & Wright, M. C. (2019). Ionic Liquids: New Emerging Pollutants, Similarities with Perfluorinated Alkyl Substances (PFASs). *Environmental Science & Technology*, *53*(18), 10539–10541. DOI: 10.1021/acs.est.9b04778 PMID: 31442027

Ouellet, M., Datta, S., Dibble, D. C., Tamrakar, P. R., & Simmons, P. I. (2011). Impact of ionic liquid pretreated plant biomass on Saccharomyces cerevisiae growth and biofuel production. *Green Chemistry*, *13*(10), 2743–2749. DOI: 10.1039/c1gc15327g

Ovejero-Perez, A., Nakasu, P. Y. S., Hopson, C., Costa, J. M., & Hallett, J. P. (2024). Challenges and opportunities on the utilisation of ionic liquid for biomass pretreatment and valorisation. *Npj Materials Sustainability*, *2*(1), 7. Advance online publication. DOI: 10.1038/s44296-024-00015-x

Ozkan, S. Z., & Karpacheva, G. P. (2018). Hybrid Materials Based on Poly-3-amine-7-methylamine-2-methylphenazine and Magnetite Nanoparticles Immobilized on Single-Walled Carbon Nanotubes. *Polymers*, *10*(5), 544–544. DOI: 10.3390/polym10050544 PMID: 30966578

Pan, S., Yao, M., Zhang, J., Li, B., Xing, C., Song, X., Su, P., & Zhang, H. (2020). Recognition of ionic liquids as High-Voltage electrolytes for supercapacitors. *Frontiers in Chemistry, 8*, 261. Advance online publication. DOI: 10.3389/fchem.2020.00261 PMID: 32432074

Park, S., & Kazlauskas, R. J. (2001). Improved preparation and use of room-temperature ionic liquids in lipase-catalyzed enantio- and regioselective acylations. *The Journal of Organic Chemistry, 66*(25), 8395–8401. DOI: 10.1021/jo015761e PMID: 11735517

Pasquini, A., Picotto, G. B., & Pisani, M. (2005). STM carbon nanotube tips fabrication for critical dimension measurements. *Sensors and Actuators. A, Physical, 123-124*, 655–659. DOI: 10.1016/j.sna.2005.02.036

Passos, H., Freire, M. G., & Coutinho, J. A. P. (2014). Ionic liquid solutions as extractive solvents for value-added compounds from biomass. *Green Chemistry, 16*(12), 4786–4815. DOI: 10.1039/C4GC00236A PMID: 25516718

Patel, M., Mishra, S., Verma, R., & Deep Shikha. (2022). *Synthesis of ZnO and CuO nanoparticles via Sol gel method and Its Characterization by using XRD and FT-IR Analysis*. DOI: 10.21203/rs.3.rs-1234162/v1

Peng, J., & Deng, Y. (2001). Ionic liquids catalyzed Biginelli reaction under solvent-free conditions. *Tetrahedron Letters, 42*(34), 5917–5919. DOI: 10.1016/S0040-4039(01)01139-X

Pereira, J., Souza, R., & Moita, A. (2024). A Review of Ionic Liquids and Their Composites with Nanoparticles for Electrochemical Applications. *Inorganics, 12*(7), 186. DOI: 10.3390/inorganics12070186

Pereira, J., Souza, R., Moita, A., & Moreira, A. (2023). Nanofluids and Ionic Fluids as Liquid Electrodes: An Overview on Their Properties and Potential Applications. *Processes (Basel, Switzerland), 11*(11), 3189. DOI: 10.3390/pr11113189

Pfruender, H., Jones, R., & Weuster-Botz, D. (2006). Water immiscible ionic liquids as solvents for whole cell biocatalysis. *Journal of Biotechnology, 124*(1), 182–190. DOI: 10.1016/j.jbiotec.2005.12.004 PMID: 16413078

Pham, P. J., Li, M., Pittman, C. U., & Li, T. (2009). Selective extraction of polyunsaturated triacylglycerols using a novel ionic liquid precursor immobilized on a mesoporous complexing adsorbent. *Biotechnology Progress, 25*(5), 1419–1426. DOI: 10.1002/btpr.219 PMID: 19882672

Pham, T. P., Cho, C. W., & Yun, Y. S. (2010). Environmental fate and toxicity of ionic liquids: A review. *Water Research*, *44*(2), 352–372. DOI: 10.1016/j.watres.2009.09.030 PMID: 19854462

Phukan, P., & Sarma, D. (2021). Synthesis of medicinally relevant scaffolds-triazoles and pyrazoles in green solvent ionic liquids. *Current Organic Chemistry*, *25*(13), 1523–1538. DOI: 10.2174/1385272825666210412160142

Pinkert, A., Marsh, K. N., Pang, S., & Staiger, M. P. (2009). Ionic liquids and their interaction with cellulose. *Chemical Reviews*, *109*(12), 6712–6728. DOI: 10.1021/cr9001947 PMID: 19757807

Pisano, R., & Durlo, A. (2023). Feynman's frameworks on nanotechnology in historiographical debate. In Condé, M. L., & Salomon, M. (Eds.), *Handbook for the historiography of science* (pp. 465–484). Springer., DOI: 10.1007/978-3-031-27510-4_26

Plechkova, N. V., & Seddon, K. R. (2008). Applications of Ionic liquids in the chemical industry. *Chemical Society Reviews*, *37*(1), 123–150. DOI: 10.1039/B006677J PMID: 18197338

Polesca, C., Passos, H., Coutinho, J. A. P., & Freire, M. G. (2022). Sustainable development of biomaterials using ionic liquids. *Current Opinion in Green and Sustainable Chemistry*, *37*, 100675. DOI: 10.1016/j.cogsc.2022.100675

Poole, C. F. (2004). Chromatographic and spectroscopic methods for the determination of solvent properties of room temperature Ionic liquids. *Journal of Chromatography. A*, *1037*(1-2), 49–82. DOI: 10.1016/j.chroma.2003.10.127 PMID: 15214660

Puiman, L., Elisiário, M. P., Crasborn, L. M. L., Wagenaar, L. E. C. H., Straathof, A. J. J., & Haringa, C. (2022). Gas mass transfer in syngas fermentation broths is enhanced by ethanol. *Biochemical Engineering Journal*, *185*, 108505. Advance online publication. DOI: 10.1016/j.bej.2022.108505

Punzalan, E., Froese, R. D., & Zimmerman, P. M. (2023). Revealing the Interactions between the Flexible Polymer and Counteranion during Propagation and Termination for Olefin Polymerization with the Ti(IV) Constrained Geometry Catalyst. *Organometallics*, *42*(22), 3236–3248. DOI: 10.1021/acs.organomet.3c00365

Puri, N., & Gupta, A. (2023). Water remediation using titanium and zinc oxide nanomaterials through disinfection and photocatalysis process: A review. *Environmental Research*, *227*, 115786. Advance online publication. DOI: 10.1016/j.envres.2023.115786 PMID: 37004858

Rahim, N. Y., Tay, K. S., & Mohamad, S. (2018). Chromatographic and spectroscopic studies on β-cyclodextrin functionalized ionic liquid as chiral stationary phase: Enantio separation of NSAIDs. *Adsorption Science and Technology*, *36*(1–2), 130–148. DOI: 10.1177/0263617416686798

Rahman, A., Chowdhury, M. A., & Hossain, N. (2022). Green synthesis of hybrid nanoparticles for biomedical applications: A review. *Applied Surface Science Advances*, *11*, 100296. DOI: 10.1016/j.apsadv.2022.100296

Ranke, J., Mölter, K., Stock, F., Bottin-Weber, U., & Hoffmann, J. (2004). Biological effects of imidazolium ionic liquids with varying chain lengths in acute Vibrio fischeri and WST-1 cell viability assays. *Ecotoxicology and Environmental Safety*, *58*(3), 396–404. DOI: 10.1016/S0147-6513(03)00105-2 PMID: 15223265

Ranke, J., Stolte, S., Störmann, R., Arning, J., & Jastorff, B. (2007). Design of sustainable chemical products—The example of ionic liquids. *Chemical Reviews*, *107*(6), 2183–2206. DOI: 10.1021/cr050942s PMID: 17564479

Ran, L., Wang, Z., Yang, B., Amiri-Margavi, A., & Alshahrani, N. (2024). Development of novel computational models based on artificial intelligence technique to predict the viscosity of ionic liquids-water mixtures. *Case Studies in Thermal Engineering*, *54*, 104076. Advance online publication. DOI: 10.1016/j.csite.2024.104076

Rastogi, S., Sethi, P., Panwar, N., Yadav, P., Verma, R., & Madan, A. (2025). *Harnessing Nanotechnology to Transform Waste Reclamation for Enhanced Efficiency and Environmental Impact.*, DOI: 10.4018/979-8-3693-7282-1.ch016

Rawat, K., & Bohidar, H. B. (2012). Universal charge quenching and stability of proteins in 1-methyl-3-alkyl (hexyl/octyl) imidazolium chloride ionic liquid solutions. *The Journal of Physical Chemistry B*, *116*(36), 11065–11074. DOI: 10.1021/jp3049108 PMID: 22891622

Ray, A., & Saruhan, B. (2021). Application of ionic liquids for batteries and supercapacitors. *Materials (Basel)*, *14*(11), 2942. DOI: 10.3390/ma14112942 PMID: 34072536

Regel-Rosocka, M., & Wisniewski, M. (2011). Ionic Liquids in Separation of Metal Ions from Aqueous Solutions. In *Applications of Ionic Liquids in Science and Technology*. InTech., DOI: 10.5772/23909

Reichardt, C., & Welton, T. (2010). *Solvents and solvent effects in organic chemistry* (4th ed.). Wiley-VCH. DOI: 10.1002/9783527632220

Reslan, M., & Kayser, V. (2018). Ionic liquids as biocompatible stabilizers of proteins. *Biophysical Reviews*, *10*(3), 781–793. DOI: 10.1007/s12551-018-0407-6 PMID: 29511969

Reynolds, J. L., Erdner, K. R., & Jones, P. B. (2002). Photoreduction of benzophenones by amines in room-temperature ionic liquids. *Organic Letters*, *4*(6), 917–919. DOI: 10.1021/ol017290o PMID: 11893185

Rezania, S., & Kordkandi, S. A. (2024). Recent advances in the adsorption of different pollutants from wastewater using carbon-based and metal-oxide nanoparticles. *Applied Sciences (Basel, Switzerland)*, *14*(24), 11492. Advance online publication. DOI: 10.3390/app142411492

Ribeiro, A. P. C., & Vieira, C. S. I., M., J., S., C., Langa, E., V. Loureno, M. J., S. Murshed, S. M., & de Castro, C. A. N. (2011). Thermal Properties of Ionic Liquids and Ionanofluids. In *Ionic Liquids: Theory, Properties, New Approaches*. InTech. DOI: 10.5772/13920

Ribeiro, L. M. O., Silva, A. E., Silva, M. C. S., & Almeida, G., & RMR. (. (2014). A study on ethanolysis and methanolysis of coconut oil for enzymatically catalyzed production of biodiesel. *Journal of Sustainable Bioenergy Systems*, *4*(4), 223–232. DOI: 10.4236/jsbs.2014.44020

Rogers, R. D., & Seddon, K. R. (2003). Ionic liquids—Solvents of the future? *Science*, *302*(5646), 792–793. DOI: 10.1126/science.1090313 PMID: 14593156

Roopan, S. M., Surendra, T. V., Elango, G., & Kumar, S. H. S. (2014). Biosynthetic trends and future aspects of bimetallic nanoparticles and its medicinal applications. *Applied Microbiology and Biotechnology*, *98*(12), 5289–5300. DOI: 10.1007/s00253-014-5736-1 PMID: 24728718

Rothe, M., Müller, E., Denk, P., & Kunz, W. (2021). Ionic liquids based on the concept of melting point lowering due to ethoxylation. *Molecules (Basel, Switzerland)*, *26*(13), 4034. Advance online publication. DOI: 10.3390/molecules26134034 PMID: 34279374

Roy, S., Speed, L.Jr, Viola, M., Luo, H., Leonard, D., & Qu, J. (2021). Oil miscible phosphonium-phosphate ionic liquid as novel antiwear and antipitting additive for low-viscosity rear axle lubricants. *Wear*, *466–467*, 203588. Advance online publication. DOI: 10.1016/j.wear.2020.203588

Rozynek, Z., Harkavyi, Y., & Giżyński, K. (2022). Fabrication of 1D particle structures outside a liquid environment using electric and capillary interactions: From fundamentals to applications. *Materials & Design, 223*, 111233. Advance online publication. DOI: 10.1016/j.matdes.2022.111233

Sagayaraj, R., Aravazhi, S., Praveen, P., & Chandrasekaran, G. (2017). Structural, morphological and magnetic characters of PVP coated ZnFe2O4 nanoparticles. *Journal of Materials Science Materials in Electronics, 29*(3), 2151–2158. DOI: 10.1007/s10854-017-8127-4

Sakhahari, K. V. (2025). Green Chemistry and Catalysis: Innovations toward a cleaner and Sustainable World. *International Journal of Pharmaceutical Research and Applications, 10*(1), 106–116. DOI: 10.35629/4494-1001106116

Sanders, J. R., Ward, E. H., & Hussey, C. L. (1986). Erratum: "Aluminum bromide-1-Methyl-3-Ethylimidazolium bromide Ionic liquids. I. Densities, viscosities, electrical conductivities, and phase transitions" [J. Electrochem. Soc., 133, 325 (1986)]. *Journal of the Electrochemical Society, 133*(7), 1526–1526. DOI: 10.1149/1.2152158

Sangwan, R., Saini, M., Verma, R., Kumar, S., Banerjee, M., & Jain, S. (2020). Synthesis of 1,8-dioxooctahydroxanthene derivatives using ionic liquids, quantum chemical studies and anticancer activity. *Journal of Molecular Structure, 1208*, 127786. DOI: 10.1016/j.molstruc.2020.127786

Santori, G., Di Nicola, G., Moglie, M., & Polonara, F. (2012). A review analyzing the industrial biodiesel production practice starting from vegetable oil refining. *Applied Energy, 92*, 109–132. DOI: 10.1016/j.apenergy.2011.10.031

Saraswat, P., Singh, R. B., Prasad, M., Misra, R., Rajput, V. D., & Ranjan, R. (2023). Applications of bio-based nanomaterials in environment and agriculture: A review on recent progresses. *Hybrid Advances, 4*, 100097. Advance online publication. DOI: 10.1016/j.hybadv.2023.100097

Sarma, B., Chen, J., Hsi, H.-Y., & Myerson, A. S. (2011). Solid forms of pharmaceuticals: Polymorphs, salts and cocrystals. *Korean Journal of Chemical Engineering, 28*(2), 315–322. DOI: 10.1007/s11814-010-0520-0

Sather, N. A., Sai, H., Sasselli, I. R., Sato, K., Ji, W., Synatschke, C. V., Zambrotta, R. T., Edelbrock, J. F., Kohlmeyer, R. R., Hardin, J. O., Mirau, P., & Stupp, S. I. (2021). 3D Printing of Supramolecular Polymer Hydrogels with Hierarchical Structure. *Small, 17*(5), 2005743. Advance online publication. DOI: 10.1002/smll.202005743 PMID: 33448102

Satyam, S., & Kumar, V. (2024). Innovations and challenges in adsorption-based wastewater remediation: A comprehensive review. *Heliyon*, *10*(9), e29573. Advance online publication. DOI: 10.1016/j.heliyon.2024.e29573 PMID: 38699034

Schirru, M., & Adler, M. (2020). An ultrasonic rheometer to measure gas absorption in ionic liquids: Design, calibration and testing. *Sensors (Basel)*, *20*(12), 1–20. DOI: 10.3390/s20123544 PMID: 32585862

Seddon, K. R. (1997). Ionic liquids for clean technology. *Journal of Chemical Technology and Biotechnology*, *68*(4), 351–356. DOI: 10.1002/(SICI)1097-4660(199704)68:4<351::AID-JCTB613>3.0.CO;2-4

Seddon, K. R., Stark, A., & Torres, M. (2000). Influence of chloride, water, and organic solvents on the physical properties of Ionic liquids. *Pure and Applied Chemistry*, *72*(12), 2275–2287. DOI: 10.1351/pac200072122275

Shahbaz, K., Mjalli, F. S., Hashim, M. A., & AlNashef, I. M. (2010). Using deep eutectic solvents for the removal of glycerol from palm oil-based biodiesel. *Journal of Applied Sciences (Faisalabad)*, *10*(24), 3349–3354. DOI: 10.3923/jas.2010.3349.3354

Shamiri, A., Chakrabarti, M., Jahan, S., Hussain, M., Kaminsky, W., Aravind, P., & Yehye, W. (2014). The Influence of Ziegler-Natta and Metallocene Catalysts on Polyolefin Structure, Properties, and Processing Ability. *Materials (Basel)*, *7*(7), 5069–5108. DOI: 10.3390/ma7075069 PMID: 28788120

Shamshina, J. L., Barber, P. S., & Rogers, R. D. (2013). Ionic liquids in drug delivery. *Expert Opinion on Drug Delivery*, *10*(10), 1367–1381. DOI: 10.1517/17425247.2013.808185 PMID: 23795613

Sharma, N. K., Vishwakarma, J., Rai, S., Alomar, T. S., AlMasoud, N., & Bhattarai, A. (2022). Green Route Synthesis and Characterization Techniques of Silver Nanoparticles and Their Biological Adeptness. *ACS Omega*, *7*(31), 27004–27020. DOI: 10.1021/acsomega.2c01400 PMID: 35967040

Sheldon, R. (2001). Catalytic reactions in ionic liquids. *Chemical Communications*, *23*(23), 2399–2407. DOI: 10.1039/b107270f PMID: 12239988

Sheldon, R. A. (2000). Atom efficiency and catalysis in organic synthesis. *Pure and Applied Chemistry*, *72*(7), 1233–1246. DOI: 10.1351/pac200072071233

Shishido, H., & Hashizume, H. (2024). Numerical analysis of the viscosity of ZrF<inf>4</inf>–BeF<inf>2</inf> molten salts for nuclear transmutation using fusion neutron sources. *Fusion Engineering and Design*, *203*, 114456. Advance online publication. DOI: 10.1016/j.fusengdes.2024.114456

Shukla, M. K., Tiwari, H., Verma, R., Dong, W., Azizov, S., Kumar, B., Pandey, S., & Kumar, D. (2023). Role and recent advancements of ionic liquids in drug delivery systems. *Pharmaceutics*, *15*(2), 702. DOI: 10.3390/pharmaceutics15020702 PMID: 36840024

Shukla, S. K., Khokarale, S. G., Bui, T. Q., & Mikkola, J. T. (2019). Ionic liquids: Potential materials for carbon dioxide capture and utilization. *Frontiers in Materials*, *6*, 42. Advance online publication. DOI: 10.3389/fmats.2019.00042

Sidat, Z., Marimuthu, T., Kumar, P., du Toit, L. C., Kondiah, P. P. D., Choonara, Y. E., & Pillay, V. (2019). Ionic liquids as potential and synergistic permeation enhancers for transdermal drug delivery. *Pharmaceutics*, *11*(2), 96. DOI: 10.3390/pharmaceutics11020096 PMID: 30813375

Sierra, A., Coleman, M. G., & Iglesias, P. (2022). Tribological Properties of Borate-Based Protic Ionic Liquids as Neat Lubricants and Biolubricant Additives for Steel-Steel Contact. *Lubricants (Basel, Switzerland)*, *10*(10), 269. Advance online publication. DOI: 10.3390/lubricants10100269

Sigma, V. USA. (2023). *Differences between cationic and anionic - 3V Sigma USA*. 3V Sigma USA. https://3vsigmausa.com/differences-between-cationic-and-anionic/

Silva, W., Zanatta, M., Ferreira, A. S., Corvo, M. C., & Cabrita, E. J. (2020). Revisiting Ionic Liquid Structure-Property Relationship: A Critical analysis. *International Journal of Molecular Sciences*, *21*(20), 7745. DOI: 10.3390/ijms21207745 PMID: 33086771

Sin Cheow, W., Xu, R., & Hadinoto, K. (2013). Towards sustainability: New approaches to nano-drug preparation. *Current Pharmaceutical Design*, *19*(35), 6229–6245. DOI: 10.2174/138161281131935002 PMID: 23469999

Singh, V., Amirchand, K. D., & Gardas, R. L. (2022). Ionic liquid-nanoparticle based hybrid systems for energy conversion and energy storage applications. *Journal of the Taiwan Institute of Chemical Engineers*, *133*, 104237. DOI: 10.1016/j.jtice.2022.104237

Siyahjani, S., Oner, S., Diker, H., Gultekin, B., & Varlikli, C. (2020). Enhanced capacitive behaviour of graphene based electrochemical double layer capacitors by etheric substitution on ionic liquids. *Journal of Power Sources*, *467*, 228353. Advance online publication. DOI: 10.1016/j.jpowsour.2020.228353

Sochi, T. (2010, December 20). *High Throughput Software for Powder Diffraction and its Application to Heterogeneous Catalysis*. ResearchGate. https://www.researchgate.net/publication/48172215_High_Throughput_Software_for_Powder_Diffraction_and_its_Application_toHeterogeneous_Catalysis

Song, J. (2021). Research Progress of Ionic Liquids as Lubricants. *ACS Omega, 6*(44), 29345–29349. DOI: 10.1021/acsomega.1c04512 PMID: 34778607

Stark, A., Ott, D., Kralisch, D., Kreisel, G., & Ondruschka, B. (2010). Ionic liquids and green chemistry: A lab experiment. *Journal of Chemical Education, 87*(2), 196–201. DOI: 10.1021/ed8000396

Stewart, R. J., Sima, M., Karz, J., & Jones, J. P. (2023). Material characterization of GPX®: A versatile in situ solidifying embolic platform technology. *Frontiers in Bioengineering and Biotechnology, 11*, 1095148. Advance online publication. DOI: 10.3389/fbioe.2023.1095148 PMID: 36726745

Sunitha, S., Kanjilal, S., Reddy, P. S., & Prasad, R. B. N. (2007). Ionic liquids as a reaction medium for lipase-catalyzed methanolysis of sunflower oil. *Biotechnology Letters, 29*(12), 1881–1885. DOI: 10.1007/s10529-007-9471-x PMID: 17634866

Suresh, & Sandhu, J. S. (2011). Recent advances in ionic liquids: Green unconventional solvents of this century: Part 1. *Green Chemistry Letters and Reviews, 4*(4), 289–310. DOI: 10.1080/17518253.2011.572295

Suresh, & Sandhu, J. S. (2011). Recent advances in ionic liquids: Green unconventional solvents of this century: Part II. *Green Chemistry Letters and Reviews, 4*(4), 311–320. DOI: 10.1080/17518253.2011.572296

Surette, J. K., Green, L., & Singer, R. D. (1996). 1-Ethyl-3-methylimidazolium halogenoaluminate melts as reaction media for the Friedel–crafts acylation of ferrocene. *Chemical Communications (Cambridge)*, (24), 2753–2754. DOI: 10.1039/CC9960002753

Swatloski, R. P., Spear, S. K., Holbrey, J. D., & Rogers, R. D. (2002). Dissolution of cellulose with ionic liquids. *Journal of the American Chemical Society, 124*(18), 4974–4975. DOI: 10.1021/ja025790m PMID: 11982358

Tadesse, H., & Luque, R. (2011). Advances on biomass pretreatment using ionic liquids: An overview. *Energy & Environmental Science, 4*(10), 3913–3929. DOI: 10.1039/c0ee00667j

Taha, M., Hassan, M., Essa, S., & Tartor, Y. (2013). Use of Fourier transform infrared spectroscopy (FTIR) spectroscopy for rapid and accurate identification of Yeasts isolated from human and animals. *International Journal of Veterinary Science and Medicine*, *1*(1), 15–20. DOI: 10.1016/j.ijvsm.2013.03.001

Taher, H., Nashef, E., Anvar, N., & Al-Zuhair, S. (2017). Enzymatic production of biodiesel from waste oil in ionic liquid medium. *Biofuels*, *10*(4), 463–472. DOI: 10.1080/17597269.2017.1316145

Takana, H., Hara, N., Makino, T., & Kanakubo, M. (2021). Effect of environmental temperature on CO<inf>2</inf> selective absorption characteristics by ionic liquid electrospray in flow system. *Journal of Electrostatics*, *114*, 103634. Advance online publication. DOI: 10.1016/j.elstat.2021.103634

Takeuchi, Y., Tarui, T., & Shibata, N. (2000). A novel and efficient synthesis of 3-fluorooxindoles from indoles mediated by Selectfluor. *Organic Letters*, *2*(5), 639–642. DOI: 10.1021/ol991400y PMID: 10814398

Tampucci, S., Guazzelli, L., Burgalassi, S., & Monti, D. (2020). pH-responsive nanostructures based on surface active fatty acid-protic ionic liquids for imiquimod delivery in skin cancer topical therapy. *Pharmaceutics*, *12*(11), 1078. DOI: 10.3390/pharmaceutics12111078 PMID: 33187215

Tang, S., Baker, G. A., & Zhao, H. (2012). Ether- and alcohol-functionalized task-specific ionic liquids: Attractive properties and applications. *Chemical Society Reviews*, *41*(10), 4030–4066. DOI: 10.1039/c2cs15362a PMID: 22456483

Thakur, A., Kumar, A., & Singh, A. (2024). Adsorptive removal of heavy metals, dyes, and pharmaceuticals: Carbon-based nanomaterials in focus. *Carbon*, *217*, 118621. Advance online publication. DOI: 10.1016/j.carbon.2023.118621

Timasheff, S. N. (1993). The control of protein stability and association by weak interactions with water: How do solvents affect these processes? *Annual Review of Biophysics and Biomolecular Structure*, *22*(1), 67–97. DOI: 10.1146/annurev.bb.22.060193.000435 PMID: 8347999

Titze-Frech, K., Ignatiev, N., Uerdingen, M., Schulz, P. S., & Wasserscheid, P. (2013). Highly selective aromatic alkylation of phenol and anisole by using recyclable Brønsted acidic ionic liquid systems. *European Journal of Organic Chemistry*, *2013*(30), 6961–6966. DOI: 10.1002/ejoc.201300579

Tomar, P., & Jain, D. (2022). A review of green solvent ionic liquids: as a future solvent. *International Journal of Advances in Scientific Research*, *13*(06), 1–16. DOI: 10.55218/JASR.202213601

Torrecilla, J. S., García, J., Rojo, E., & Rodríguez, F. (2009). Estimation of toxicity of ionic liquids in leukemia rat cell line and acetylcholinesterase enzyme by principal component analysis, neural networks and multiple linear regressions. *Journal of Hazardous Materials*, *164*(1), 182–194. DOI: 10.1016/j.jhazmat.2008.08.022 PMID: 18805639

Transmission Electron Microscopy. (2024). Nanoscience Instruments. https://www.nanoscience.com/techniques/transmission-electron-microscopy/

Trost, B. (1991). The Atom Economy—A Search for Synthetic Efficiency. *Science*, *254*(5037), 1471–1477. DOI: 10.1126/science.1962206 PMID: 1962206

Troter, D. Z., Todorović, Z. B., Đokić-Stojanović, D. R., Stamenković, O. S., & Veljković, V. B. (2016). Application of ionic liquids and deep eutectic solvents in biodiesel production: A review. *Renewable & Sustainable Energy Reviews*, *61*, 473–500. DOI: 10.1016/j.rser.2016.04.011

Trujillo-Rodríguez, M. J., Pino, V., & Míró, M. (2020). High-throughput microscale extraction using ionic liquids and derivatives: A review. *Journal of Separation Science*, *43*(9–10), 1890–1907. DOI: 10.1002/jssc.202000045 PMID: 32074395

Ullah, A., Rehman, U. U., Ahmad, R., & Rahman, F. (2024). Bioinspired green synthesis of copper, nickel, and hybrid nanoparticles using Myristica Fragrans seeds: Biomedical applications and beyond. *Materials Research Express*, *11*(8), 085401. DOI: 10.1088/2053-1591/ad68cf

United Nations General Assembly. (2015). *Transforming our world: The 2030 Agenda for Sustainable Development*. United Nations. https://sdgs.un.org/2030agenda

Urzúa, J. I., Valenzuela, M. L., Cavieres, J., & Inestrosa-Izurieta, M. J. (2023). Ionic liquid mixtures as energy storage materials: A preliminary and comparative study based on thermal storage density. *RSC Advances*, *13*(28), 19412–19419. DOI: 10.1039/D3RA02901H PMID: 37383686

Vafaeezadeh, M., & Alinezhad, H. (2016). Brønsted acidic ionic liquids: Green catalysts for essential organic reactions. *Journal of Molecular Liquids*, *218*, 95–105. DOI: 10.1016/j.molliq.2016.02.017

Vafaeezadeh, M., & Hashemi, M. M. (2014). Efficient fatty acid esterification using silica supported Brønsted acidic ionic liquid catalyst: Experimental study and DFT modeling. *Chemical Engineering Journal*, *250*, 35–41. DOI: 10.1016/j.cej.2014.04.001

Valli Sowbhagyam, D. (2024). Ionic Liquids as Green Solvents: A Comprehensive Review. *International Research Journal on Advanced Engineering Hub (IRJAEH), 2*(02), 220–224. DOI: 10.47392/IRJAEH.2024.0035

Van der Bruggen, B., Canbolat, Ç. B., Lin, J., & Luis, P. (2017). *The Potential of Membrane Technology for Treatment of Textile Wastewater.*, DOI: 10.1007/978-981-10-5623-9_13

Vancov, T., McIntosh, S., Curtis, J., & Yu, J. (2012). Use of ionic liquids in converting lignocellulosic material to biofuels. *Renewable Energy, 45*, 1–6. DOI: 10.1016/j.renene.2012.02.033

Vargas-Bernal, R. (2020). Introductory Chapter: Hybrid Nanomaterials. *IntechOpen EBooks.* DOI: 10.5772/intechopen.92012

Varma, R. S., & Namboodiri, V. V. (2001). An expeditious solvent-free route to ionic liquids using microwaves. *Chemical Communications (Cambridge)*, (7), 643–644. DOI: 10.1039/b101375k

Visser, A. E., Swatloski, R. P., Reichert, W. M., Griffin, S. T., & Rogers, R. D. (2000). Traditional extractants in nontraditional solvents: Groups 1 and 2 extraction by crown ethers in room-temperature ionic liquids. *Industrial & Engineering Chemistry Research, 39*(10), 3596–3604. DOI: 10.1021/ie000426m

Walden, P. (1914). Molecular weights and electrical conductivity of several fused salts. *Bulletin de l'Academie Imperiale des Sciences de St. Petersburg*, 405, 42.

Wang, D., Wang, R., Chen, S., Gao, J., Cai, C., Zheng, Y., Liu, X., Qu, B., Chen, N., & Zhuo, D. (2024). Low viscosity and highly flexible stereolithographic 3D printing resins for flexible sensors. *Materials & Design, 243*, 113052. Advance online publication. DOI: 10.1016/j.matdes.2024.113052

Wang, J., Zhang, Y., Guo, X., Tang, J., Chen, Z., Ha, M. N., Cui, P., & Ke, Q. (2024). Manganese oxide-based catalysts for the sustainable synthesis of value-added chemicals through oxidation processes: A critical review and perspectives for the future. *Green Chemistry, 26*(5), 2365–2383. DOI: 10.1039/D3GC04117D

Wang, L., Li, H., & Li, P. (2009). Task-specific ionic liquid as base, ligand and reaction medium for the palladium-catalyzed Heck reaction. *Tetrahedron, 65*(1), 364–368. DOI: 10.1016/j.tet.2008.10.042

Wang, L., & Zhang, X. (2017). The antimicrobial activity of nanoparticles: Present situation and prospects for the future. *International Journal of Nanomedicine, 12*, 1227–1249. DOI: 10.2147/IJN.S121956 PMID: 28243086

Wang, Y., Zhang, W., Luo, C., Wu, X., & Yan, G. (2016). Superparamagnetic FeCo@ SnO2 nanoparticles on graphene-polyaniline: Synthesis and enhanced electromagnetic wave absorption properties. *Ceramics International, 42*(10), 12496–12502. DOI: 10.1016/j.ceramint.2016.05.038

Wasserscheid, P., & Keim, W. (2000). Ionic liquids—New "Solutions" for transition metal catalysis. *Angewandte Chemie, 39*(21), 3772–3789. DOI: 10.1002/1521-3773(20001103)39:21<3772::AID-ANIE3772>3.0.CO;2-5 PMID: 11091453

Wasserscheid, P., & Welton, T. (Eds.). (2002). *Ionic liquids in synthesis.* Wiley-VCH., DOI: 10.1002/3527600701

Wei, P., Pan, X., Chen, C., Li, H., Yan, X., Li, C., Chu, Y., & Yan, B. (2021). Emerging impacts of ionic liquids on eco-environmental safety and human health. *Chemical Society Reviews, 50*(24), 13609–13627. DOI: 10.1039/D1CS00946J PMID: 34812453

Welton, T. (1999). Room-temperature Ionic liquids. Solvents for synthesis and catalysis. *Chemical Reviews, 99*(8), 2071–2084. DOI: 10.1021/cr980032t PMID: 11849019

Welton, T. (2004). Ionic liquids in catalysis. *Coordination Chemistry Reviews, 248*(21–24), 2459–2477. DOI: 10.1016/j.ccr.2004.04.015

Welton, T. (2015). Solvents and sustainable chemistry. *Proceedings - Royal Society. Mathematical, Physical and Engineering Sciences, 471*(2183), 20150502. DOI: 10.1098/rspa.2015.0502 PMID: 26730217

Welton, T. (2018). Ionic liquids: A brief history. *Biophysical Reviews, 10*(3), 691–706. DOI: 10.1007/s12551-018-0419-2 PMID: 29700779

Wheeler, C., West, K. N., Eckert, C. A., & Liotta, C. L. (2001). Ionic liquids as catalytic green solvents for nucleophilic displacement reactions. *Chemical Communications (Cambridge)*, (10), 887–888. DOI: 10.1039/b101202a

Wilkes, J. S. (2002). A short history of Ionic liquids—From molten salts to neoteric solvents. *Green Chemistry, 4*(2), 73–80. DOI: 10.1039/b110838g

Wilkes, J. S., & Zaworotko, M. J. (1992). Air and water sTable 1-ethyl-3-methylimidazolium based Ionic liquids. *Journal of the Chemical Society. Chemical Communications*, (13), 965. DOI: 10.1039/c39920000965

Wong, E. L. S., Vuong, K. Q., & Chow, E. (2021). Nanozymes for environmental pollutant monitoring and remediation. *Sensors (Basel), 21*(2), 408. Advance online publication. DOI: 10.3390/s21020408 PMID: 33430087

Wu, F., Zhao, Y., Wang, Y., Xu, Y., Tang, M., Wang, Z., Han, B., & Liu, Z. (2022). Hydrogen-bonding and acid cooperative catalysis for benzylation of arenes with benzyl alcohols over ionic liquids. *Green Chemistry*, *24*(8), 3137–3142. DOI: 10.1039/D1GC04485K

Wu, Q., Chen, H., Han, M., Wang, D., & Wang, J. (2007). Transesterification of cottonseed oil to biodiesel catalyzed by highly active ionic liquids. *Industrial & Engineering Chemistry Research*, *46*(24), 7955–7960. DOI: 10.1021/ie070678o

Wu, X., Zhu, Q., Chen, Z., Wu, W., Lu, Y., & Qi, J. (2021). Ionic liquids as a useful tool for tailoring active pharmaceutical ingredients. *Journal of Controlled Release*, *338*, 268–283. DOI: 10.1016/j.jconrel.2021.08.032 PMID: 34425167

Xing, H., Wang, T., Zhou, Z., & Dai, Y. (2005). Novel Brønsted-acidic ionic liquids for esterifications. *Industrial & Engineering Chemistry Research*, *44*(11), 4147–4150. DOI: 10.1021/ie0488703

Xue, Y., Zhang, X.-G., Lu, Z.-P., Xu, C., Xu, H.-J., & Hu, Y. (2022). Enhancing the Catalytic Performance of Candida antarctica Lipase B by Chemical Modification With Alkylated Betaine Ionic Liquids. *Frontiers in Bioengineering and Biotechnology*, *10*, 850890. Advance online publication. DOI: 10.3389/fbioe.2022.850890 PMID: 35265607

Xu, J. M., Wu, Q., Zhang, Q. Y., Zhang, F., & Lin, X. F. (2007). Basic ionic liquid as catalyst and reaction medium: A rapid and facile protocol for aza-Michael addition reactions. *European Journal of Organic Chemistry*, *2007*(11), 1798–1802. DOI: 10.1002/ejoc.200600999

Xu, L., Li, W., He, J., Liu, Z., & Zhou, Y. (2011). Assessment of a dry and a wet route for the production of biofuels from microalgae: Energy balance analysis. *Bioresource Technology*, *102*(8), 5113–5122. DOI: 10.1016/j.biortech.2011.01.066 PMID: 21334889

Yang, F., & Feng, P. (2020). Densities and viscosities of ionic liquid with organic solvents. *Applied Sciences (Basel, Switzerland)*, *10*(23), 1–18. DOI: 10.3390/app10238342

Yang, F., Zheng, Q., Wang, B., Tan, H., & Wang, X. (2024). Improvement of ionic liquid properties for the absorption refrigeration cycle. *Energy Storage and Saving*, *3*(4), 352–361. DOI: 10.1016/j.enss.2024.10.005

Yang, H., Chen, L., Zhou, Y., & Wang, H. (2017). New algal lipid extraction procedure using an amphiphilic amine solvent and ionic liquid. *Biomass and Bioenergy*, *100*, 108–115. DOI: 10.1016/j.biombioe.2017.03.018

Yang, H., Luo, X.-F., Matsumoto, K., Chang, J.-K., & Hagiwara, R. (2020). Physicochemical and electrochemical properties of the (fluorosulfonyl)(trifluoromethylsulfonyl)amide ionic liquid for Na secondary batteries. *Journal of Power Sources*, *470*, 228406. Advance online publication. DOI: 10.1016/j.jpowsour.2020.228406

Yang, Z., Huang, Z.-L., Hu, Z.-L., Zhang, K.-P., & Wang, P.-W. (2010). Both hydrolytic and transesterification activities of *Penicillium expansum* lipase are significantly enhanced in ionic liquid [BMIm] [PF6]. *Journal of Molecular Catalysis. B, Enzymatic*, *63*(1–2), 23–30. DOI: 10.1016/j.molcatb.2009.11.014

Yannai, X., Ge, S., Rennert, G., Gruener, N., & Fares, F. A. (1996). 3,3′-Diindolylmethane induces apoptosis in human cancer cells. *Biochemical and Biophysical Research Communications*, *228*(1), 153–158. DOI: 10.1006/bbrc.1996.1631 PMID: 8912651

Yaqoob, A. A., Parveen, T., Umar, K., & Mohamad Ibrahim, M. N. (2020). Role of nanomaterials in the treatment of wastewater: A review. *Water (Basel)*, *12*(2), 495. Advance online publication. DOI: 10.3390/w12020495

Yoo, C. G., Pu, Y., & Ragauskas, A. J. (2017). Ionic liquids: Promising green solvents for lignocellulosic biomass utilization. *Current Opinion in Green and Sustainable Chemistry*, *5*, 5–11. DOI: 10.1016/j.cogsc.2017.03.003

Yuan, W., Yang, X., He, L., Xue, Y., Qin, S., & Tao, G. (2018). Viscosity, conductivity, and electrochemical property of dicyanamide ionic liquids. *Frontiers in Chemistry*, *6*, 59. Advance online publication. DOI: 10.3389/fchem.2018.00059 PMID: 29600245

Yudaev, P. A., & Chistyakov, E. M. (2022). Ionic Liquids as Components of Systems for Metal Extraction. *ChemEngineering*, *6*(1), 6. DOI: 10.3390/chemengineering6010006

Yu, T., Cai, Q., Lian, G., Bai, Y., Zhang, X., Zhang, X., Liu, L., & Zhang, S. (2021). Mechanisms behind high CO<inf>2</inf>/CH<inf>4</inf> selectivity using ZIF-8 metal organic frameworks with encapsulated ionic liquids: A computational study. *Chemical Engineering Journal*, *419*, 129638. Advance online publication. DOI: 10.1016/j.cej.2021.129638

Yu, Y., & Chen, Y. (2021). Density prediction of ionic liquids at different temperatures using the average free volume model. *ACS Omega*, *6*(23), 14869–14874. DOI: 10.1021/acsomega.1c00547 PMID: 34151068

Zakrewsky, M., Lovejoy, K. S., Kern, T. L., Miller, T. E., Le, V., Nagy, A., Goumas, A. M., Iyer, R. S., Del Sesto, R. E., Koppisch, A. T., Fox, D. T., & Mitragotri, S. (2014). Ionic liquids as a class of materials for transdermal delivery and pathogen neutralization. *Proceedings of the National Academy of Sciences of the United States of America*, *111*(37), 13313–13318. DOI: 10.1073/pnas.1403995111 PMID: 25157174

Zalewski, M., Krawczyk, T., Siewniak, A., & Sobolewski, A. (2021). Carbon dioxide capture using water-imidazolium ionic liquids-amines ternary systems. *International Journal of Greenhouse Gas Control*, *105*, 103210. Advance online publication. DOI: 10.1016/j.ijggc.2020.103210

Zang, H., Yao, S., Luo, Y., Tang, D., & Song, H. (2018). Complex-precipitation using functionalized chiral ionic liquids with L-proline anion and chromatographic analysis for enantio separation of racemic amino acids. *Chirality*, *30*(10), 1182–1194. DOI: 10.1002/chir.23011 PMID: 30188593

Zhang, W., & Zhang, S. (2024). Catalysts: Combinatorial heterogeneous catalysis. In *Encyclopedia of Condensed Matter Physics* (pp. 729–737). Elsevier. DOI: 10.1016/B978-0-323-90800-9.00201-8

Zhang, B., Guo, Y., Lin, L., Qiao, D., & Xie, F. (2023). For efficient treatment of starch using aqueous ionic liquid at room temperature. *Sustainable Materials and Technologies*, *36*, e00592. Advance online publication. DOI: 10.1016/j.susmat.2023.e00592

Zhang, C., Malhotra, S. V., & Francis, A. J. (2014). Toxicity of ionic liquids to Clostridium sp. and effects on uranium biosorption. *Journal of Hazardous Materials*, *264*, 246–253. DOI: 10.1016/j.jhazmat.2013.11.003 PMID: 24316798

Zhang, H., & Zhang, X. (2022). MIL series of metal organic frameworks (MOFs) as novel adsorbents for heavy metals in water: A review. *Journal of Hazardous Materials*, *429*, 128271. Advance online publication. DOI: 10.1016/j.jhazmat.2022.128271 PMID: 35093745

Zhang, K. P., Lai, J. Q., Huang, Z. L., & Yang, Z. (2011). *Penicillium expansum* lipase-catalyzed production of biodiesel in ionic liquids. *Bioresource Technology*, *102*(3), 2767–2772. DOI: 10.1016/j.biortech.2010.11.057 PMID: 21138789

Zhang, L., Cui, Y., Zhang, C., Wang, L., Wan, H., & Guan, G. (2012). Biodiesel production by esterification of oleic acid over Brønsted acidic ionic liquid supported onto Fe-incorporated SBA-15. *Industrial & Engineering Chemistry Research*, *51*(50), 16590–16596. DOI: 10.1021/ie302419y

Zhang, R., Hallström, S., Mao, H., Kjellqvist, L., & Chen, Q.ZHANG. (2021). Accurate viscosity prediction for molten slags: A new model and database. *ISIJ International, 61*(5), 1379–1388. DOI: 10.2355/isijinternational.ISIJINT-2020-586

Zhang, S., & Wang, C. (2023). Precise Analysis of Nanoparticle Size Distribution in TEM Image. *Methods and Protocols, 6*(4), 63. DOI: 10.3390/mps6040063 PMID: 37489430

Zhang, X., Zhang, J., Yin, J., Liu, X., Qiu, W., He, J., Jiang, W., Zhu, L., Li, H., & Li, H. (2025). Mo-MOF-based ionanofluids for highly efficient extraction coupled catalytic oxidative desulfurization. *Separation and Purification Technology, 353*, 128289. DOI: 10.1016/j.seppur.2024.128289

Zhang, Y., Bakshi, B. R., & Demessie, E. S. (2008). Life cycle assessment of an ionic liquid versus molecular solvents and their applications. *Environmental Science & Technology, 42*(5), 1724–1730. DOI: 10.1021/es0713983 PMID: 18441827

Zhang, Y., Wu, B., Xu, H., Liu, H., Wang, M., He, Y., & Pan, B. (2016). Nanomaterials-enabled water and wastewater treatment. *NanoImpact, 3–4*, 22–39. DOI: 10.1016/j.impact.2016.09.004

Zhao, D., Liao, Y., & Zhang, Z. (2007). Toxicity of ionic liquids. *Clean (Weinheim), 35*(1), 42–48. DOI: 10.1002/clen.200600015

Zhao, H., Xia, S., & Ma, P. (2005). Use of ionic liquids as 'green' solvents for extractions. *Journal of Chemical Technology and Biotechnology, 80*(10), 1089–1096. DOI: 10.1002/jctb.1333

Zhao, H., Zhang, C., & Crittle, T. D. (2013). Choline-based deep eutectic solvents for enzymatic preparation of biodiesel from soybean oil. *Journal of Molecular Catalysis. B, Enzymatic, 85–86*, 243–247. DOI: 10.1016/j.molcatb.2012.09.003

Zhen, B., Jiao, Q., Wu, Q., & Li, H. (2014). Catalytic performance of acidic ionic liquid-functionalized silica in biodiesel production. *Journal of Energy Chemistry, 23*(1), 97–104. DOI: 10.1016/S2095-4956(14)60122-4

Zheng, Z., Xu, Q., Guo, J., Qin, J., Mao, H., Wang, B., & Yan, F. (2016). Structure–antibacterial activity relationships of imidazolium-type ionic liquid monomers, poly(ionic liquids) and poly(ionic liquid) membranes: Effect of alkyl chain length and cations. *ACS Applied Materials & Interfaces, 8*(19), 12684–12692. DOI: 10.1021/acsami.6b03391 PMID: 27145107

Zhou, J. H., Wu, L. M., Wang, J., & Zhang, L. (2011). Preparation of biodiesel from Tung oil catalyzed by sulfonic-functional Brønsted acidic ionic liquids. *Advanced Materials Research, 314–316*, 1459–1462. . DOI: 10.4028/www.scientific.net/AMR.314-316.1459

Zhou, S., Liu, L., Wang, B., Xu, F., & Sun, R. C. (2012). Biodiesel preparation from transesterification of glycerol trioleate catalyzed by basic ionic liquids. *Chinese Chemical Letters*, *23*(4), 379–382. DOI: 10.1016/j.cclet.2012.01.034

Zhu, H., & O'Dell, L. A. (2021). Nuclear magnetic resonance characterisation of Ionic liquids and organic Ionic plastic crystals: Common approaches and recent advances. *Chemical Communications*, *57*(46), 5609–5625. DOI: 10.1039/D1CC02151F PMID: 34048516

Zhuo, Y., Cheng, H.-L., Zhao, Y.-G., & Cui, H.-R. (2024). Ionic Liquids in Pharmaceutical and Biomedical Applications: A Review. *Pharmaceutics*, *16*(1), 151. DOI: 10.3390/pharmaceutics16010151 PMID: 38276519

Zullo, V., Iuliano, A., & Guazzelli, L. (2021). Sugar-Based Ionic Liquids: Multifaceted Challenges and Intriguing Potential. *Molecules (Basel, Switzerland)*, *26*(7), 2052. DOI: 10.3390/molecules26072052 PMID: 33916695

Zvereva, E. E., Katsyuba, S. A., Dyson, P. J., & Aleksandrov, A. V. (2017). Leaching from Palladium Nanoparticles in an Ionic Liquid Leads to the Formation of Ionic Monometallic Species. *The Journal of Physical Chemistry Letters*, *8*(15), 3452–3456. DOI: 10.1021/acs.jpclett.7b01212 PMID: 28692296

About the Contributors

Syam Sundar Lingala obtained his doctorate degree in the field of Mechanical Engineering (Energy Systems) from the Jawaharlal Nehru Technological University-Hyderabad, Hyderabad, India in 2010 and Post-Doctoral Research Fellow from the University of Aveiro, Portugal, M. Tech in Thermal Engineering from the Jawaharlal Nehru Technological University-Hyderabad, Hyderabad, India in 2003 and B.E. in Mechanical Engineering from the Andhra University, India in 1998. His research interests include energy, thermal power plants, modeling, and optimization of forced convection heat transfer from tube banks and micro-channel heat sinks, thermal system optimization using entropy generation minimization and genetic algorithm, mixed convection, Nanofluids, and ionic liquids. Dr. Syam Sundar Lingala has guided 02 PhDs, and 04 Master thesis and 05 PhDs are on-going. Dr. Syam Sundar Lingala is the top 100 thousand or in the top 2% of the most cited researchers in the world throughout their career and in their scientific field, according to a Stanford University study signed by the team led by John P.A. Ioannidis. He has more than 100 publications in refereed International Journals and 10 book chapters.

Muhammad Waqar Ashraf is Professor & Dean of College of Sciences & Human Studies at Prince Mohammad Bin Fahd University. His research interests include Analytical Chemistry, Organo-metallic Chemistry, Sustainability & Waste Effluents Treatment technologies. He is author of over 80 research publications and a book. Dr. Ashraf has successfully carried out funded research projects related to environmental issues in the Kingdom, funded by King Abdulaziz City of Science & Technology (KACST) and SABIC. He has presented his research work at various renowned international conferences held at England, Italy, Canada, Greece and the Netherland. He acts as a reviewer for KACST and other reputed scientific journals. He has been member of editorial boards and chaired conference organizational committees. Dr.Ashraf is an active member of Environmental Technology & Management Association (ETMA) and American Chemical Society (ACS) Saudi

Arabian Chapter. Currently, he is Assistant to Vice President for Academic Affairs. He is the member of various university committees and Secretary General of the Scientific Council.

***

Kim Andrews holds a PhD in Chemistry (Organic Chemistry) from Purdue University in West Lafayette Indiana, USA, (December 2007). A MSc. in Chemistry (Gas Phase Organic Chemistry) from University of Tennessee in Knoxville Tennessee, USA (May 1996) and a BSc. in Chemistry from Mercer University in Macon Georgia, USA (May 1993). Her PhD graduate studies from Purdue were focused on Organo-Metallic Chemistry with emphasis in asymmetrical porphyrin synthesis and their ability to intercalate into DNA as potential photodynamic therapeutic agents. After completing her PhD graduate studies, she worked for a fortune 500 chemical company which lead to filing and obtaining European and American patent applications. Her work led to leading high profile scientific projects in Germany where she worked and lived during her assignments. In addition to the fore mentioned experience Dr. Kim has also worked as a senior site manufacturing chemist and she later took on a leadership role for another specialty chemical company as a principal scientist to help expand the company product portfolio. Upon receiving her MSc. in Chemistry, she worked for a fortune 500 pharmaceutical company where she served as a process product improvement specialist and technical service product lead specialist in manufacturing. Moreover, her work experience also includes working at a national laboratory where she supported the treatment of radioactive wastewater as a contributing team member. Dr. Kim's research interest include Porphyrin synthesis and intercalation into B form DNA for medicinal applications, Identifying and generating substrates for encapsulation chemistry, Waste water treatment using chromatography, Commercial biomedical and industrial coatings applications. At PMU she teaches Introductory Chemistry and Chemistry for Engineers I.

Tejal Bele is an accomplished academic and research scholar with a deep passion for pharmacology, currently pursuing her Ph.D. at Savitribai Phule University, Pune, with an M.Pharm in Pharmacology, where she was awarded a Gold Medal for her academic excellence. With over six years of experience in both research and teaching, she has contributed significantly to advancing the field of pharmacology. Her work is focused on preclinical evaluation of potential neuropharmacological agents and utilizing cutting-edge computational methods such as network pharmacology, molecular docking and simulation to predict the interaction between therapeutic molecules and biological targets. As a researcher, she had the privilege of presenting her work at an international & national conference and received the Best Oral Presentation Award. She has published several research papers in peer-reviewed journals and is actively involved in research activities. Also, she guides students on their research projects, encouraging them to pursue innovative research and contributing to their academic growth.

Kashvi Chaturvedi, is studying B. Tech, specialisation in Artificial Intelligence and Data Science at Ajeenkya DY Patil University. She started solidifying her programming skills by completing various certification courses like Data Structures in C, Tableau functions and Operating Systems along with doing few projects using languages like Python, C, C++ and creating Web Page using HTML, CSS and Java Script. She had also attended an Arduino-Based Automation course at Dolphin Labs to enhance her software knowledge. There, she learned how to work with sensors, microcontrollers, and circuit design. She has also written multiple review papers on topics such as Geographic Information Systems, Comparative Analysis of BJT Configurations, Swiss Bank vs. European Central Bank: Data Analysis. She is also researching on and learning many topics such as Small Modular Reactors, Bibliometric Analysis, UNSDGs etc. Because of her interest in latest technologies, she also participated in Global Hyperloop Competition hosted by IIT Madras representing her college as team Vac-Vectoor where she learnt more about various subtopics from the mentors.

Susanta Das is a seasoned educator with a proven record of teaching and mentoring diverse students, working at multiple institutions, and providing service to the various initiatives of organizations. He received his Ph.D. and M.A. degrees from Western Michigan University (WMU), USA, and M.Sc. degree from Banaras Hindu University (BHU), India, all in Physics. He continued his research as a Marie-Curie post-doctoral fellow at Stockholm University, Stockholm, Sweden on beam diagnostics for the DESIREE (Double ElectroStatic Ion Ring ExpEriment) in the project DITANET (Diagnostic Techniques for particle Accelerators – a Marie-Curie initial training NETwork), at the Indian Institute of Science Education and Research-Kolkata, India on high-pressure physics as a Project Scientist-B, and at the University of Electro-Communications, Tokyo, Japan on ion-surface interactions as a Post-doctoral fellow. Throughout his career, Dr. Das worked and collaborated with many researchers, Ph.D., master, and visiting students from different countries. He further visited many countries to discuss research and an international conference participant (UK, Italy, Germany, Greece, Belgium, Bulgaria, Romania, Brazil etc.). He worked at several institutions before assuming his current role at Ajeenkya D Y Patil University. He received the Marie-Curie post-doctoral fellowship in Sweden, Science Academies' Summer Research Fellowship in India, Gowen Frostic Doctoral

Fellowship, Department Graduate Research and Creative Scholar Award by WMU, and Leo R. Parpart Doctoral Fellowship by Dept. of Physics, WMU, among many others, throughout his academic journey. He co-authored several research articles in WoS/Scopus indexed international journals, conference proceedings, and scholarly book chapters and co-edited several books published by renowned international publishing houses (IGI-Global, Springer Nature, Nova Science Publishers, Wiley, Scrivener-Wiley, Notion Press). At present, Dr. Das continues to delve into cutting-edge fields, with a keen interest in nanotechnology, UN SDGs, and data science. His multifaceted contributions, spanning teaching, research, and administrative leadership, showcase a dedicated professional who is instrumental in advancing the vision and mission of the institutions he serves.

Shraddha Devarshi is a Doctor of Pharmacy graduate from BVDU's Poona College of Pharmacy in 2016. During this tenure, she has worked on a project titled quality of life of Beta-thalassemia patients. She has been awarded Gold medal for her academic excellence. She is currently working as Assistant Professor in Indira University, School of Pharmacy, Pune. She has 9 years of teaching experience. She has 5 research articles, 2 case reports, and 1 book chapter published in national and international journals with total citation of 10 and hi-index of 2. She has also presented her work in national and international conferences. She guides Pharm.D. students in their project work, demonstrating her keen interest in research. She had delivered guest lecture on Writing SOAP notes, how to teach Pharmacotherapeutics effectively?, Do's & Don't of Project & Clerkship. She was also a panelist at Clinical Pharmacy Summit 2024 focused on bridging gap between academia and clinical practice. Her areas of research include drug utilisation pattern studies, quality of life studies, and health services. Her areas of expertise in Cardiology, Pulmonology, infectious diseases, Pharmacotherapeutics, Clinical Research, Clinical Pharmacy, Patient Counselling. Years of Experience: 9 years Qualification: Doctor of Pharmacy. Subject Expertise: Pharmacotherapeutics, Clinical Research, Clinical Pharmacy. Google scholar: 2bMmSjEAAAAJ Orcid: 0000-0002-6725-8917

Rutuja Kamble is a dynamic and dedicated professional with 15 years of experience. Her contributions encompass teaching, with a deep passion for inspiring and mentoring young minds, as well as fostering the growth and development of the institution. She has published several papers in national and international journals, presented numerous posters at various national and international platforms, and received Special Appreciation, secured First Rank, and been awarded Best Presenter at both national and international levels. Actively engaged in continuous professional development, she regularly participates in Faculty Development Programmes to further enhance her expertise and teaching practices. Her research interests include the synthesis, purification, and characterization of organic compounds, as well as online in silico studies and docking studies

E V Ramana is a researcher at the University of Aveiro, Portugal. He holds a masters degree in applied electronics and PhD in Physics from India.He was a post-doc in South Korea for 2 years and 7 years in Portugal. Since 2019, he has been serving as a researcher. His main research interests are materials for energy storage, conversion and refrigeration.

Gourisankar Roymahapatra is currently working as an Associate Professor of Chemistry at the School of Applied Science and Humanities, Haldia Institute of Technology (HIT) Haldia, India. He completed his Ph.D. from Jadavpur University, Kolkata (2014) with N-Heterocyclic Carbene Chemistry as a major field under the mentoring of Dr. Joydev Dinda of HIT Haldia, and Prof. Ambikesh Mahapatra of Jadavpur University. Before joining to Global Institute of Science and Technology (GIST Haldia), as a Lecturer in Chemistry in the year 2011, he worked at MCC PTA Chem. Corp. Pvt. Ltd India (MCPI), Haldia, (2003 ~ 2011) as a Senior Chemist. He joined HIT Haldia in 2015 as an Assistant Professor of Chemistry. His area of research interest includes NHC complexes, organometallics, catalysis, antibiotics, anti-carcinogenic, gas adsorption, super alkali, computational chemistry and DFT. For his contribution in research, he got the 'Distinguished Young Scientist Award in Chemistry -2014'; 'Bharat Gaurabh Award – 2018'; and 'Indian Chemical Society Research Excellent Award 2021'. Dr. Roymahapatra is a Life Fellow of the Indian Chemical Society (FICS), and a Life Fellow of the Institutes of Chemistry

Sunil Sankathala is pursuing B. Tech Computer Science and Engineering at Ajeenkya D Y Patil University, with a diverse range of interests that bridge technology, sustainability, and innovation. His academic focus lies in computer science fundamentals like programming, data structures, algorithms, and software development. He actively supplements this with hands-on experience in web development and robotics. Beyond the core curriculum, He has a strong research inclination in emerging and impactful fields. His current research interests include Artificial Intelligence in Healthcare, bioplastics, nano bioplastics, e-waste management, and solid waste management—areas where he believes technology can make a meaningful difference in solving real-world problems. He has been publishing research articles and book chapters in reputed national and international journals and books. He is also part of an ambitious Hyperloop project, which involves interdisciplinary collaboration and cutting-edge engineering aimed at revolutionizing transportation. Working on this project has enhanced his skills in teamwork, innovation, and systems thinking. In the tech domain, he has developed proficiency in web development using technologies such as HTML, CSS, JavaScript, and React. His interests in robotics further reflect building smart, responsive systems that interact with the physical world. Throughout his course, he has helped strengthen problem-solving and collaborative skills. He is proficient in languages such as Python, C, C++, MySQL, Ubuntu, Machine Learning and familiar with frameworks and tools like React, Git. He actively participates in hackathons, coding competitions, and tech events to challenge himself and learn from peers. He is also a quick learner and highly adaptable, with an eagerness to stay updated with the latest trends in the tech world. He aspires to become a skilled software engineer and contribute to building innovative solutions that have a positive impact on society. The long-term goal is to work in a dynamic environment where knowledge, grow professionally, and make meaningful contributions to the field of technology. Driven by a spirit of exploration and sustainability, he aims to contribute to technological solutions that address both human and environmental needs. The long-term goal is to become a versatile engineer and researcher, working at the intersection of computer science, environmental innovation, and human well-being.

Aditya Shrivastav is an aspiring technologist currently pursuing a Bachelor of Technology (B. Tech.) in Computer Science and Engineering at Ajeenkya D Y Patil University. He holds solid programming expertise in C, C++, and Python, alongside strong command over data and business tools such as Microsoft Excel and Tally software. Aditya is a member of Team Vac-Vectoor Hyperloop and participated with the team in the Global Hyperloop Competition organized by IIT Madras, contributing to cutting-edge transportation solutions. He actively engages in hackathons and various technical events to expand his skills and contribute to innovative problem-solving. He has authored and co-authored book chapters in the domains of solid waste management, bioplastics, nanotechnology applications, and artificial intelligence, showcasing his multidisciplinary approach and research acumen. Aditya is also working on an innovative project on predictive e-waste management. The project is targeted toward IT departments, schools, and colleges, aiming to support sustainable practices and resource optimization. In parallel with his academic journey, Aditya is serving as a faculty intern at Disha Computer Institute, where he teaches C, C++, basic Python, basic Java, Excel formulas, and Tally software. His dedication and teaching proficiency earned him a Letter of Appreciation and the opportunity to conduct a workshop on Advanced Excel at G. H. Raisoni College of Arts, Commerce and Science. With a strong foundation in technology, a passion for sustainability, and hands-on experience in both teaching and project development, Aditya Shrivastav is committed to driving impact through research, education, and innovation.

Pallavi Singh Chouhan is a highly skilled academician and researcher with 8 years 10 months of professional experience. For the past 8 years, she has been serving as an Assistant Professor in the Indira University, School of Pharmacy, Pune. Additionally, she has gained valuable industry experience as a QA executive in the Packaging and Manufacturing department. She is a university approved guide and involved in mentoring M. Pharm Quality Assurance students through their research projects and dissertations. Her research interests lie in the application of Quality by Design (QBD), analytical method development, stability testing in pharmaceuticals, and Pharmaceutical Validation. She has filed a patent and has authored one research and 3 review publications.

Amruta Walvekar is an academician and researcher with 9 years of professional experience, including 8 years at Indira University, School of Pharmacy, Pune. She is Passionate about mentoring students, fostering academic excellence, and accomplishing institutional goals. She has contributed extensively to teaching, delivering well-structured and engaging lectures in pharmaceutical chemistry related subjects. She has authored 2 research and 2 review papers in reputed journals with citation index of 9 and h-index of 2. She has contributed to 1 book chapter which is indexed in international publication in the field of pharmaceutical science. She has been actively involved and participated in poster and oral presentation at various conferences to forecast innovative ideas. She has keen interest to attend various seminars and FDPs organized by various organization at national and international levels. She has guided under graduate and post graduate students on academic projects, career planning, personal development, emphasizing practical knowledge and innovative thinking.

Index

A

B

C

D

E

F

G

I

L

M

N

O

P

Q

R

S

T

V

X

www.ingramcontent.com/pod-product-compliance
Lightning Source LLC
LaVergne TN
LVHW061547290926
849162LV00004B/933